Caffeine
Chemistry, Analysis, Function and Effects

Food and Nutritional Components in Focus

Series Editors:
Professor Victor R Preedy, *School of Medicine, King's College London, UK*

Titles in the Series:
1: Vitamin A and Carotenoids: Chemistry, Analysis, Function and Effects
2: Caffeine: Chemistry, Analysis, Function and Effects

How to obtain future titles on publication:
A standing order plan is available for this series. A standing order will bring delivery of each new volume immediately on publication.

For further information please contact:
Book Sales Department, Royal Society of Chemistry, Thomas Graham House, Science Park, Milton Road, Cambridge, CB4 0WF, UK
Telephone: +44 (0)1223 420066, Fax: +44 (0)1223 420247
Email: booksales@rsc.org
Visit our website at http://www.rsc.org/Shop/Books/

Caffeine
Chemistry, Analysis, Function and Effects

Edited by

Victor R Preedy
School of Medicine, King's College London, UK
Email: victor.preedy@kcl.ac.uk

RSCPublishing

Food and Nutritional Components in Focus No. 2

ISBN: 978-1-84973-367-0
ISSN: 2045-1695

A catalogue record for this book is available from the British Library

Published by The Royal Society of Chemistry,
Thomas Graham House, Science Park, Milton Road,
Cambridge CB4 0WF, UK

Registered Charity Number 207890

For further information see our web site at www.rsc.org

Printed and bound in Great Britain by CPI Group (UK) Ltd, Croydon, CR0 4YY, UK

Preface

In the past three decades there have been major advances in our understanding of the chemistry and function of nutritional components. This has been enhanced by rapid developments in analytical techniques and instrumentation. Chemists, food scientists and nutritionists are, however, separated by divergent skills, and professional disciplines. Hitherto this transdisciplinary divide has been difficult to bridge.

The series *Food and Nutritional Components in Focus* aims to cover in a single volume the chemistry, analysis, function and effects of single components in the diet or its food matrix. Its aim is to embrace scientific disciplines so that information becomes more meaningful and applicable to health in general.

The series *Food and Nutritional Components in Focus* covers the latest knowledge base and has a structured format.

Caffeine has 4 major sections, namely

Caffeine in Context
Chemistry and Biochemistry
Analysis
Function and Effects

The first section covers caffeine in relation to nutrition and as an ingredient in sugar sweetened beverages. The chemistry and biochemistry section covers basic features, crystal structures, complexes, syntheses of labeled caffeine, and diffusion characteristics in aqueous media. The section on analysis includes liquid chromatography–mass spectrometry, multiple injection capillary electrophoresis, automated flow methods and immunoassay, quantification of self-reported caffeine use, teas and coffees, phenolic compounds and

Food and Nutritional Components in Focus No. 2
Caffeine: Chemistry, Analysis, Function and Effects
Edited by Victor R Preedy
© The Royal Society of Chemistry 2012
Published by the Royal Society of Chemistry, www.rsc.org

xanthines. Finally, the section on function and effects includes caffeine in relation to the brain, cognitive performance, neuroprotection, sleep deprivation, exercise performance, ventilation and pulmonary function, young children, type 2 diabetes, and apoptosis.

Each chapter transcends the intellectual divide with a novel cohort of features namely by containing:

- *Summary points*
- *Key Facts* (areas of focus explained for the lay person)
- *Definitions of words and terms*

It is designed for chemists, food scientists and nutritionists, as well as health care workers and research scientists. Contributions are from leading national and international experts including contributions from world renowned institutions.

Victor R Preedy
King's College London

Contents

Caffeine in Context

Food and Nutritional Components in Focus No. 2
Caffeine: Chemistry, Analysis, Function and Effects
Edited by Victor R Preedy
© The Royal Society of Chemistry 2012
Published by the Royal Society of Chemistry, www.rsc.org

The Chemistry of Caffeine

Analysis

Function and Effects

Caffeine in Context

CHAPTER 1

Caffeine and Nutrition: an Overview

RUBEM CARLOS ARAUJO GUEDES*[1],
MÁRLISON JOSÉ LIMA DE AGUIAR[1] AND
CILENE REJANE RAMOS ALVES-DE-AGUIAR[2]

[1] Departamento de Nutrição, Universidade Federal de Pernambuco, BR-50670901, Recife, PE, Brazil; [2] Departamento de Psicologia, Universidade Federal de Pernambuco, BR-50670901, Recife, PE, Brazil
*E-mail: rguedes@ufpe.br

1.1 Introduction: Caffeine Consumption and its Effects on the Organism

Caffeine is extracted on a large scale from the plant *Coffea arabica* (Figure 1.1 (A)), which originated in Ethiopia and spread to other regions of the world between the twelfth and fifteenth centuries A.D. The daily consumption of caffeine (1,3,7-trimethylxanthine; Figure 1.1 (B)) is common in modern society. In most parts of the world, people largely consume caffeine regardless of age and economic status. The chemical is present in our diet in the form of beverages like coffee (Figure 1.1 (C)), tea, soft drinks and energy drinks, chocolate, and other foods, and also in medicines (Table 1). All of these products are characterized by containing substances from the group of the xanthines (caffeine, theophylline, and theobromine), the most potent of which is caffeine. This substance can be ingested through infusions, medications, or caffeine-laced soft drinks (McKim, 1996) and acts on brain function to produce both positive and negative effects.

The frequent consumption of drugs like caffeine, as well as malnutrition induced by the dietary deficiency of protein, can disrupt the behavioral and electrophysiological organization of brain function. This disruption is especially severe if occurring early in life, during the period of intense brain

Food and Nutritional Components in Focus No. 2
Caffeine: Chemistry, Analysis, Function and Effects
Edited by Victor R Preedy
© The Royal Society of Chemistry 2012
Published by the Royal Society of Chemistry, www.rsc.org

development (Guedes, 2011). The capacity that normal organisms have of focusing attention to optimally perform a task, as well as the production of normal brain electrical activity, are basic neural functions that caffeine and the nutritional status of the organism can influence. The relevance of this theme and the lack of studies on the caffeine–nutrition interface highlight the need for systematic investigation with clinical and experimental approaches. In this chapter, we present experimental results on the behavioral phenomenon designated as "latent inhibition (LI) and on the brain electrophysiological phenomenon known as "cortical spreading depression" (CSD). We demonstrate how LI and CSD can be used in studies, comparing their features in malnourished and well-nourished rats.

Although the relationships between caffeine and normal and abnormal function of the brain are detailed in specific sections of this book, we briefly comment here on the two experimental models mentioned above that we have used to explore the theme "caffeine and nutrition." With the use of the LI behavioral model, it is possible to study the effects of behaviorally active substances like caffeine and to observe disturbances in behavioral parameters (Bakshi *et al*, 1995; Aguiar *et al*, 2011). Recent studies have targeted verifying the effect of caffeine on attention deficit and hyperactivity disorder and suggest that caffeine acts on cognitive performance by improving the state of concentration in novel environments (Caballero *et al*, 2011; Mahoney *et al*, 2011).

Caffeine at moderate doses seems to produce several behavioral effects in humans. Among these, we can highlight (i) increased alertness and reduced fatigue, especially under conditions of little stimulation, such as working at night, and (ii) improved performance on tasks that involve vigilance, or long-lasting responses when the alertness has been reduced (Smith, 2002). Furthermore, caffeine enables users to have control over its consumption; *i.e.,* users seem to use caffeine when they need to benefit from its positive effects on, for instance, mental performance and fatigue (Dagan and Doljansky, 2006; Giesbrecht *et al*, 2010).

The excessive consumption of caffeine can produce negative effects in the organism. Under these conditions, attenuation of the effect of sedatives, increased anxiety, and worsening of the symptoms of anxiety disorders have been reported (Pan and Chen, 2007). The action of caffeine on cognition and memory still requires further investigation; it seems to depend on the type of task used. In activities involving operational skills (operating machinery, car driving), caffeine intake appears to be beneficial. In the execution of complex cognitive tasks (involving intelligence, memory, and learning), data are not yet clear. Studies reveal that the time of day in which these tasks are carried out seems to strongly influence mental performance (*e.g.,* Adan *et al* 2008).

With regard to the effects of caffeine abstinence, abstinence symptoms have been suggested to be related to the uncontrolled and excessive use of compounds containing caffeine. This intake seems to produce some withdrawal symptoms, which include headache, stress, fatigue, decreased alertness, depression and anxiety, and changes in EEG activity (Sigmon *et al* 2009). Recently, Heatherley (2011) has discussed the possibility that improvements in alertness and driving performance after caffeine intake would represent withdrawal reversal rather than a net effect of caffeine. All of these effects may be related to the action of caffeine as a psychostimulant of the central nervous system in humans, in an age-independent way (Mahoney *et al* 2011; McKim, 1996).

Several studies indicate that the action of caffeine on the organism of mammals is also dose-dependent. To produce stimulant effects in rodents, caffeine should be administered in doses ranging from 10 to 40 mg kg^{-1} (Fredholm *et al* 1999; McKim, 1996). In animal models, several behavioral effects of caffeine have been described, including (i) increased spontaneous locomotor activity in the open-field paradigm, (ii) difficulty in the discrimination of stimuli in visual discrimination models, and (iii) decreased perception under negative stimuli, *e.g.*, electric shock through the conditioned avoidance model (Fredholm *et al* 1999; McKim 1996; Mahoney 2011). Other evidence indicates that caffeine could produce an effect known as conditioned place preference, which suggests a reinforcing action of this drug in animals (Brockwell *et al* 1991). However, doses above 40 mg kg^{-1} can produce a state of insomnia, irritability, headache, vertigo, and tinnitus in humans (Fredholm *et al* 1999; McKim 1996). Recent studies have reported that caffeine-containing compounds can elicit chronic energetic effects, mainly by increasing alertness and decreasing sleep in adolescents and children when taken in doses above 50 mg per day (Warzark *et al* 2011).

The most consumed caffeine-containing products are listed in the left column of Table 1, and their content ranges of caffeine are presented on the right column on a "per 100 mL", or "per 100 g" or "per tablet" basis, respectively, in the case of liquids (coffee, tea and energy drinks), solids (chocolate bar), or medication (tablets) products. The sometimes wide range of caffeine content depends on variations in analytical methods, in the varieties of the caffeine-source plants, and the methods of product preparation (mainly in the case of coffee and tea beverages preparation). For more details, the reader is advised to see the reports by Barone and Roberts (1996), and Stavric et al,

Table 1.1 Quantities of caffeine found in most consumed products.*

Product	Quantity of caffeine
Coffee	11 to 250 mg 100 mL^{-1}
Tea	2 to 540 mg 100 mL^{-1}
Chocolate bar	21 to 76 mg 100 g^{-1}
Soft drinks	10 to 20 mg 100 mL^{-1}
Energy drinks	30 to 112 mg 100 mL^{-1}
Medication tablets	30 to 200 mg per tablet

*adapted from the site http://www.mayoclinic.com/health/caffeine/AN01211

Figure 1.1   Origin, chemical structure, and consumption of caffeine. (A) The coffee
plant, *Coffea arabica*, with some green seeds that will mature and be
roasted and ground to generate the coffee powder used to prepare the
coffee infusion, which is rich in caffeine. (B) The chemical structure of the
caffeine molecule, 1,3,7-trimethylxanthine. (C) Consumption of small
amounts of coffee infusion is believed to improve attention, and drinking
coffee has become a pleasant social habit in almost all cultures. In this
picture, one of the authors (RCAG) is drinking a small cup of coffee
during a break from writing this text (unpublished pictures from the
authors' lab).

(1988). For recommendations about consumption of caffeine-containing
energy drinks, see Higgins et al (2010).

1.2   Pharmacodynamics and Pharmacokinetics of Caffeine

After ingestion, caffeine is rapidly absorbed in the gastrointestinal tract in both
humans and in laboratory animals (Magkos and Kavouras 2005). When plasma
caffeine concentrations after caffeine oral intake are compared with those after
intravenous administration, they vary over time with comparable features,
suggesting similar pharmacokinetics in the two routes of administration (Arnaud
1993). Concerning its action on the central nervous system, caffeine seems to pass
freely from blood to brain and also exerts a dose-dependent, protective effect
against the blood–brain barrier disruption found in degenerative diseases of the
central nervous system, like Alzheimer's disease and Parkinson's disease (Chen
et al 2010). The stimulant effects of caffeine on the central nervous system are
primarily the result of its role in the blockade of adenosine receptors (mostly on
the high-affinity A_1 and A_{2A} receptors), the inhibition of cAMP phosphodiester-

ase activity, the intracellular mobilization of calcium, and the binding to benzodiazepine receptors (Chen *et al* 2010; Lee and Chung 2010).

When administered orally or intravenously in humans or rats, caffeine acts with reinforcing properties in response to electrophysiological and behavioral stimulation (Griffiths and Mumford 1995). However, the bitter taste of caffeine can interfere with administration, especially in behavioral studies, as experiments in rats have shown that the animals freely ingest only low doses of caffeine (Heppner *et al* 1986). Thus, the immediate consequence of consuming caffeine orally would be a trend to generate aversion (because of its bitter taste). Other studies show that caffeine intake in the form of medication can cause nausea and gastric irritation, especially in children (*e.g.*, McKim, 1996). Furthermore, caffeine can interfere with a subsequent reinforcing effect in animal models of oral self-administration. Similar findings have been observed in humans; the reinforcing effect of caffeine varies with dose, *i.e.*, low and medium doses maintain the behavior of oral self-administration and high doses can even produce aversion (Griffiths and Mumford 1995). Other routes, while less explored, may also be used for caffeine administration, such as intramuscular injection and application in the form of suppositories.

Caffeine is absorbed primarily by the stomach and intestine (absorption reaches 99% in the intestinal tract; see Fredholm *et al.*, 1999, for details), with higher affinity for lipids compared to water; the peak plasma of caffeine occurs 15–45 minutes after ingestion, and the half-life is approximately 5–6 hours (Smith 2002). Caffeine is distributed through the blood, with 10 to 30 percent being transported by proteins (McKim 1996). In addition to its relative facility in passing the blood–brain barrier, caffeine has a lipophilic character and, for this reason, is also found in all organs and can be present in breast milk (Arnaud 1993; Fredholm *et al* 1999; McKim 1996). The excretion of the products of caffeine metabolism is performed by the liver (Fredholm *et al* 1999), and only 2% is excreted by the kidneys without being metabolized (Arnaud 2011).

The selective antagonism of A_1 and A_{2A} receptors that is involved in the mechanism of action of caffeine has been the focus of recent investigations regarding behavioral outcomes (Randall *et al* 2011). Authors have shown that increased locomotor activity and high frequency of lever pressing in rats result from the caffeine stimulating effect mediated by blockade of A_{2A} receptors, whereas the anxiogenic effect is probably related to action on the A_1 receptor antagonist. However, additional investigation is required to better clarify the specific action of adenosine receptors on behavioral alterations.

1.3 Caffeine–Nutrition Interaction: Effects on Physiological Processes

In this part of the chapter, we address the possible interaction between caffeine and changes in the nutritional status of the organism. This discussion is based on data from our lab on an experimental model of attention designated as LI, and on an electrophysiological phenomenon designated as CSD as described in the following text.

1.4 Caffeine–Nutrition Interaction: Latent Inhibition Model of Attention

Attention is defined as a basic psychological process that is the focus of our consciousness in only a limited aspect of the totality of our experience. According to the cognitive interpretation, attention is seen as a "filtering" process that controls the afferent passage of information through the sensory system. In behavioral analysis, however, the complex information processing required for simple acts of perception, language, and thought involves selection of relevant stimuli, so that the behavior becomes controlled by a narrow range of stimuli that are related to important consequences. Therefore, deficient filtering of information from the outside world results in an inability to ignore irrelevant stimuli. Because of these characteristics, attention is considered a selective process (Alves *et al* 2002). An animal model of simulation of a cognitive process that uses the manipulation of environmental variables for the behavioral study of selective attention is the LI model. In LI, the animal is first repeatedly exposed to a neutral stimulus, without consequence, and thereafter is submitted to a conditioning paradigm in which the neutral stimulus now functions as a conditioned stimulus. The pre-exposure to this conditioned stimulus impairs the subsequent conditioning (Aguiar *et al* 2011).

In our experiments, LI was assessed using a conditioned taste-aversion paradigm. Briefly, we measured the effectiveness of sucrose pre-exposure in preventing subsequent acquisition of a conditioned taste aversion to sucrose. The LI procedure consisted of three phases, conducted on five successive days, at the same time of the day during the morning. During phase 1 (pre-exposure to sucrose, days 1 to 3), animals were individually placed in the experimental chamber (Figure 1.2) and given access to either 50 mL of a 5% sucrose solution (sucrose pre-exposed group, PE) or 50 mL of tap water (non–pre-exposed group, NPE) for 30 minutes. On day 4 (phase 2; "conditioning" day), all animals were given access to 50 mL of a 5% sucrose solution for 30 minutes, immediately followed by an intraperitoneal injection of LiCl (50 mg kg^{-1} in 1 mL kg^{-1} distilled water) as an aversive stimulus to produce the conditioned taste aversion. Finally, on day 5 (phase 3; testing day), each animal was placed again in the experimental chamber and given simultaneous access to both 5% sucrose and water for 30 minutes. In PE and NPE animals, LI, expressed as the sucrose suppression ratio (SSR), was assessed by comparing the amount of sucrose *versus* water consumed on day 5, according to the following formula: SSR = [mL sucrose consumed/(mL sucrose consumed + mL water consumed)]. Both caffeine and saline rats were subdivided into PE and NPE groups, and these substances were administered to the respective groups 20 minutes prior to the pre-exposure and conditioning phases.

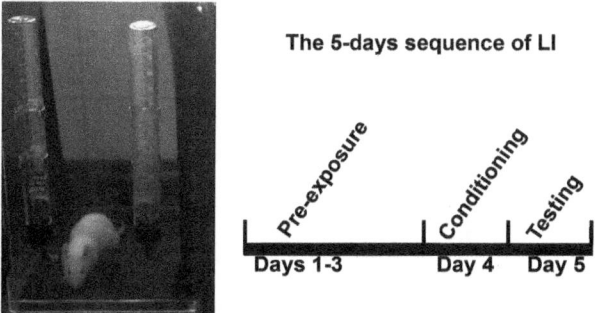

Figure 1.2 The 5-day sequence of latent inhibition (LI). Left, the experimental chamber to run the latent inhibition (LI) procedure, in a conditioned taste-aversion paradigm. Right, the sequence of the three phases of the LI experiment, described as follows: in *phase 1* (pre-exposure to sucrose, days 1 to 3), animals were placed in this chamber and had access to either a 5% sucrose solution (sucrose pre-exposed group, PE) or tap water (non-pre-exposed group, NPE) for 30 min. On *phase 2* ("conditioning"; day 4), all animals were given access to the sucrose solution for 30 min, immediately followed by an intraperitoneal injection of LiCl (50 mg kg^{-1}) as an aversive stimulus to produce the conditioned taste aversion. Finally, on *phase 3* ("testing"; day 5), each animal had simultaneous access to both sucrose and water for 30 min. LI, expressed as the "sucrose suppression ratio" (SSR), was assessed by comparing the amount of sucrose *versus* water consumed on day 5, according to the following formula: SSR = [mL sucrose consumed/(mL sucrose consumed + mL water consumed)] (unpublished picture from the authors' lab).

In well-nourished rats, previous experiments by others showed that the administration of 10 mg kg^{-1} caffeine could not modify selective attention, as evaluated with the LI model (Bakshi *et al* 1995). We recently performed a dose-response curve analysis for caffeine and showed that the caffeine effect depends on the dose employed. A dose of 30 mg kg^{-1}, but not treatment with 15 mg kg^{-1} caffeine, was effective in modulating LI (Aguiar *et al* 2011; see also our unpublished data in Figure 1.3). The results can be explained by the central action of caffeine particularly on the adenosine A$_{2A}$ and dopamine D$_2$ receptors (Fredholm 1999; McKim 1996). The effects of caffeine on selective attention are suggested to depend on the increased dopamine D$_2$ transmission following the inhibition of adenosine A$_{2A}$ receptors by caffeine (Aguiar *et al* 2011). Other studies also support the involvement of dopaminergic activity in the mechanisms of attention. D$_2$-antagonistic drugs that are effective in the treatment of psychotic illnesses like schizophrenia also improve selective attention, which is impaired in this disease, and have been tested in the LI model (Weiner and Arad 2009). It is worth mentioning that the biological basis of schizophrenia involves selective attention impairment (Moser *et al* 2000), but no one knows if, and how, nutritional disturbances would modulate it. According to these authors, because of its characteristics, LI is a sensitive and reliable model for the study of selective attention and is considered an

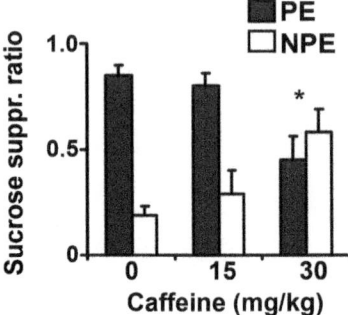

Figure 1.3   The dose-dependent effect of caffeine on LI in well-nourished rats. In the
absence of caffeine (left pair of bars), the rats non–pre-exposed to sucrose
(NPE; white bar) presented a suppression of sucrose drinking, as
compared with the pre-exposed group (PE; black bar). Caffeine
antagonized this response when injected at a dose of 30 mg kg^{-1} (right
pair of bars), but not at 15 mg kg^{-1} (middle pair of bars). Data are
expressed as mean $\pm$ standard error of the mean. The asterisk indicates
that the 30 mg kg^{-1} group is significantly different from the other two
groups (ANOVA plus Holm-Sidak test) (unpublished data from the
authors' lab).

interesting animal model of schizophrenia. In fact, studies in humans and
laboratory animals show that dopaminergic agents can disrupt attention as
measured by LI (Alves *et al* 2002; Moser *et al* 2000; Schimidt-Hansen *et al*

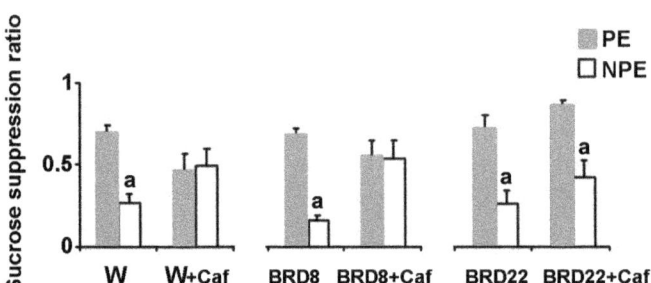

Figure 1.4   The caffeine–nutrition interaction in LI. The left and middle panels of this
figure show that caffeine antagonized LI in both well-nourished and
early-malnourished rats (fed respectively a commercial chow diet with
22% protein and a protein-deficient diet with 8% protein mostly from
vegetable sources). This effect of caffeine on LI was not seen in another
group of malnourished rats (right panel), fed the deficient diet with
protein content (but not protein quality) that had been increased to 22%
by augmenting the proportion of foodstuffs of vegetable origin. This
result suggests that the amino acid imbalance of this diet may modulate
the caffeine effects on latent inhibition. Data are expressed as mean $\pm$
standard error of the mean. The lower-case letter (a) indicates NPE values
for SSR that are significantly different from the corresponding PE values.
Data are from a previous publication of the authors (Aguiar *et al* 2011).

2009). These data collectively support the hypothesis that caffeine acts on LI *via* the dopaminergic system, suggesting amphetamine-like effects for caffeine (Fredholm *et al* 1999). Data also support the speculation that the caffeine-induced modulation of cerebral dopamine receptors could be a very reasonable explanation, at least in part, for caffeine effects on selective attention (Aguiar *et al* 2011).

We have recently shown that caffeine antagonizes LI in both well-nourished and early-malnourished rats (fed respectively a commercial chow diet with 22% protein and a protein-deficient diet with 8% protein, mostly from vegetable sources). This effect of caffeine on LI was not seen in another group of malnourished rats fed the deficient diet with protein content (but not the protein quality) that had been increased to 22% by increasing the proportion of foodstuffs of vegetable origin (Figure 1.4). These findings suggest that the amino acid imbalance of this diet may modulate caffeine effects on latent inhibition (Aguiar *et al* 2011).

1.5   Caffeine–Nutrition Interaction: Effects on Cortical Spreading Depression

The numerous functions of the brain are produced directly or indirectly as the consequence of the electrical activity generated by brain cells. Recording spontaneous brain electrical activity, both in humans and in laboratory animals, constitutes an interesting way of obtaining important information about brain functioning. One of the most commonly used techniques for this goal is the EEG recording, which consists of measuring the differences in electrical potential between two electrodes located on different regions of the scalp. As a noninvasive technique, EEG is largely employed as a useful approach that aids in diagnosis of several neurological disorders (Guedes 2011).

Considering that caffeine can influence brain electrical activity (Tchekalarova *et al* 2010a) and seizure susceptibility (Tchekalarova *et al* 2010b), we investigated the effect of both caffeine and nutritional deficiency on the electrophysiological phenomenon known as CSD. CSD has been electrophysiologically characterized as a reversible and slowly propagating "wave" of reduction of cortical electrical activity. Measuring the time spent by a CSD wave in crossing the distance between two recording electrodes enables estimation of the CSD velocity of propagation along the cortical tissue (Figure 1.5). Thus, this approach is a reasonable and simple method for estimating brain CSD susceptibility under clinically relevant conditions known to influence brain excitability (see Guedes 2011, for details). The velocity with which CSD propagates in the cortical tissue is much lower (in the order of mm min^{-1}) than the propagation velocity of neuronal action potentials (in the order of meters s^{-1}). Classical migraine, brain ischemia, and epilepsy are three important human diseases that are currently postulated to have some relationship to CSD in terms of common mechanisms (Guedes 2011).

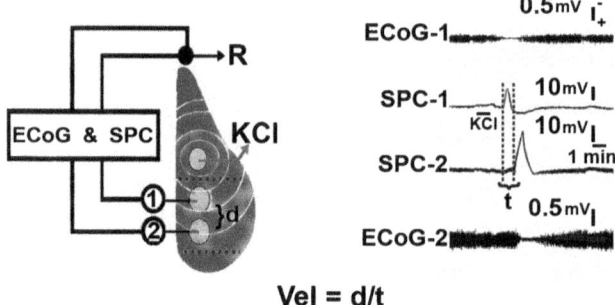

$$Vel = d/t$$

Figure 1.5   The description of cortical spreading depression (CSD). The left panel represents an upper view of the right cerebral hemisphere of the rat showing the CSD eliciting location (where KCl was applied) and the recording points (1 and 2) where the recording electrodes were placed, separated by the distance, **d**. After 1min KCl stimulation, CSD was elicited, propagated concentrically, and the EEG depression and the slow potential change (SPC) typical of CSD were recorded at the remote points 1 and 2, against a common reference electrode (R) placed on the nasal bones (recordings shown in the right panel). The CSD velocity of propagation is calculated based on the time, **t** (in the right panel) spent by the CSD wave to pass the interelectrode distance, **d** (in the left panel). The negative-up polarity indicated in the upper trace of the right panel applies also to the other three traces (unpublished picture from the authors' lab).

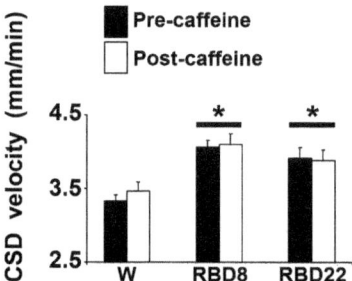

Figure 1.6   CSD velocity in caffeine-treated well-nourished and malnourished rats. After 1–2 h of CSD recording (in which we measured the "baseline" CSD velocity), we injected intraperitoneally 30 mg kg^{-1} caffeine and the CSD recording continued for more 2 h. We evaluated the effects on CSD propagation by comparing, in the same animal, the CSD velocities before and after caffeine. The post-caffeine CSD velocities were comparable to the pre-caffeine values, indicating no significant effect of the acute administration of this drug on CSD propagation. However, ANOVA (plus the Holm-Sidak test) confirmed the previously described (Guedes 2011) significant effect of nutrition: the asterisks indicate that malnutrition increased CSD velocities as compared to the well-nourished controls ($F_{[2, 31]} = 14.439$; $P<0.001$). Data are expressed as mean $\pm$ standard error of the mean. Data are from the authors' previous publication (Aguiar *et al* 2011).

We subjected both well-nourished and early-malnourished adult rats to the electrophysiological recording of CSD. The application of a small cotton ball (1–2 mm in diameter) soaked in 2% KCl (approximately 270 mM) for 1 min to a point of the frontal cortical surface was a very effective stimulus in eliciting one CSD episode, which propagated and was recorded at two parietal recording points. During the recording session, we injected intraperitoneally 30 mg kg^{-1} caffeine and evaluated the effects on CSD propagation by comparing, in the same animal, the CSD velocities before and after caffeine. The post-caffeine CSD velocities were comparable to the pre-caffeine values (Figure 1.6), indicating no significant effect of the acute administration of this drug on CSD propagation. However, ANOVA confirmed the previously described (Guedes 2011) significant effect of nutrition, indicating that malnutrition increased CSD velocities, as compared to the well-nourished controls (F [2, 31] = 14.439; P<0.001).

1.6 Concluding Remarks

An imbalance of some nutrients such as proteins in the diet or consumption of psychostimulant drugs such as caffeine can cause changes in behavioral parameters such as mood, perception, motivation, and attention, as well as modifications in brain electrophysiological features like EEG wave patterns and the brain's susceptibility to generating and propagating epileptic electrical activity. In this chapter, we described evaluations of the widely accepted idea that caffeine can influence behavioral and electrophysiological properties in the brain and extended the discussion to assessment of the interaction between caffeine and nutritional deficiency. Our recent experimental results (Aguiar *et al* 2011), presented and discussed here, constitute some evidence for a novel impairing behavioral effect of caffeine on the phenomenon designated as LI (latent inhibition), a process considered to be related to attention. In addition, we also demonstrated a failure of caffeine to affect the electrophysiological phenomenon known as CSD (cortical spreading depression). The type of nutritional imbalance early in life seems to influence the caffeine effect on LI, as suggested by the differential effects resulting from the two protein-deficient diets (see Figure 1.4). The mechanism involved in at least some of the caffeine-induced brain effects is probably the interaction between this drug and adenosine A_{2A} and dopamine D_2 receptors (Fredholm *et al* 1999). In fact, studies on LI demonstrate that dopaminergic agents can harm attention (Alves *et al* 2002; Bakshi *et al* 1995). Furthermore, caffeine increases rotation behavior following dopamine receptor activation, and the effects of caffeine on LI are similar to those of amphetamine (Fredholm *et al* 1999). Thus, data collectively support our speculation that caffeine-induced modulation of dopaminergic receptors is involved, at least in part, in the effects on LI.

By comparing the effect of caffeine on LI and the absence of effect on CSD, we suggest that the action of caffeine on the brain, by being more selective than the actions of malnutrition, more effectively influences the dopaminergic processes of *subcortical* nuclei involved in the LI phenomenon, in comparison to the cortical processes underlying CSD. This suggestion receives support from our own data documenting a lack of action of dopaminergic drugs on CSD in the rat brain (see the final part of the discussion in Aguiar *et al.*, 2011). The suggestion of a dopamine-based modulation of caffeine at the subcortical but not at the cortical level deserves further investigation.

Finally, we must comment on the complex task of extrapolating experimental data from rats to humans. This task has obvious limitations and requires considerable caution. Despite such pitfalls, the main overall conclusion from the converging pieces of evidence presented here is that the development of experimental models devoted to investigating this theme is highly desirable and is surely valid in the case of behavioral and electrophysiological models. In this context, we are convinced that the LI and CSD models are valuable for understanding the relationship among caffeine consumption, nutrition, and development and function of the organism. The analysis of LI and CSD features under conditions varying from moderate–eventual to intense–habitual caffeine intake, as well as under nutritional deficiency, constitutes an interesting experimental approach for investigating the neurophysiological effects of nutritional and non-nutritional conditions. An understanding of the mechanisms by which such conditions affect the phenomena of LI and–or CSD may be useful in the development of more effective treatment strategies for diseases presumably associated with those phenomena, such as attention–disturbance-related diseases in the case of LI and epilepsy and migraine in the case of CSD.

Summary Points

- The daily consumption of caffeine can produce positive or negative impact on physiologic processes in normal human beings, and this constitutes a worldwide medical- and public health matter of concern.
- In the present chapter, data from experiments in the rat on the effect of caffeine on one behavioral and one electrophysiological phenomenon, designated respectively as latent inhibition [LI] and cortical spreading depression [CSD], have been discussed.
- Considering that malnutrition early in life can disrupt the organization of the developing organism, we investigated and discussed the interaction between caffeine and nutritional deficiency on the LI and CSD phenomena.
- The administration of caffeine during LI conditioning impaired selective attention in rats in a dose-dependent way, with the dose of 30 mg kg^{-1} caffeine being effective, but not the treatment with 15 mg kg^{-1}.
- Caffeine antagonized LI in both well-nourished and early-malnourished rats, respectively fed a commercial chow diet with 22% protein and a qualitatively and quantitatively protein-deficient diet (with 8% protein mostly of vegetable sources).

- In the deficient diet condition, the increase of the protein quantity (but not of its quality) abolished the LI effect, suggesting that the amino acid imbalance of this diet may modulate the caffeine effects on LI.
- We evaluated the effects on CSD propagation by comparing, in the same animal, the CSD velocities before- and after 30 mg kg^{-1} caffeine. In contrast to the LI findings, the post-caffeine CSD velocities were comparable to the pre-caffeine values, indicating no significant effect of the acute administration of this drug on CSD propagation.
- In malnourished rats, analysis of variance (ANOVA) confirmed the previously described significant effect of nutrition, indicating that malnutrition increased CSD velocities, as compared to the well-nourished controls.
- Our results constitute evidence for a novel impairing behavioral effect of caffeine on the LI phenomenon, which is considered a process related to attention. In addition, we also demonstrated a failure of caffeine in affecting the CSD electrophysiological phenomenon.
- We suggest that caffeine may modulate dopaminergic subcortical receptors participating in attention processes, but does not interact at the cortical level, in a way that would affect cortical spreading depression.

Key Facts of Latent Inhibition (LI)

Aspect of LI	Fact	Highlighting point
Exposure of the animal to a neutral stimulus *before* a conditioning procedure	The neutral stimulus to which the animal was *pre-exposed* (PE) will be used later, in the same animal, as a conditioned stimulus	Compared to non–pre-exposed animals, the PE subject will display *poor conditioning*
Pre-exposed stimulus needed to produce LI	Usually a *neutral* stimulus	The stimulus is initially *irrelevant* for the animal
Brain *mechanisms* involved in LI	Evidence points to important involvement of *dopaminergic* neurons in LI	Dopaminergic *drugs* can affect the LI-phenomenon
LI involvement in brain *physiological* processes	Evidence indicates that the LI phenomenon is related to processes determining *selectivity* of attention	LI represents an output of a *filtering attention* process that implies in learning to ignore irrelevant stimuli
LI and brain functional disorders	Attentional processes that produces LI in normal subjects are dysfunctional in *schizophrenia* patients	LI production in animals has been considered an experimental model of schizophrenia

Key Facts of Cortical Spreading Depression (CSD)

Aspect of CSD	Fact	Highlighting point
Cortical cell organization	CSD needs a certain number of neural cell bodies to occur	CSD is a "*social*" or "*cooperative*" cellular phenomenon in the neural tissue
Stimulus needed to elicit CSD	Any kind of rapid energy variation (chemical, electrical, mechanical, *etc.*) applied to the tissue for a certain time	CSD can be elicited by several, *nonspecific* stimuli
Tissue recovery from CSD	The electrical activity of the depressed region recovers in 5–10 min after depression	CSD is a completely *reversible* phenomenon
Functionally distinct regions of the brain	CSD apparently propagates in the same way from a motor to a sensory region, and *vice versa*	Brain functional limits do not influence CSD propagation
Relation of CSD to Phylogenesis	CSD has been observed in *all* animal species so far studied, from fishes to mammals, including the *human* species	CSD seems to be a very *general phenomenon* of the nervous system
CSD velocity of propagation	In the order of a few mm min^{-1}	In contrast to action potentials, CSD is a paradoxically much slower propagating phenomenon of the nervous tissue
Is CSD a physiological or a pathological phenomenon?	Some CSD features speak in favor of a brain protecting role, but there is also evidence linking CSD to certain neurological diseases	Understanding the underlying CSD mechanisms will help in knowing brain functioning under normal and–or under some neurological disorders

This table was adapted from the "Table 148.1" entitled "Key features of relevant aspects of CSD" recently published by one of the present authors [see Guedes, 2011].

Definitions and Explanations of Key Terms

Latent Inhibition (LI): is an experimental model in which the animal is first repeatedly exposed to a neutral stimulus, and thereafter submitted to a conditioning paradigm in which the neutral stimulus now functions as a conditioned stimulus. The pre-exposure to this conditioned stimulus impairs the subsequent conditioning. In our experiments, LI was assessed using a conditioned taste-aversion paradigm, in which the pre-exposure to sucrose was the neutral stimulus.

Electroencephalogram (EEG): is the recording of the spontaneous brain electrical activity that all living mammals present, even when they sleep. This can be electronically recorded on chart-paper, ink-writer machines, or digitally in computer-based devices. EEG is currently used as a helpful diagnostic test in the neurological clinic, aiding in diagnoses of certain diseases like epilepsy.

Epilepsy: is a neurological disease resulting from an exacerbated and uncontrolled activity of a certain brain neuronal population. The main clinical signs of a generalized epileptic seizure are loss of consciousness and uncontrolled muscle contractions (motor seizure).

Migraine: is a neurological disorder characterized by chronic episodes of headaches that in the most severe cases can cause intense pain that lasts for hours or, infrequently, even days. In a particular type of migraine, the "migraine with aura", headache is preceded by visual symptoms (denominated "aura") that include the sensation of seeing multiple light spots, geometric designs of very simple form, zigzag lines and flashes of light, that slowly move across the visual field.

Cortical Spreading Depression (CSD): is a reversible electrophysiological brain response provoked by adequate stimulation of a point of the brain tissue. It consists of reduction ("depression") of the EEG electrical activity that concentrically and slowly propagates from the stimulated cortical point to remote areas. It is a totally reversible response: the depressed EEG activity returns to the normal pattern in a few minutes. Current knowledge of the phenomenon points to a link between CSD and human neurological diseases like epilepsy, migraine, and brain ischemia.

Benzodiazepine receptors: are structures (probably of protein nature) in the neural cell membrane that favors or promotes the action of the inhibitory neurotransmitter gamma-amino butyric acid (GABA). The reduction in the activity of this brain inhibitory process may predispose the individual to epileptic seizures.

Caffeine half-life: is defined as the time necessary for the plasma caffeine level to fall to 50% of its initial concentration.

Blood-brain barrier: is defined as a functional barrier existing at the interface between the walls of the blood vessels irrigating the central nervous system and the extra cellular space of the nervous tissue. Experimental evidence demonstrates that this functional barrier controls the otherwise deleterious passage of certain substances from the blood to the brain.

Attention deficit and hyperactivity disorder: is a neurobehavioral disorder that is clinically characterized by attention problems and hyperactivity coexisting in the same patient. It is possible, albeit infrequent, that each behavior occurs alone. The symptoms usually start before seven years of age.

Cognitive tasks: are those tasks that a normal person can perform employing processes of memory and learning, demonstrating a normal capacity of using intelligence. For more details, see Chapter 15 *Caffeine and Cognitive Performance*, by Prof. A. Adan.

Open field paradigm: is designed to measure qualitative and quantitative aspects of general locomotor exploring activity in rodents. It consists of a circular or squared box with surrounding walls high enough to prevent escape. The floor is divided into squares. The number of squares crossed in a certain time interval is an index of locomotor activity. The modern apparatuses have infrared beams that can automatically quantify the animal's locomotor activity. The positive or negative effects of pharmacological agents can be assessed by changes in the open field measurements.

Abbreviations

cAMP	Cyclic adenosine monophosphate
ANOVA	Analysis of variance
CSD	Cortical spreading depression
EEG	Electroencephalogram
LI	Latent inhibition
NPE	Non–pre-exposure (to sucrose, in the LI paradigm)
PE	Pre-exposure (to sucrose, in the LI paradigm)
SSR	Sucrose suppression ratio

Acknowledgements

The authors thank the Brazilian agencies CAPES (Procad/2007), CNPq (INCT de Neurociencia Translacional–No. 573604/2008-8), MS/SCTIE/DECIT (No. 17/2006), Facepe (APQ0975-4.05/08), and IBN-Net/Finep (No. 4191) for financial support. R.C.A. Guedes is a Research Fellow from CNPq (No. 301190/2010-0).

References

Adan, A., Prat, G., Fabbri, M. and Sànchez-Turet, M., 2008. Early effects of caffeinated and decaffeinated coffee on subjective state and gender differences. Progress in Neuro-Psychopharmacology & Biological Psychiatry. 32: 1698–1703.

Aguiar, M. J. L., Alves-de-Aguiar, C. R. R. and Guedes, R. C. A., 2011. Caffeine/nutrition interaction in rat brain: Influence of latent inhibition and cortical spreading depression. European Journal of Pharmacology. 650: 268–274.

Alves, C. R. R., DeLucia, R, and Silva, M. T. A., 2002. Effects of fencamfamine on latent inhibition. Progress of Neuro-Psychopharmacology Biology and Psychiatry. 26: 1089–1093.

Arnaud, M. J., 1993. Metabolism of caffeine and other components of coffee. In: Garattini, S. (ed.) Caffeine, Coffee and Health. Raven Press, New York, pp. 43–95.

Arnaud, M. J., 2011. Pharmacokinetics and metabolism of natural methylxanthines in animal and man. In: Fredholm, B. B. (ed.), Methylxanthines. Handbook of Experimental Pharmacology 200. Springer, Berlin, pp. 33–91.

Bakshi, V. P., Geyer, M. A., Taaid, N. and Swerdlow, N. R., 1995. A comparison of the effects of amphetamine, strychnine and caffeine on prepulse inhibition and latent inhibition. Behavioural Pharmacology. 6: 801–809.

Barone, J. J. and Roberts, H. R., 1996. Caffeine consumption. Food Chemistry and Toxicology. 34: 119–129.

Brockwell, N. T., Eikelboom, R. and Beninger, R. J., 1991. Caffeine-induced place and taste conditioning: production of dose-dependent preference and aversion. Pharmacology, Biochemistry and Behavior. 38: 513–517.

Caballero, M., Núñez, F., Ahern, S., Cuffí, M.L., Carbonell, L., Sánchez, S., Fernández-Dueñas, V. and Ciruela, F., 2011. Caffeine improves attention deficit in neonatal 6-OHDA lesioned rats, an animal model of attention deficit hyperactivity disorder (ADHD). Neuroscience Letters. 494: 44–48.

Chen, J. F., Yu, L., Shen, H-Y, He, J-C, Wang, X. and Zheng, R., 2010. What knock-out animals tell us about the effects of caffeine. Journal of Alzheimer's Disease. 20: S17–S24.

Dagan, Y. and Doljansky, J. T., 2006. Cognitive performance during sustained wakefulness: A low dose of caffeine is equally effective as modafinil in alleviating the nocturnal decline. Chronobiology International. 23: 973–983.

Fredholm, B. B., Bättig, K., Holmén, J., Nehlig, A. and Zvartau, E. E., 1999. Actions of caffeine in the brain with special reference to factors that contribute to its widespread use. Pharmacological Reviews. 51: 83–133.

Giesbrecht, T., Rycroft, J. A., Rowson, M. J. and De Bruin, E. A., 2010. The combination of L-theanine and caffeine improves cognitive performance and increases subjective alertness. Nutritional Neurosciences. 13: 283–290.

Griffiths, R. R. and Mumford, G. K., 1995. Caffeine-a drug of abuse? In: Bloom, F. E. and Kupfer, D. J. (ed.). Psychopharmacology: The Fourth Generation of Progress. Raven Press, New York, pp. 1699–1713.

Guedes, R. C. A., 2011. Cortical spreading depression: a model for studying brain consequences of malnutrition. In: Preedy, V. R., Watson, R. R. and Martin, C. R. (ed.). Handbook of Behavior, Food and Nutrition, Springer, Berlin, pp. 2343–2355.

Heatherley, S. V. 2011. Caffeine withdrawal, sleepiness, and driving performance: What does the research really tell us? Nutritional Neuroscience. 14: 89–95.

Heppner, C. C., Kemble, E. D. and Cox, W. N., 1986. Effects of blood deprivation on caffeine consumption in male and female rats. Pharmacology Biochemistry and Behavior. 24: 1555–1559.

Higgins, J. P., Tuttle, T. D., Higgins, C. L., 2010. Energy Beverages: Content and Safety. Mayo Clinic Proceedings. 85:1033–1041

Lee, S. W. and Chung, S. S., 2010. A review of the effects of vitamins and other dietary supplements on seizure activity. Epilepsy & Behavior. 18: 139–150.

Magkos, F. and Kavouras, S. A., 2005. Caffeine Use in Sports, Pharmacokinetics in Man, and Cellular Mechanisms of Action. Critical Reviews in Food Science and Nutrition. 45: 535–562.

Mahoney, C. R., Brunyé, T. T., Giles, G., Lieberman, H. R. and Taylor, H. A., 2011. Caffeine-induced physiological arousal accentuates global processing biases. Pharmacology, Biochemistry and Behavior. 99: 59–65.

Mckim, W. A. 1996. Caffeine and the methylxanthines. In: Mckim, W. A. (ed.) Drugs and Behavior: an introduction to behavioral pharmacology. Prentice Hall, New Jersey, pp 191–210.

Moser, P. C., Hitchcock, J. M., Lister, S. and Moran, P. M., 2000. The pharmacology of latent inhibition as an animal model of schizophrenia. Brain Research Reviews. 33: 275–307.

Pan, H. Z. and Chen, H. H., 2007. Hyperalgesia, low-anxiety, and impairment of avoidance learning in neonatal caffeine-treated rats. Psychopharmacology. 191:119–125.

Randall, P. A., Nunes, E. J., Janniere, S. L., Stopper, C. M., Farrar, A. M., Sager, T. N., Baqi, Y., Hockemeyer, J., Müller, C. E. and Salamone, J. D., 2011. Stimulant effects of adenosine antagonists on operant behavior: differential actions of selective A_{2A} and A_1 antagonists. Psychopharmacology (in press).

Schmidt-Hansen, M., Killcross, A. S. and Honey, R. C., 2009. Latent inhibition, learned irrelevance, and schizotypy: assessing their relationship. Cognitive Neuropsychiatry. 14: 11–29.

Sigmon, S. C., Herning, R. I., Better, W., Cadet, J. L. and Griffiths, R. R., 2009. Caffeine withdrawal, acute effects, tolerance, and absence of net beneficial effects of chronic administration: cerebral blood flow velocity, quantitative EEG, and subjective effects. Psychopharmacology. 204: 573–585.

Smith, A., 2002. Effects of caffeine on human behavior. Food and Chemical Toxicology. 40: 1243–1255.

Stavric, B., Klassen, R., Watkinson, B., Karpinski, K., Stapley, R. and Fried, P., 1988. Variability in caffeine consumption from coffee and tea: possible significance for epidemiological studies. Food and Chemical Toxicology. 26: 111–118.

Tchekalarova, J., Kubová, H. and Mareš, P., 2010a. Postnatal period of caffeine treatment and time of testing modulate the effect of acute caffeine on cortical epileptic afterdischarges in rats. Brain Research. 1356: 121–129.

Tchekalarova, J., Kubová, H. and Mareš, P., 2010b. Effects of early postnatal caffeine exposure on seizure susceptibility of rats are age- and model-dependent. Epilepsy Research. 88: 231—238.

Warzak, W. J., Evans. S, Floress, M. T., Gross, A. C. and Stoolman, S., 2011. Caffeine consumption in young children. Journal of Pediatrics. 158: 508–509.

Weiner, I. and Arad, M., 2009. Using the pharmacology of latent inhibition to model domains of pathology in schizophrenia and their treatment. Behavioural Brain Research. 204: 369–386.

CHAPTER 2

Caffeine as an Ingredient in Sugar Sweetened Beverages

LYNN J. RIDDELL[1], DHOUNGSIRI SAYOMPARK[1,2], PENNY OLIVER[1] AND RUSSELL S. J. KEAST[1]

[1] Centre for Physical Activity and Nutrition, Sensory Science Group, Deakin University, Burwood, Victoria, Australia; [2] Faculty of Science and Technology, Rajamangala University of Technology Tawan-ok, Chonburi, Thailand
*E-mail: russell.keast@deakin.edu.au

2.1 Introduction: Caffeine Consumption Patterns

Caffeine is the most widely used psychoactive drug in the world, with more than 80% of the US population classed as regular consumers (Garrett and Griffiths 1998). An analysis of the Continuing Survey of Food Intakes by Individuals (CSFII) in the US indicates that 87% of US population over 2 years of age consumed caffeine daily and the average intake in caffeine consumers was 193 mg per day or 1.2 mg kg^{-1} per day (Frary *et al* 2005). SSB were the primary source of caffeine in children and adolescents under 18 years of age and provided between 50–64% of the daily caffeine intake. For adults 18–34 years, SSB provided 30% of total daily caffeine, dropping to 11% for adults 34 years and older (Frary *et al* 2005). The total daily intake of caffeine observed in the CSFII is slightly lower that than observed in the 1995 National Nutrition Survey of Australian adults who reported consuming on average 270 mg of caffeine per day. Caffeine intakes amongst children, aged 2 to 14 years,

Food and Nutritional Components in Focus No. 2
Caffeine: Chemistry, Analysis, Function and Effects
Edited by Victor R Preedy
© The Royal Society of Chemistry 2012
Published by the Royal Society of Chemistry, www.rsc.org

were reported as 17 mg per day. It is suggested that cola flavored SSB provide around 62% of this intake (Desbrow *et al* 2004).

Is the popularity of caffeinated foods mere coincidence? Is the flavor coffee, chocolate, tea and cola soft drinks such that without caffeine they would still be widely consumed? Or is the popularity of caffeine containing foods due to the influence of caffeine in the body?

2.2 Caffeine in the Body

Within 20 minutes of ingestion, 90% of the caffeine (Figure 2.1) consumed is removed from the stomach (Heckman *et al* 2010), the majority to be processed by the liver to enable absorption into the bloodstream (Heckman *et al* 2010). Once caffeine enters the bloodstream it exhibits effects on the central nervous system and metabolism. Half of the dose of caffeine is eliminated from the body of healthy humans in approximately 4–5 hours (Mandel 2002). The acute effects of caffeine on the human body have been identified as stimulating both mental function and the sympathetic neural system. Caffeine is a non-selective adenosine receptor antagonist (competitive inhibitor), primarily inhibiting adenosine receptors A_1 and A_{2a}. Adenosine A_1 receptors are most abundantly found in the hippocampus, the cerebral cortex, the cerebellar cortex and the thalamic nuclei in the brain, whereas the adenosine A_{2a} receptors are found in basal ganglia areas of the brain where it has a role in regulation of dopamine release. Adenosine has an inhibitory effect on the central nervous system and therefore caffeine acting as a competitive inhibitor at adenosine receptors A_1 and A_{2a} effectively acts as a stimulant (Daly and Fredholm 1998). Thus the stimulatory effects of caffeine on the body are mainly due to this inhibition of adenosine (Mandel 2002) (Figure 2.2).

Improvements in concentration and alertness upon consumption of caffeine are two benefits in mental functions that have been well documented, even amongst the sleep deprived. These effects have been identified with

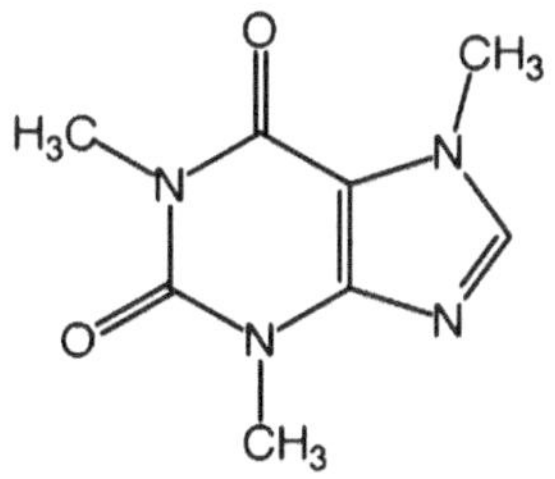

Figure 2.1 Structure of caffeine (1,3,7-trimethylxanthine). This figure shows the chemical structure of caffeine. Caffeine is an alkaloid which means it contains nitrogen and hydrogen ions; it also has a methyl group attached to the first, third and seventh nitrogen group giving it the chemical name of 1,3,7-trimethylxanthine.

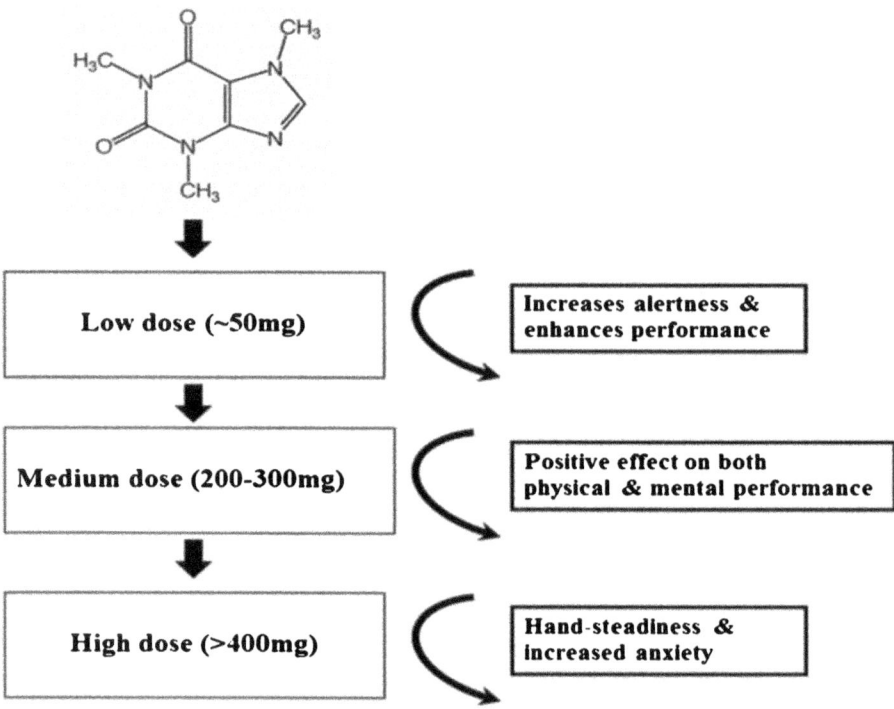

Figure 2.2 The impact of increasing doses of caffeine on the body. This figure describes some of the body's responses to taking in caffeine. At lower doses the effects are stimulatory and a positive impact on performance and alertness. As the dose increases however, anxiety levels have been shown to increase and overstimulation occurs and can be seen in symptoms such as hand shaking.

administered doses as low as 32 mg, however with larger doses of 500 mg cognitive functioning has been shown to deteriorate (Heckman *et al* 2010). Although it has not been well documented, doses of 1000–2000 mg of caffeine per day have been linked in some cases to anxiety.

2.2.1 Withdrawal from Caffeine

In a critical review of the withdrawal symptoms associated with caffeine, Juliano and Griffiths (2004) documented the evidence from 57 experimental studies and 9 retrospective survey studies. Valid withdrawal symptoms identified were the onset of headaches, a decrease in feeling alert, difficulty concentrating, fatigue, drowsiness, lower energy levels, a depressed mood, a decrease in feelings of contentment, irritability, and a decrease in feelings of a clear head. Withdrawal symptoms most commonly occurred within 12–24 hours following abstinence from caffeine and may last anywhere between two and nine days. Upon a re-dose of caffeine, withdrawal effects are rapidly

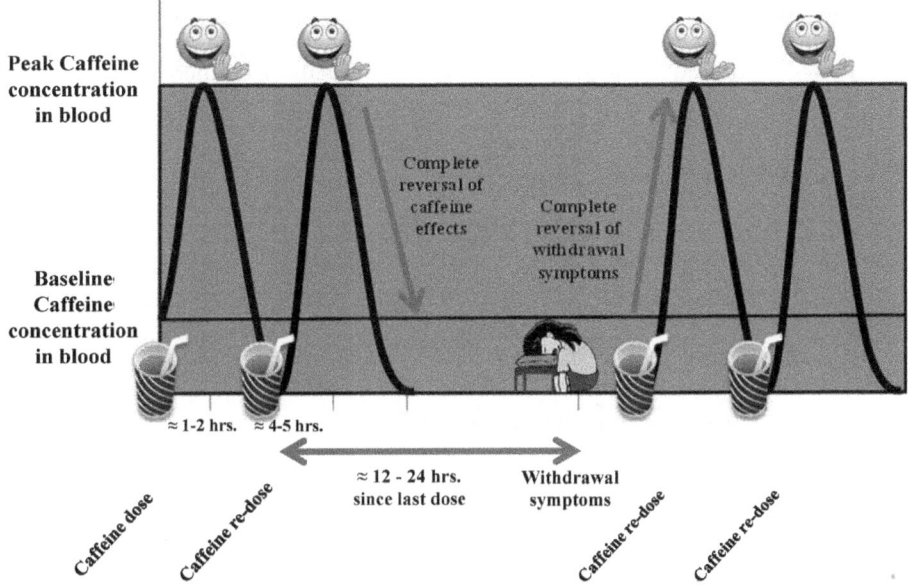

Figure 2.3 Caffeine's cycle of dependence. This figure describes the cycle of caffeine dependence, moving between positive effects of caffeine following ingestion and the onset of symptoms of withdrawal which are reversed by consuming caffeine again.

reversed in their entirety. Reinforcing effects of caffeine have been demonstrated through studies that investigated abstinence of caffeine in caffeine consumers, whereby an increase in preference for caffeinated beverages was identified (Hughes *et al* 1998; Juliano and Griffiths 2004). This evidence supports some of the motivations of caffeine consumers to continue to consume caffeinated beverages or foods. Re-dosing in caffeine consumers may occur for the acute positive effects caffeine has on the body, to lead to a reversal of withdrawal symptoms associated with caffeine abstinence, or for taste preferences for caffeine (Figure 2.3).

2.2.2 Is Caffeine a Drug of Dependence?

In a study of 162 caffeine consumers, Hughes *et al.* reported 56% had a strong desire or unsuccessful attempt to quit caffeine consumption, 50% spent a great deal of time obtaining caffeine, 28% used more than intended, and 18% suffered withdrawal symptoms (Hughes *et al* 1998). Although different mechanisms of action are involved, drugs of dependence have a similar site of action to that of caffeine. For this reason, and its ability to reverse withdrawal effects, caffeine has been thought by some to be a drug of dependence. Tolerance to the effects of caffeine has been demonstrated in chronic users indicating its drug like properties. Caffeine users have also been shown to have a physical dependence to the substance due to the onset of withdrawal symptoms upon abstinence from caffeine. These properties, commonly witnessed in addictive substances, fuel the argument for caffeine as a drug of dependence. However such contentions are controversial and The International Food Information Council Foundation (IFIC Foundation) believes caffeine does not meet criteria to be classified as an addictive substance.

What follows is a review of the role of caffeine in SSB, including as a flavoring agent, and as a factor driving consumption of SSB, and the implications of increased SSB consumption on development of overweight and obesity.

2.3   Caffeine as a Flavor Agent

When added to foods or beverages, caffeine has the ability to impart bitter taste or modify flavor through cognitive level *via* mixture suppression, or peripheral level *via* interference with taste transduction (Peri *et al* 2000). The method by which caffeine (or any compound) can modify flavor is a direct result of its concentration (Figure 2.4). If the concentration of caffeine is in the supra-threshold range, then the bitter taste will modify flavor by a cognitive effect labeled mixture suppression. In mixture suppression theory, the intensity of a mixture of two flavors is less than the sum of the individual intensities. Caffeine may also influence flavor at sub-threshold concentrations because it interacts with taste receptors modifying taste transduction at the peripheral level, independent of any taste perception (Keast and Roper 2007). In addition to activating bitter taste, caffeine may have an influence on sweetness as it has been shown to modify the taste transduction signals of a variety of sweeteners. Presumably caffeine can translocate through cellular membranes and access intra-cellular mechanisms associated with taste.

The addition of caffeine to popular SSB is ostensibly as an essential flavor component, contributing to the complex flavor experience these drinks deliver (PepsiCo 1981; Griffiths and Vernotica 2000). The role of caffeine as a flavoring or flavor enhancer in SSB has been questioned previously by Griffiths and Vernotica (Griffiths and Vernotica 2000) and again more recently by Keast and Riddell (Keast and Riddell 2007; Riddell and Keast 2007). Griffiths *et al.* (Griffiths and Vernotica 2000) tested 25 adults, who were caffeine dependent and reported preferring caffeine-containing beverages over non-caffeinated varieties. Subjects were asked to detect differences in beverage flavor and composition, including detecting differences between diet and regular SSB as well as detecting caffeine in SSB at varying concentrations (0.05–1.6 mg^{-1} mL) (Griffiths and Vernotica 2000). The two most popular brands of commercial cola flavored beverages, Pepsi and Coke, contain caffeine at concentrations at or below 0.1mg mL^{-1}. At this concentration, only two subjects could detect caffeine as a flavor, beverages with caffeine concentrations above 0.2 mg mL^{-1} could be detected by half of the participants

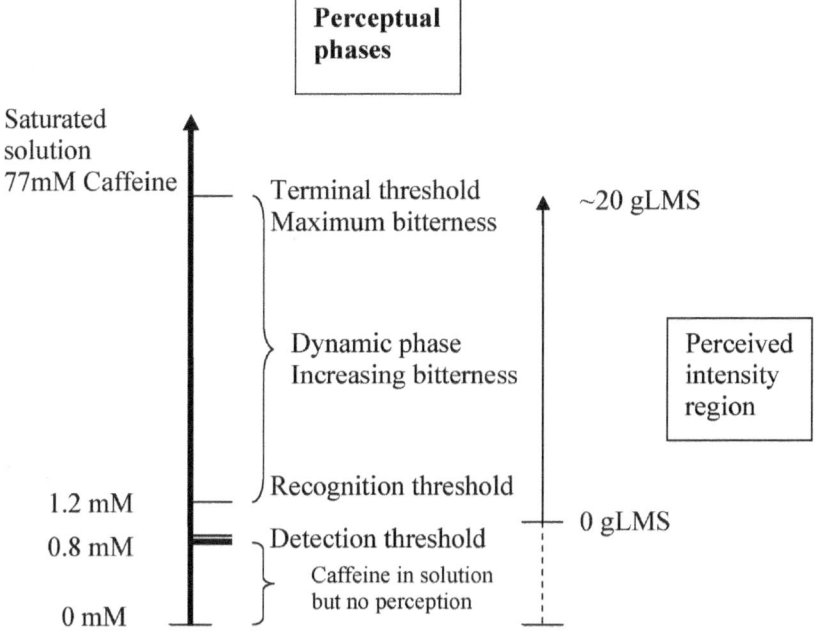

Figure 2.4 Schematic illustration of the relationship between caffeine and gLMS. Schematic illustration of the relationship between chemical concentration of caffeine and perceived intensity using general Labeled Magnitude Scale (gLMS). The left hand side of the bold black y-axis represents chemical concentration from 0 millimolar (0 mM) solution to the detection threshold (0.8 mM caffeine), recognition threshold (1.2 mM) and a saturated solution (77 mM caffeine). The right hand side of the bold black y-axis represents the perceptual relationship to increasing concentration. The far right vertical axis represents the gLMS scale from no perception to maximum bitterness.

(Griffiths and Vernotica 2000). Presently there is only one cola beverage, 'Jolt Cola' that provides caffeine at this concentration (Griffiths and Vernotica 2000). Keast and Riddell (Keast and Riddell 2007) similarly found the concentration of caffeine in popular cola beverages to be below taste threshold. Their study differed in design to that of Griffiths *et al.*; it relied on 30 subjects who were trained to detect the five distinct taste qualities (sweet – sucrose, salty – NaCl, bitter – quinine, sourness – citric acid, and umami – glutamic acid monosodium salt and inosine 5'-phosphate) and preference or liking of caffeinated beverages were not a component of subject selection. After training, subjects were asked to drink both diet and non-diet caffeinated and non-caffeinated varieties of soft drinks. Of the 30 highly trained subjects none were able to discriminate between the caffeinated and non-caffeinated beverages (Keast and Riddell 2007). The authors suggest that SSB activate multiple sensory systems resulting in many sensory messages being sent to the brain almost simultaneously, thus, any flavor effects caffeine may have in a simple solution such as water, were effectively masked by this sensory overload (Keast and Riddell 2007). Combined, these findings question the role of caffeine within these SSB. If caffeine is not detected as a flavor agent within complex SSB, why, then is it a common ingredient in popular drinks? Caffeinated beverages such as coffee, tea, cocoa, and SSB are hedonically pleasant to drink, but some of the pleasure may be due to reversal of onset of withdrawal symptoms (Yeomans *et al* 2000). In effect, the liking of the caffeinated drink is partially due to making the consumer feel better, not simply the flavor of the beverage. In addition, when caffeine is co-consumed with a source of sugar (glucose), there appears to be a synergy of cognitive modulating effects. The physiological and psychological influence of caffeine on consumers may help ensure repeat purchase of products containing caffeine. Combined, these factors suggest that caffeine may be added to modify consumer behavior, rather than enhance flavor (Griffiths and Vernotica 2000; Keast and Riddell 2007; Riddell and Keast 2007).

2.4   Flavor Preferences

The development of flavor preferences and aversions encompass taste, odor and texture of foods or beverages and occur when individuals associate post-ingestive consequences of a food or beverage with particular flavor cues. If post-ingestive consequences or intrinsic reinforcing factors of foods occur post-ingestion, they can become associated with particular food and beverages, resulting in consumers preferring those foods over others (Myers and Sclafani 2006). These positive associations with specific foods can disrupt energy balance and energy intake as individuals select reinforcing foods and drinks leading to larger and more frequent doses being required to produce the desired effects (Myers and Sclafani 2006). SSB have several modes of positive reinforcement which may result in preferences for these drinks over others (Vartanian *et al* 2007). The high sugar content, intrinsic energy, marketing claims, environmental situations in which these drinks are consumed, as well as their caffeine content, may all contribute. Early preference for sweetened foods like SSB may also calibrate individuals to high levels of sweetness and generalize to other food choices (Vartanian *et al* 2007). The positive effects of caffeine post-ingestion are thought to reinforce subsequent consumption of caffeine containing foods (Tinley *et al* 2003). A study by Yeomans *et al* reported that pairing a novel beverage with caffeine, increased the liking of that beverage in a caffeine deprived state, once the caffeine was removed from the beverage, the liking decreased (Yeomans *et al* 2000). The authors suggest that increased preference for caffeinated beverages may be due to the positive post-ingestive consequences of caffeine. These reinforcing effects of caffeine alongside avoidance of withdrawal strongly suggest caffeination of foods and beverages can increase preference and consumption of these beverages, leading to higher energy intakes (Yeomans *et al* 2000).

Table 1 Caffeine content in selected SSB.

Drink	Amount of Caffeine (mg)
Coca Cola® (375 mL)	33.9 ± 0.9
Diet Coke (375 mL)	46.3 ± 1.7
Coke Zero (375 mL)	35.8 ± 2.6
Pepsi® (375 mL)	38.9 ± 1.0
Diet Pepsi (375 mL)	36.7 ± 0.6
Pepsi One (375 mL)	57.1 ± 3.3

This table details the caffeine content of commonly available cola beverages. Adapted from Chou and Bell (2007)

2.5   Caffeine in Sugar Sweetened Beverages

Caffeine is found naturally in coffee, cocoa (chocolate), and black teas, and its stimulant effect in these foods has been known for hundreds of years. Caffeine is now included as an ingredient in a wider variety of beverages due to its use as a food additive, and SSB and energy-drinks often contain caffeine in their formulation. Caffeine is recognized by the American Food and Drug Administration (FDA) as being a multipurpose Generally Recognized As Safe (GRAS) food substance. Variation exists in permissibility of the use of caffeine in beverages internationally with countries differing in both the allowable concentration of caffeine and the beverages in which it can be added. In the US, the permissible limit of caffeine in cola-type beverages is 200 mg L^{-1}, whereas the Australian Food Standards Code restricts the addition of caffeine to cola type SSB and related beverages to less than 145 mg kg^{-1} (approximately 54.5 mg per 375 mL) (FSANZ 2008). The level of caffeine found in common SSB is described in Table 1.

2.6   SSB Consumption

The analysis of the American NHANES surveys suggests that SSB consumption has tripled over the last thirty years (1971–1994) (Troiano *et al* 2000) and SSB now provide 33% of added sugars in their diets (Troiano *et al* 2000). Mean annual consumption of SSB in the US is estimated to be 1.2 drinks per person, per day adding 800 kJ per day. The amount children and adolescents consume is thought to be higher, averaging 2 drinks and adding 1470 kJ per day (Nestle 2000). Similar increases have also been observed in Australia with intakes doubling over a 30 year period with figures suggesting increases from 47 L per person per year in 1969, to 119 L per person per year in 1999 (Statistics 1998). Similarly, the contribution of free-sugars from SSB has doubled between the 1985 and 1995 national surveys. As the consumption of one standard serving of SSB adds 630 kJ and 42 grams of sugar, it is no surprise to discover that regular consumers have higher energy intakes (up to10%) (Striegel-Moore *et al* 2006) and larger portions of their intakes are contributed by beverages (Troiano *et al* 2000).

2.7   SSB and Body Weight

The association between SSB, weight change and obesity is still not clear, but the bulk of the evidence indicates a positive association between increased SSB consumption and body weight. Recent reviews of the literature have investigated this association (Malik *et al* 2006; Vartanian et al. 2007). The reviews included cross sectional, prospective cohorts, and clinical studies. The pooled evidence from these studies indicates there is indeed a positive correlation between consumption of SSB, higher energy intakes and weight gain (Malik *et al* 2006; Vartanian *et al* 2007).

2.7.1   Cross Sectional Studies

One of the largest cross sectional studies used results from three NHANES between the years 1970–1994, using a representative sample of 10 371 children and adolescents aged 2 to19 years in the United States. The authors concluded that beverages contributed up to 24% of total energy amongst this age range and SSB were the major source of liquid energy (Troiano *et al* 2000). Overweight and obese subjects consumed 2% more energy from beverages than normal weight subjects and amongst each age group SSB contributed a greater amount of energy to total energy intake in overweight and obese subjects in both males and females (2–5 y: 3.1% compared to 2.4%; 6–11y: 5.4% compared to 4%; and 12–19y: 10.3% compared to 7.6%) (Troiano *et al* 2000). A large, repeat cross sectional study conducted in 1986, 1990, 1994 and 1999 in Sweden used 5 915 persons to document trends in food choices and prevalence of overweight and obesity, measured by waist circumference (Krachler *et al* 2006). Their findings suggest hamburgers, fried potatoes and SSB repeatedly show up as positive determinants to higher waist circumferences and occurrences of overweight and obesity (Krachler *et al* 2006).

2.7.2   Prospective Studies and Weight Change

A large prospective cohort from the United States which followed 16 771 of 9 to 14 year olds over a 2 year period found a significant association between SSB consumption and BMI score (Berkey *et al* 2004). Another study which followed 2 379 girls aged 9 to 10 years at baseline for 10 years found that SSB consumption increased 3 fold over the course of the study and was associated with changes in BMI and higher energy intakes (Striegel-Moore *et al* 2006). The authors found that each 100 g serving of SSB contributed an additional 82 calories (344.4 kJ) and a significant increase in BMI ($p < 0.05$) (Striegel-Moore *et al* 2006). This finding is interesting as 100 g of SSB contains only 41 calories (172.2 kJ). The authors have noted that in addition to excess energy, SSB may encourage other eating habits resulting in higher caloric intakes (Striegel-Moore *et al* 2006).

2.7.3 Intervention Trials of SSB and Body Weight

Intervention trials targeting obesity suggest reducing consumption of SSB can also reduce energy intakes, promote weight loss and BMI changes, and decreases the prevalence of overweight and obesity. James *et al*, (2004) and Ebbeling *et al*, (2006) recently conducted intervention trials on children between approximately 7 to 18 years. Both studies used a variety of strategies to reduce consumption of SSB, with positive impacts on BMI change in both intervention groups.

2.7.4 Studies Finding no Association between SSB and Weight Change

Despite these large studies suggesting a link between SSB, energy intakes and weight change there are several studies which do not support these findings. A cross sectional study on 6–7 year old children from various cities in Spain found SSB to be associated with greater energy intakes, but not with higher BMI values (Rodriguez-Artalejo *et al* 2003). As this study is cross sectional and the effects of positive energy balance on body weight may occur over time, the higher energy intakes seen in regular consumers of SSB could be argued to eventually lead to weight changes. (Rodriguez-Artalejo *et al* 2003). Blum *et al* also found no association between SSB and weight gain during a two year prospective study involving 166, 8 to 10 years olds (Blum *et al* 2005). Limitations of this study include its small sample size, small number of food questionnaires (only two conducted during the two years) and no food frequency questionnaires conducted on weekends.

2.8 Linking Caffeine in SSB and Development of Overweight and Obesity

It is unlikely that caffeine is the only mechanism leading to overconsumption of SSB and weight gain. It is more likely that this is but one component in a complex mix of subtle and not so subtle influences. In addition to the intrinsic energy these drinks provide, their low satiation, high glycemic index, as well as aggressive marketing are all implicated in their excess consumption. The satiation effects of liquid energy is thought to be less than those provided from solid energy with studies showing a decreased ability to down regulate energy intake after consuming energy in liquid form compared with solid form (DiMeglio and Mattes 2000). The glycemic index (GI) of SSB may be another mechanism driving up energy intake as high GI foods and beverages such as SSB may promote excess energy consumption due to the differing responses carbohydrate foods have on blood sugar. High GI foods produce rapid responses in insulin levels post-ingestion, causing both glucose and fatty acid concentrations in the blood to decrease. This response may signal low fuel availability in the body and promote further food intake, despite recent ingestion of energy.

Both liking and wanting are complex components of the food reward system and evidence suggests two unique brain substrates underlie the distinction between liking and wanting. Liking has been shown to be affected by the opioid system, and wanting by the dopamine system (Berridge 1996). Liking primarily encompasses the hedonic pleasure experienced on the palette following the ingestion of a food (Berridge 1996). SSB are hedonically pleasant to drink, with sweetness being the primary taste innately liked (Harris 2008; Beauchamp and Mennella 2009). Caffeine on the other hand is a bitter substance, but in SSB the concentration is often below bitter taste thresholds, therefore yields no bitterness (Keast and Riddell 2007; Keast 2008). The addition of caffeine to SSB, while having no direct influence on taste (and therefore liking), may influence wanting of the SSB. The mechanisms underlying wanting arise from a person's appetite, craving or motivation to consume a food (Berridge 1996). Evidence suggests dopamine is expressed as a result of caffeine ingestion (Fredholm *et al* 1999). As the dopamine system appears to play a role in stimulating food consumption, perhaps repeated caffeine ingestion promotes an increased desire to consume caffeinated foods or beverages. The hedonic pleasure experienced from the ingestion of sweet substances together with the motivation to consume caffeine may therefore have a synergistic effect thereby increasing consumption of caffeinated SSB.

2.9   Would Decreasing Caffeine in SSB Impact Body Weight?

Egger and Swinburn have suggested use of the epidemiological triad as a framework to model the causes and contributors of obesity (Figure 2.5) (Egger *et al* 2003). The epidemiological triad considers the host, the environment, the agent and its vectors. In this instance, SSB can be considered vectors for increased energy intake; the evidence presented indicates that this is fairly widely accepted. However, the inclusion of caffeine in SSB formulation could also conceivably be considered an additional vector for weight gain as its inclusion may be at least partially responsible for excess, repeat consumption. Interventions which target high volume vectors have the potential to bring about large changes (Egger *et al* 2003). Thus decreasing SSB consumption *via* restrictions on advertising and availability (*i.e.* the school vending machine) is frequently called for by public health professionals. We argue that by decreasing the amount of caffeine in SSB, or at a minimum, restricting any further approval of its permissibility as a flavor agent, may also potentially impact body weight by minimizing the impact of caffeine on overconsumption of SSB.

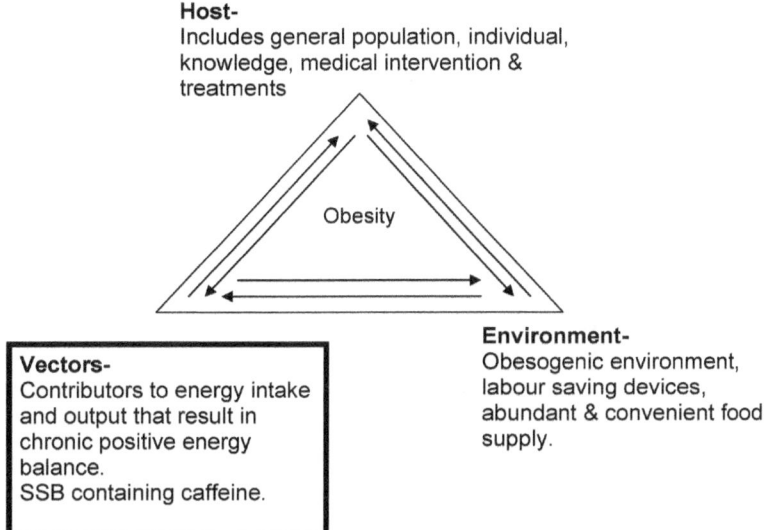

Host-
Includes general population, individual, knowledge, medical intervention & treatments

Obesity

Vectors-
Contributors to energy intake and output that result in chronic positive energy balance.
SSB containing caffeine.

Environment-
Obesogenic environment, labour saving devices, abundant & convenient food supply.

Figure 2.5　The epidemiological triad and obesity. This figure provides a conceptual framework to describe the interplay between different determinants of obesity. This figure has been modified from Egger and Swinburn (Egger *et al* 2003).

2.10　Conclusion

Caffeine has been a controversial additive in beverages for over 100 years. The overwhelming evidence shows that, in general, humans tolerate doses up to 400 mg per day with no adverse effects. Caffeine has been shown to promote consumption through a cycle of dependence and this may become a problem if daily caffeine re-dosing is *via* high energy SSB. It is conceivable that the presence of caffeine within SSB drives repeat consumption of SSB through the caffeine dependence pathway and *via* the development of wanting of the beverage *via* the dopamine pathway influencing appetite. SSB have been identified as a high volume vector in the development of overweight and obesity and strategies to lower the consumption of these beverages has the potential to reduce weight gain, ultimately reducing the proportion of individuals within a population with elevated BMI levels. As studies have shown the removal of caffeine from novel beverages can decrease liking of those beverages, it can be suggested removal of caffeine from SSB could decrease consumption of these drinks (Yeomans *et al* 2000). As intervention trials have indicated that SSB reduction is a means of reducing obesity prevalence, methods to reduce consumption need to be considered. The permissibility of an addictive compound such as caffeine to SSB as a flavor agent needs to be challenged in light of the international public health crisis surrounding excessive body weight.

Summary Points

- Caffeine is hedonically pleasant to consume and can result in feelings of alertness and energy and may enhance wanting of caffeinated Sugar Sweetened Beverages.
- Reversal of symptoms of caffeine withdrawal may be an important motivator for repeat consumption of caffeinated Sugar Sweetened Beverages.
- The majority of Sugar Sweetened Beverages consumed worldwide contain caffeine as an additive.
- Despite caffeine being approved for use as a flavor additive, consumers cannot detect any flavor activity of caffeine in the complex flavor matrix of Sugar Sweetened Beverages.
- Increasing Sugar Sweetened Beverages consumption has been linked with increased body weight and rates of overweight and obesity.
- The presence of caffeine within Sugar Sweetened Beverages may result in repeat overconsumption of these obesogenic beverages.
- Removal of caffeine from Sugar Sweetened Beverages formulation may result in decreased consumption, decreased energy intake and decreased weight gain in individuals.

Key Facts:

Key Facts of Caffeine:

- Caffeine is an alkaloid naturally present in plants used to make beverages such as tea or coffee.
- Caffeine acts as a stimulant to the central nervous system, the circulatory and respiratory systems and acts as a diuretic.
- Caffeine is also added to some sugar sweetened beverages and is approved as Generally Regarded as Safe (GRAS) as a flavor agent.

Key Facts of National Health and Nutrition Examination Survey (NHANES)

- NHANES is an ongoing survey of the nutritional and health status of people living within the United States of America.
- NHANES began a series of surveys in the 1960s and in the late 1990s the survey became continuous with a rolling intake of people and timed released of data from the survey.
- The survey data are an important means to track the health of Americans and changes that occur over time.
- It provides invaluable data tracking changes in body weight and dietary intake, including beverage intake, over time.

Definitions

Caffeine: an alkaloid compound that acts as a stimulant of the central nervous system. Found naturally in certain foods and beverages and added to sugar sweetened beverages.

Cross sectional studies: Studies that involve measuring an exposure at a fixed time point in a defined population.

Flavor preferences: The preference for one flavor over another flavor, or the development of liking the flavor over multiple tastings.

Intervention trials: Studies that measure the impact of a change in exposure on a predefined outcome, generally a marker of disease development.

Overweight: a condition of excess body weight, defined within as a body mass index of 25–29.9.

Obesity: a condition of excess body weight, defined within as a body mass index in excess of 30.

Post-ingestive consequences: physiological and psychological consequences experienced following ingestion of a food or beverage.

Prospective or cohort studies: studies that measure the impact of exposure on disease progression over a period of time. Population groups studied are generally healthy and free of disease.

Sugar Sweetened Beverages: manufactured carbonated beverages sweetened with sugar, excludes carbonated water, fruit juices and artificially sweetened beverages.

Abbreviations

CSFII	Continuing Survey of Food Intake by Individuals
BMI	Body Mass Index
GRAS	Generally Regarded as Safe
FDA	Food and Drug Administration
gLMS	general Linear Magnitude Scale
GI	Glycemic Index
NHANES	National Health and Examination Survey
SSB	Sugar Sweetened Beverages

References

Nielsen, A. C. 2008. "Convenience Report" Available at: http://au.acnielsen.com/site/documents/ConvenienceReport2008-extob.pdf. Accessed April 2011.

Australia Bureau of Statistics (ABS), 1998. Apparent Consumption of Food Stuffs 1996–97. Australia Bureau of Statistics. Australian Government, Canberra.

Beauchamp, G. K. and Mennella, J. A., 2009. Early flavor learning and its impact on later feeding behavior. Journal of Pediatric Gastroenterology and Nutrition. 48 Suppl 1: S25–30.

Benowitz, N. L., Jacob, P. 3rd, Mayan H. and Denaro C., 1995. Sympathomimetic effects of paraxanthine and caffeine in humans. Clinical Pharmacology and Therapeutics. 58(6): 684–91.

Berkey, C. S., Rockett, H. R., Field, A. E., Gillman M. W. and Colditz, G. A., 2004. Sugar-added beverages and adolescent weight change. Obesity Research. 12(5): 778–88.

Berridge, K. C. 1996. Food reward: brain substrates of wanting and liking. Neuroscience and Biobehavioral Reviews. 20(1): 1–25.

Blum, J. W., Jacobsen, D. J. and Donnelly, J. E., 2005. Beverage consumption patterns in elementary school aged children across a two-year period. Journal of the American College of Nutrition. 24(2): 93–8.

Chou, K.-H. and Bell, L., 2007. Caffeine content of prepackaged national-brand and private-label carbonated beverages. Journal of Food Science. 72(6): C337–42.

Daly, J. W. and Fredholm, B. B., 1998. Caffeine–an atypical drug of dependence. Drug Alcohol Dependance. 51(1–2): 199–206.

Desbrow, B., Girot, P.-M. and Somerset, S., 2004. The dynamics of caffeine intake in Australia: analysis of the National Nutrition Survey 1995. Dietitians of Association of Australia 22nd National Conference.

DiMeglio, D. P. and Mattes, R. D., 2000. Liquid versus solid carbohydrate: effects on food intake and body weight. International Journal of Obesity and Related Metabolic Disorders. 24(6): 794–800.

Ebbeling, C. B., Feldman, H. A., Osganian, S. K., Chomitz, V. R., Ellenbogen, S. J. and Ludwig, D. S., 2006. Effects of decreasing sugar-sweetened beverage consumption on body weight in adolescents: a randomized, controlled pilot study. Pediatrics. 117(3): 673–80.

Egger, G., Swinburn, B. and Rossner, S., 2003. Dusting off the epidemiological triad: could it work with obesity? Obesity Reviews. 4(2): 115–9.

Frary, C. D., Johnson, R. K. and Wang, M. Q., 2005. Food sources and intakes of caffeine in the diets of persons in the United States. Journal of the American Dietetic Association. 105(1): 110–3.

Fredholm, B. B., Battig, B. B., Holmen, J., Nehlig, A. and Zvartau, E. E., 1999. Actions of caffeine in the brain with special reference to factors that contribute to its widespread use. Pharmacological Reviews. 51(1): 83–133.

FSANZ., 2008. Australia New Zealand Food Standards Code Incorporating amendments up to and including Amendment 95. Available at: http://www.foodstandards.gov.au/foodstandards/foodstandardscode.cfm. Accessed 1 April 2011.

Garrett, B. E. and Griffiths, R. R., 1998. Physical dependence increases the relative reinforcing effects of caffeine versus placebo. Psychopharmacology (Berl). 139(3): 195–202.

Griffiths, R. and Vernotica, E., 2000. Is caffeine a flavoring agent in cola soft drinks? Archives of Family Medicine. 9: 727–734.

Harris, G. 2008. Development of taste and food preferences in children. Current Opinion in Clinical Nutrition and Metabolic Care. 11(3): 315–9.

Heckman, M. A., Weil, J. and Gonzalez de Mejia, E., 2010. Caffeine (1,3,7-trimethylxanthine) in foods: a comprehensive review on consumption, functionality, safety, and regulatory matters. Journal of Food Science. 75(3): R77–87.

Hughes, J. R., Oliveto, A. H., Liguori, A. H., Carpenter, J. and Howard, T., 1998. Endorsement of DSM-IV dependence criteria among caffeine users. Drug Alcohol Dependence. 52(2): 99–107.

James, J., Thomas, P., Cavan, P. and Kerr, D., 2004. Preventing childhood obesity by reducing consumption of carbonated drinks: cluster randomised controlled trial. (BMJ) British Medical Journal. 328(7450): 1237.

James, J. E. and Keane, M. A., 2007. Caffeine, sleep and wakefulness: implications of new understanding about withdrawal reversal. Human Psychopharmacology. 22(8): 549–58.

Juliano, L. M. and Griffiths, R. R., 2004. A critical review of caffeine withdrawal: empirical validation of symptoms and signs, incidence, severity, and associated features. Psychopharmacology (Berl). 176(1): 1–29.

Keast, R. and Riddell, L., 2007. Caffeine as a flavour additive in soft-drinks. Appetite. 49: 255–259.

Keast, R. and Roper, J., 2007. A complex relationship among chemical concentration, detection threshold, and suprathreshold intensity of bitter compounds. Chemical Senses. 32(3): 245–253.

Keast, R. S. J. 2008. Modification of the bitterness of caffeine. Food Quality and Preference. 19: 465–472.

Krachler, B., Eliasson, M., Stenlund, H., Johansson, I., Hallmans, G. and Lindahl, B., 2006. Reported food intake and distribution of body fat: a repeated cross-sectional study. Nutrition Journal. 22(5): 34.

Malik, V. S., Schulze, M. B. and Hu, F. B., 2006. Intake of sugar-sweetened beverages and weight gain: a systematic review. American Journal of Clinical Nutrition. 84(2): 274–288.

Mandel, H. G. 2002. Update on caffeine consumption, disposition and action. Food and chemical toxicology: An International Journal Published for the British Industrial Biological Research Association. 40(9): 1231–4.

Myers, K. P. and Sclafani, A., 2006. Development of learned flavor preferences. Developmental Psychobiology. 48(5): 380–8.

Nestle, M. 2000. Soft drink "pouring rights": marketing empty calories to children. Public Health Reports. 115(4): 308–319.

Olsen, N. J. and Heitmann, B. L., 2009. Intake of calorically sweetened beverages and obesity. Obesity Reviews. 10(1): 68–75.

PepsiCo, I. 1981. The physical or technical effect of caffeine in cola beverages. Vol III, Appendix XII of comments of the National Soft Drink Association submitted to the department of Health and Human Services

Food and Drug Administration in response to the proposal to delete caffeine in cola-type beverages from the list of generally regarded as safe and to issue an interium food additive regulation governing its future use. FDA Docket No. 80N–0418

Peri, I., Mamrud-Brains, H., Rodin, S., Krizhanovsky, V., Shai, Y., Nir, S. and Naim, M., 2000. Rapid entry of bitter and sweet tastants into liposomes and taste cells: implications for signal transduction. [In Process Citation]. The American Journal of Physiology. 278(1): C17–C25.

Riddell, L. and Keast, R., 2007. Is caffeine in soft drinks really necessary? Medical Journal of Australia. 187(11): 655.

Rodriguez-Artalejo, F., Garcia, E. L., Gorgojo, L., Garces, C., Royo, M. A., Martin Moreno, J. M., Benavente, M., Macias, A. and De Oya, M., 2003. Consumption of bakery products, sweetened soft drinks and yogurt among children aged 6–7 years: association with nutrient intake and overall diet quality. British Journal of Nutrition. 89(3): 419–29.

Striegel-Moore, R., Thompson, D., Affenito, S., Franko, D., Obarzanek, E., Barton, B., Schreiber, G., Daniels, S., Schmidt, M. and Crawford, P., 2006. Correlates of beverage intake in adolescent girls: the National Heart, Lung, and Blood Institute Growth and Health Study. Journal of Pediatrics. 148(2): 183–187.

Tinley, E. M., Yeomans, M. R. and Durlach, P. J., 2003. Caffeine reinforces flavour preference in caffeine-dependent, but not long-term withdrawn, caffeine consumers. Psychopharmacology (Berl). 166(4): 416–23.

Troiano, R., Briefel, R., Carroll, M. and Bialostosky, K., 2000. Energy and fat intakes of children and adolescents in the united states: data from the national health and nutrition examination surveys. American Journal of Clinical Nutrition. 72(5(supp)): 1343S–1353S.

Vartanian, L., Schwartz, M. and Brownell, K., 2007. Effects of soft drink consumption on nutrition and health: a systematic review and meta-analysis. American Journal of Public Health. 97(4): 667–675.

Yeomans, M., Jackson, A., Lee, M., Steer, B., Tinley, E., Durlach, P. and Rogers, P., 2000. Acquisition and extinction of flavour preferences conditioned by caffeine in humans. Appetite. 35(2): 131–141.

The Chemistry of Caffeine

CHAPTER 3

The Chemistry of Caffeine

JOSÉ JOÃO CARVALHO[1], FRANZISKA EMMERLING[2]
AND RUDOLF J. SCHNEIDER*[2]

[1] Chemicals Registration and Regulatory Affairs, DR. KNOELL CONSULT
GmbH, Dynamostrasse 19, D-68165 Mannheim, Germany; [2] BAM Federal
Institute for Materials Research and Testing, Richard-Willstaetter-Str. 11,
D-12489 Berlin, Germany
*E-mail: rudolf.schneider@bam.de

In Lieu of an Introduction: Caffeine Trivia

"Mm! How sweet the coffee tastes, more delicious than a thousand kisses!"
Johann Sebastian Bach
    (in the aria "Be quiet, do not chat" of the operetta "Kaffeekantate").

Bach wrote a full cantata dedicated to coffee (Coffee Cantata) (Bach 1732–
1734) in a Leipzig coffee house, as a statement against a movement in
Germany in this time to prevent women from drinking coffee as it was thought
to make them sterile (Renneberg *et al* 2011). From the frequent visits of Bach
to Zimmermann's coffee house in Leipzig until the isolation of caffeine, 90
years passed. Another Johann (Wolfgang von Goethe), not a composer but a
poet, is believed to have contributed decisively to the discovery of this chemical
by the young scholar Friedlieb Ferdinand Runge. Goethe asked Runge, who
had been performing experiments with *Belladonna* extracts in Jena, to find out
what substance there was in coffee to create such an effect on people (Weiberg
and Bealer 2001).

Food and Nutritional Components in Focus No. 2
Caffeine: Chemistry, Analysis, Function and Effects
Edited by Victor R Preedy
© The Royal Society of Chemistry 2012
Published by the Royal Society of Chemistry, www.rsc.org

On the other side, Frederick II, whose 300th birthday is celebrated in Berlin in 2012, tried to ban coffee from his territory in 1781, being quoted (Weiberg and Bealer 2001) "Everybody is using coffee; this must be prevented. His Majesty was brought up on beer, and so were both his ancestors and officers. Many battles have been fought and won by soldiers nourished on beer, and the King does not believe that coffee-drinking soldiers can be relied upon to endure hardships in case of another war." (Wikiquotes). The correlation between the thousands of coffee shops today all over the city of his former court, and the absence of war, has unfortunately not been scientifically studied to date.

From the time when laws were passed banning its use, until our day when it has become the legally most accepted drug of the world, caffeine has come a long way in chemistry, medicine, societal habits and legal regulation. Human societies have had a major task to accomplish since the industrial revolution: to pack our daily activities into a much-too-short 24-hour shift. And obviously a lot of them consider caffeine as helping them to attain this task. For science, it can be deducted from an analysis of the annual publication numbers how vital caffeine research still is (Figure 3.1). The large positive increment in the 90s is related to the discovery of the adenosine and adenosine triphosphate receptors and the interest in agonists–antagonists based on caffeine which was then being produced and studied.

In order not to stir up the battles of coffee friends and foes this chapter only reports facts and figures of caffeine as a chemical.

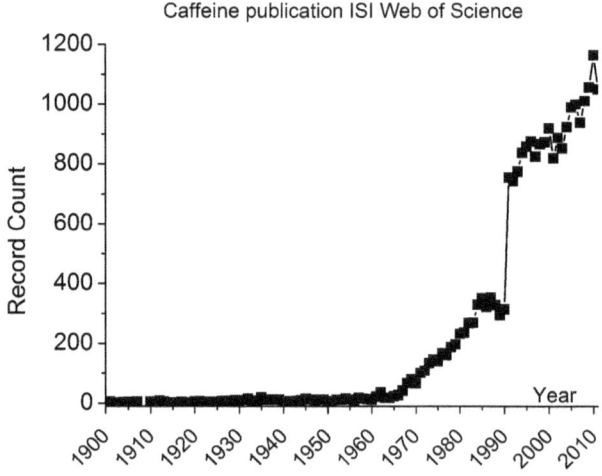

Figure 3.1   Publications on caffeine (ISI Web of Science. Analysis of all records from the database ISI Web of Science® (24 700 publications) in different fields; chemistry: 12%; pharmacology: 20%; neurosciences: 14%). Search term: "caffeine" (search performed on 8 February 2012).

3.1 Nomenclature and Generic Properties

Caffeine is an organic molecule made up of the four most common elements: carbon, hydrogen, nitrogen and oxygen and has the molecular formula $C_8H_{10}N_4O_2$. Caffeine is the common name for the chemical trimethylxanthine (Figure 3.2), the systematic names of which include 1,3,7-trimethylxanthine (IUPAC); 3,7-dihydro-1,3,7-trimethyl-1H-purine-2,6-dione (IUPAC); 1,3,7-trimethyl-2,6-dioxopurine; caffeine; coffeinum; thein; theine; guaranine; methyltheobromine; methyltheophylline; mateine (O'Neil 2006). It is registered in the Chemical Abstract Service (CAS) as 3,7-dihydro-1,3,7-trimethyl-1H-purine-2,6-dione with the registration number 58-08-2 (CAS 2010). Other identifiers are listed in Table 3.1.

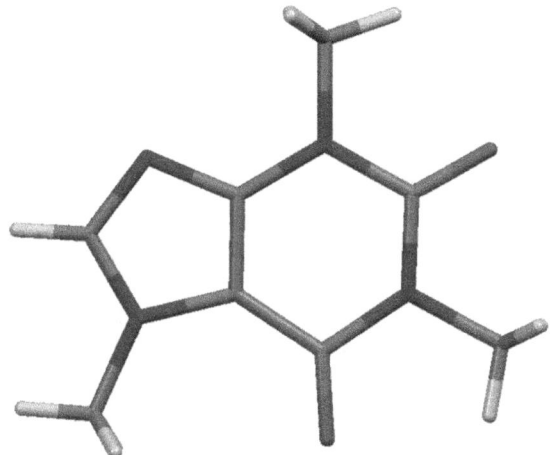

Figure 3.2 Caffeine, the chemical. 1,3,7-Trimethyl-2,6-dioxopurine (caffeine).

Table 3.1 Chemical identity and registration codes of caffeine.

Molecular Formula	$C_8H_{10}N_4O_2$
Molecular Weight	194.1906 g mol^{-1}
Chemical Composition	Carbon (49.48%)
	Hydrogen (5.19%)
	Nitrogen (28.85%)
	Oxygen (16.48%) (O'Neil, 2006)
CAS No.	58-08-2
Beilstein Registry Number	17705
EC Number	200-362-1
MDL Number	MFCD00005758
PubChem Substance ID	24892984
HSDB number	36

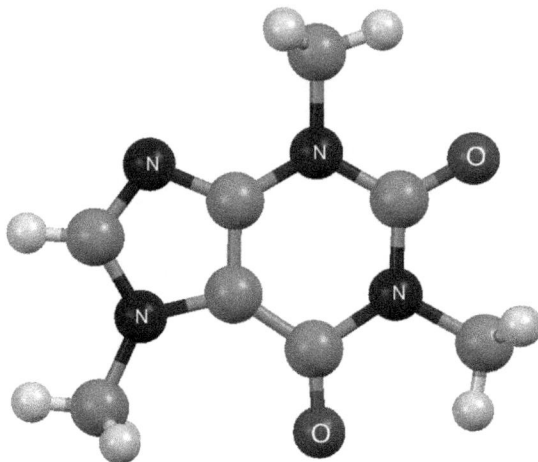

Figure 3.3   Molecular structure of caffeine. Structural formula of caffeine; atom colour coding: dark grey = carbon, light grey = hydrogen. The xanthine ring is a double ring composed of one six- and one five-membered heterocycle.

Caffeine is a naturally occurring alkaloid, and the principal member of the methylated xanthine alkaloids, commonly found in tea leaves, coffee beans, mate leaves, guarana paste and kola (cola) nuts. Caffeine can be obtained by both extraction of natural products (green coffee beans, tea leaves (also their dusts and wastes), kola nuts, *etc.*) using organic solvents (HSDBank 2011) or obtained as a by-product of the production of caffeine-free coffee and other decaffeinated consumer products. Semi-synthetic processes involve the methylation of various xanthines, including theophylline and theobromine. Moreover, *de novo* synthesis using dimethylurea and malonic acid as building blocks is feasible but less common (Burdock 2009).

Caffeine belongs to the group of purine alkaloids, often referred to as methylxanthines, which include theophylline, theobromine, and paraxanthine, among others (Carvalho *et al* 2010). The basic structure of these compounds is the purine nucleus – an aromatic heterocycle which is also the base of DNA and RNA bases like adenine and guanidine (Table 3.2).

The different xanthines (see Table 3.2) are classified by number and position of the methyl groups: caffeine (three methyl groups, on carbon 1, 3 and 7); theophylline (two methyl groups, on carbon 1 and 3); theobromine (two methyl groups, on carbon 3 and 7); paraxanthine (two methyl groups, on carbon 1 and 7); and xanthine itself without methyl groups.

3.2   Physicochemical Properties

Caffeine's physicochemical properties are listed in Table 3.3. At room temperature, caffeine is a white fleecy powder, similar to corn starch, or

Table 3.2 Caffeine, xanthines, and the purine structure.

Compound	CAS no.	Structure
Caffeine 1,3,7-Trimethylxanthine	58-08-2	
Theophylline 1,3-Dimethylxanthine	58-55-9	
Theobromine 3,7-Dimethylxanthine	83-67-0	
Paraxanthine 1,7-Dimethylxanthine	611-59-6	
Xanthine	69-89-6	
Purine	120-73-0	

Table 3.3 Physicochemical properties of caffeine.

Color and form	White powder, prismatic crystals (PubChem 2007)
Odor and taste	Odorless with bitter taste (O'Neil 2006)
Physical state	Solid (O'Neil 2006)
Boiling point	Not available; 178 °C (sublimates)[a] (O'Neil 2006)
Melting point	238 °C (O'Neil 2006)
	236 – 236.5 °C (European Pharmacopoeia 2008)
Specific density (18 °C)	1.23 (O'Neil 2006)
Vapor pressure	20 hPa at 89 °C (Sigma-Aldrich 2008)
pH of a 1% solution in water ($\sim$ 10 g L^{-1})	6.9 (O'Neil 2006)
	5.5 – 6.5 at 20 °C (Sigma-Aldrich 2008)
Solubility in water (2.17 g 100 mL^{-1})	1 g dissolves in:
	46 mL water at 20 °C
	5.5 mL water at 80 °C
	1.5 mL boiling water (O'Neil 2006)
Solubility in organic solvents	1 g dissolves in:
	66 mL ethanol
	22 mL ethanol at 60 °C
	50 mL acetone
	5.5 mL chloroform
	530 mL diethyl ether
	100 mL benzene
	22 mL boiling benzene
	Soluble in pyridine, ethyl acetate; slightly soluble in petroleum ether (O'Neil 2006).
Octanol–water partition coefficient	log K_{ow} = −0.07 (HSDBank 2011)
Dissociation constant	pKa = 10.4 (10 °C) (PubChem 2007)
Stability	Decomposed by concentrated alkaline solutions (HSDBank 2011).

[a]Fast sublimation is obtained at 160–165 °C under 1 mm Hg pressure at 5 mm distance.

appears in long prismatic crystals (O'Neil 2006). It is odorless and the pure product has a slightly bitter taste (HSDBank 2011). When subjected to temperature increase, *e.g.* in order to determine the melting point, caffeine will not melt but sublimates, *i.e.* similarly to dry ice it passes directly from the solid state to the gas phase without passing through a liquid phase. It sublimes at 178 °C and fast sublimation is obtained between 160–166 °C (O'Neil 2006). The melting point (mp) of caffeine is reported to be 238 °C (O'Neil 2006; HSDBank 2011). Other values (234–239 °C) can be found in the literature and reference books if the substance is previously dried in an oven at 100–105 °C (European Pharmacopoeia 2008).

Caffeine is equally soluble in oil and water, expressed by its estimated octanol–water partition coefficient (log K_{ow}) close to zero (Sigma-Aldrich 2008). It is soluble in chloroform (182 g L^{-1}) but also in water at room temperature and very soluble in boiling water (667 g L^{-1}). The solubility in water can also be increased by adding alkali benzoates, cinnamates, citrates or salicylates to the aqueous solution (WHO-IARC 1991). Its pK$_a$ is 10.4. A solution containing 1 gram of caffeine in 100 mL water will have a pH of 6.9.

The pure product is claimed to be stable but will decompose in the presence of alkali (WHO-IARC 1991) and by heating (HSDBank 2011). Polymerization of caffeine does not occur. No data is available concerning caffeine reactivity towards other substances, however it has been reported as non-corrosive to glass by the manufacturers in the product MSDS – Material Safety Data Sheet (Sciencelab 2010). The product must be stored in airtight containers (HSDBank 2011).

3.3 Spectral Data

Caffeine presents a maximum absorption in the UV-Vis range at 276 nm, which is the wavelength used in most official UV-Vis analytical methods (Ohnsmann *et al* 2002; Belay *et al* 2008). The spectrum obtained by infrared spectroscopy (FT-IR) shows bands between 1800 and 1500 cm^{-1}, an additional band at 1050 cm^{-1}, and the most intense bands at 1710 (carbonyl group), 1659 (carbonyl group) and 1554 cm^{-1}, respectively (Garrigues *et al* 2000).

The mass spectra of caffeine after electron impact (EI) ionization and collision-induced dissociation, under vacuum conditions, will present the typical molecular ion at *m/z* 194.0 and fragments at *m/z* 108.9; 67.0; 82.0; 137.0; 165.0 (Verenitch and Mazumder 2008). Ionization at atmospheric pressure using an electrospray source (ESI) will generate typically *m/z* of 195.0 (molecular ion plus a proton, [M+H]$^+$) and *m/z* 138.0 and *m/z* 110.0 as most intense fragment ions after collision in a quadrupole cell (Carvalho 2011).

3.4 Technical Substance, Identification and Impurities

Caffeine can be obtained in many commodities and forms, related to its various uses. For medical and cosmetic purposes, specifications are described for either the anhydrous form or the monohydrate by the national pharmacopoeias, as well as the European Pharmacopoeia (European Pharmacopoeia 2008). The anhydrous form of caffeine must obey the following specifications: contain not less than 98.5% and not more than the equivalent of 101.5% (m/m, by dry weight) of the active ingredient, *i.e.* caffeine.

The first identification is made by the melting point (must be within 234 °C and 239 °C), by infrared spectrophotometry referenced to a caffeine certified reference material (CRM) (*e.g.* Fluka TraceCERT® 56396) or by the test of loss on drying (0.5% max. weight loss when drying 1.000 g in an oven at 105 °C for 1 hour).

For the second identification, three tests are described in the European Pharmacopoeia including a selective precipitation reaction and a specific reaction on xanthines (European Pharmacopoeia 2008).

The most relevant impurity described for caffeine technical material is theophylline. Other relevant impurities are *N*-[6-amino-1,2-dimethyl-

2,4(1H,3H)-dioxopyrimidin-5-yl]formamide and isocaffeine (1,3,9-trimethyl-3,9-dihydro-1H-purine-2,6-dione) (European Pharmacopoeia 2008).

3.5 Crystal Structure

With respect to its structure, caffeine is experiencing a renaissance in the current scientific discussion (Leiterer *et al* 2008). Even though caffeine has such

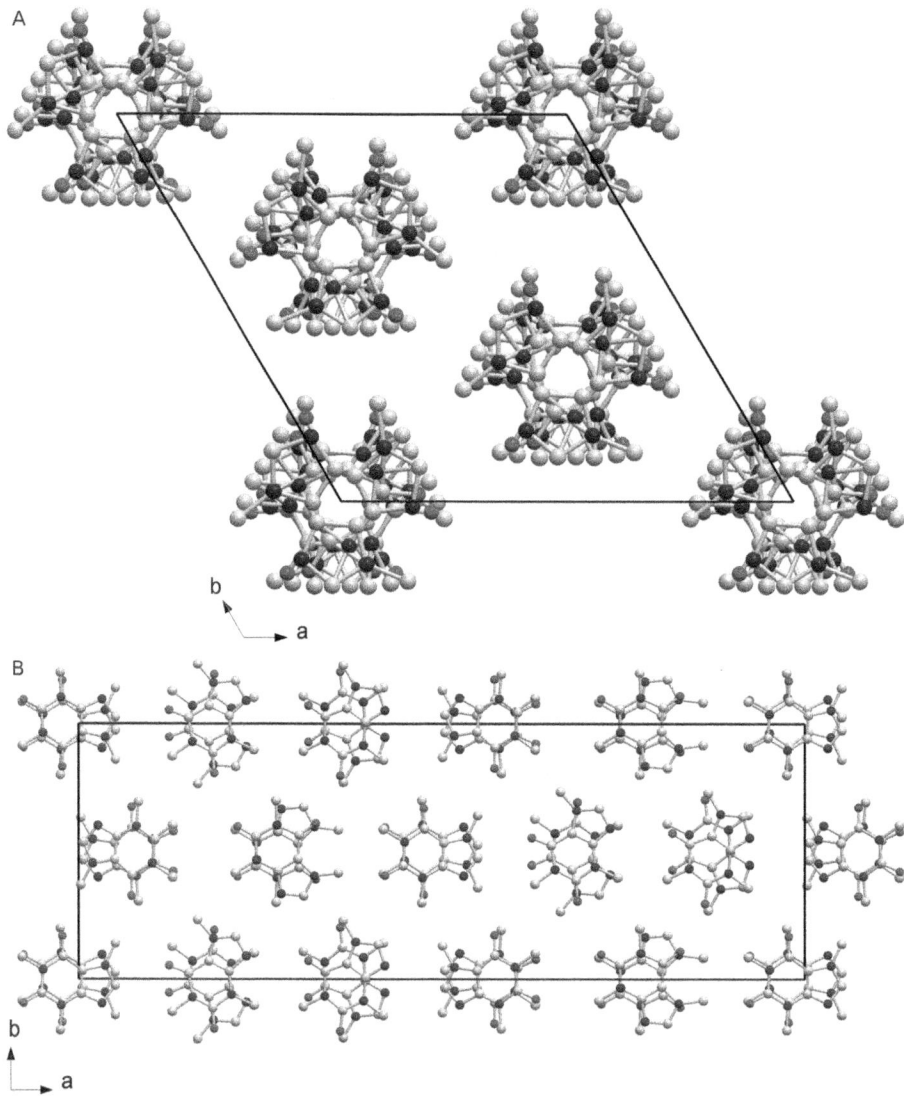

Figure 3.4   Crystal structures of caffeine polymorphs. Crystal structure of α-caffeine (A) and β-caffeine (B) polymorphs.

Figure 3.5   Hexagonal caffeine crystallites. ESEM images of deposits with hexagonal symmetry and magnified image of a single hexagon. Deposition was carried out on silicon surface using dichloromethane as solvent.

a wide use in food and even as a pharmaceutical agent, the crystal structures of the polymorphs (α, β) were only recently characterized (Derollez *et al* 2005; Lehmann and Stowasser 2007). The room temperature modification (β) crystallizes in the monoclinic space group *Cc* and contains five independent caffeine molecules in the asymmetric unit, whereas the high temperature α-form (*R-3c*) is dynamically disordered (Figure 3.4). Moreover, the alpha form is metastable at room temperature and is known to transform to the beta form at different rates depending on temperature and humidity. Under defined conditions a unique, surface mediated morphology of caffeine crystals with nearly perfect hexagonal shape can be observed (Figure 3.5) (Sarfraz *et al* 2012). When caffeine is crystallized from aqueous solutions, the formation of the monohydrate is observed, the crystal structure of which was first determined by Sutor in 1958 (Sutor 1958). A synthetic derivate of caffeine also crystallized in a monoclinic space group ($P2_1/c$) (Carvalho *et al* 2007). The monohydrate converts quantitatively to the anhydrous β-modification at room temperature (Pirttimaki *et al* 1993; Griesser *et al* 1999). This modification undergoes a phase transition when heated above 150 °C, yielding the α-polymorph (Derollez *et al* 2005). After cooling to room temperature, the α-modification is metastable.

3.6   Hazard Classification

According to European Union regulations – REACH Regulation 1272/2008 (EU 2008) – caffeine and caffeine mixtures are classified to fall into category 4 ("warning") regarding acute toxicity, which means an acute toxicity estimation (ATE) between 300 mg and 2000 mg kg^{-1} body weight when ingested and between 1000 to 2000 mg kg^{-1} body weight with dermal exposure (ECHA

2012). The International Agency for Research on Cancer (World Health Organization) classified caffeine in group 3, meaning not carcinogenic to humans and stating in its report "there is inadequate evidence for the carcinogenicity in humans and experimental animals" (WHO-IARC 1991).

Summary Points

- This chapter focuses on the chemical and physical properties of caffeine.
- Caffeine is a derivate of purine, more specifically, of the xanthine heterocyclic structure substituted with 3 methyl groups.
- There are recent discoveries: caffeine crystallizes in an alpha and a beta polymorph and in nearly perfect hexagons.

List of abbreviations

ATE	Acute Toxicity Estimation
CAS No.	Chemical Abstract Service Registration Number
cm^{-1}	Wavenumber (reciprocal of the wavelength)
CRM	Certified Reference Material
DNA	Deoxyribonucleic Acid
e.g.	Latin: exempli gratia (for example)
ESEM	Environmental Scanning Electron Microscope
etc.	Latin: et cetera (and the rest)
FT-IR	Fourier-Transform Infrared (spectroscopy)
HSDB	Hazardous Substances Databank
ID	Identification Code
i.e.	Latin: id est (that is)
IUPAC	International Union of Pure and Applied Chemistry
K_{ow}	Octanol–Water Partition Coefficient (measure for lipophilicity)
LC-MS/MS	Liquid Chromatography-Tandem Mass Spectrometry
mp	Melting Point
MSDS	Material Safety Data Sheet
(m/m)	Mass Fraction: the fraction of the mass of one substance in the mass of the total mixture
pK_a	Acid Dissociation Constant
RNA	Ribonucleic Acid
RT	Room Temperature, usually 20 °C
UV-Vis	Spectral region of Ultraviolet-Visible light

References

Bach, J.S. 1732–1734. The coffee cantata (libretto by Christian F. Henrici): Schweigt stille, plaudert nicht. Leipzig.

Belay, A., Ture, K., Redi, M. and Asfaw, A., 2008. Measurement of caffeine in coffee beans with UV/Vis spectrometer. Food Chemistry. 108: 310–315.

Burdock, G. A. E., 2009. Fenaroli's Handbook of Flavor Ingredients. 6th Edition. CRC Press, Boca Raton, FL., 229–239.

Carvalho, J.J. 2011. Immunochemical and chromatographic methods for two anthropogenic markers of contamination in surface waters: caffeine and coprostanol. PhD Thesis, Humboldt-Universitaet zu Berlin, e-doc service: http://edoc.hu-berlin.de/docviews/abstract.php?id=38981

Carvalho, J.J., Emmerling, F. and Schneider, R.J., 2007. 7-(5-Carboxypentyl)-1,3-dimethyl-xanthine monohydrate. Acta Crystallographica, Section E: Structure Reports Online. 63: O3718–U2211.

Carvalho, J.J., Weller, M.G., Panne, U. and Schneider, R.J., 2010. A highly sensitive caffeine immunoassay based on a monoclonal antibody. Analytical and Bioanalytical Chemistry. 396: 2617–2628.

CAS, 2010. Chemical Abstracts Service - online database: www.cas.org. American Chemical Society.

Derollez, P., Correia, N.T., Danede, F., Capet, F., Affouard, F., Lefebvre, J. and Descamps, M., 2005. Ab initio structure determination of the high-temperature phase of anhydrous caffeine by X-ray powder diffraction. Acta Crystallographica, Section B: Structural Science. 61: 329–334.

ECHA, 2012. European Chemicals Agengy database for registration of chemical subtances under Regulation (EC) No. 1272/2008 of the European Parliament and of the Council of 16 December 2008. Free access to registration data: http://echa.europa.eu/web/guest

EU, 2008. Regulation (EC) No. 1272/2008 of the European Parliament and of the Council of 16 December 2008 on classification, labelling and packaging of substances and mixtures. Official Journal of the European Communities. L353/ 31.12.2008: 81.

European Pharmacopoeia, 2008. Caffeine monograph. European Directorate for the Quality of Medicines and Healthcare - The Council of Europe.

Garrigues, J.M., Bouhsain, Z., Garrigues, S. and de la Guardia, M., 2000. Fourier transform infrared determination of caffeine in roasted coffee samples. Fresenius Journal of Analytical Chemistry. 366: 319–322.

Griesser, U.J., Szelagiewicz, M., Hofmeier, U.C., Pitt, C. and Cianferani, S., 1999. Vapor pressure and heat of sublimation of crystal polymorphs. Journal of Thermal Analysis and Calorimetry. 57: 45–60.

HSDBank, 2011. The Hazardous Substances Data Bank. United States National Library of Medicine: ID 36.

Lehmann, C.W. and Stowasser, F., 2007. The crystal structure of anhydrous beta-caffeine as determined from X-ray powder-diffraction data. Chemistry - A European Journal. 13: 2908–2911.

Leiterer, J., Emmerling, F., Panne, U., Christen, W. and Rademann, K., 2008. Tracing coffee tabletop traces. Langmuir. 24: 7970–7978.

O'Neil, M.J., 2006. The Merck Index - An Encyclopedia of Chemicals, Drugs, and Biologicals. 14th Edition. Whitehouse Station, NJ: Merck and Co., Inc.

Ohnsmann, J., Quintas, G., Garrigues, S. and de la Guardia, M., 2002. Determination of caffeine in tea samples by Fourier transform infrared spectrometry. Analytical and Bioanalytical Chemistry. 374: 561–565.

Pirttimaki, J., Laine, E., Ketolainen, J. and Paronen, P., 1993. Effects of Grinding and Compression on Crystal-Structure of Anhydrous Caffeine. International Journal of Pharmaceutics. 95: 93–99.

PubChem, 2007. Online database. National Center for Biotechnology Information (NCBI).

Renneberg, R., Bofinger, M. and Chow, M.F., 2011. Who cloned my cat? Fun adventures in biotechnology. Chapter 47 - Goethe and the Caffeine Pan. Stanford Publishing Pte. Ltd., Singapore.

Sarfraz, A., Simo, A., Fenger, R., Christen, W., Rademann, K., Panne, U. and Emmerling, F., 2012. Morphological Diversity of Caffeine on Surfaces: Needles and Hexagons. Crystal Growth Design. DOI: dx.doi.org/10.1021/cg101358q

Sciencelab, 2010. Caffeine Material Safety Data Sheet (MSDS). Sciencelab.com, Inc., Houston, Texas.

Sigma-Aldrich, 2008. Caffeine Material Safety Data Sheet (MSDS). Sigma-Aldrich Chemie GmbH, Steinheim.

Sutor, D.J., 1958. The Structures of the Pyrimidines and Purines. VI. The Crystal Structure of Theophylline. Acta Crystallographica. 11: 83–87.

Verenitch, S. S. and Mazumder, A., 2008. Development of a methodology utilizing gas chromatography ion-trap tandem mass spectrometry for the determination of low levels of caffeine in surface marine and freshwater samples. Analytical and Bioanalytical Chemistry. 391: 2635–2646.

Weiberg, B.A. and Bealer, B.K., 2001. The world of caffeine - the science and culture of the world's most popular drug, 1st edition. Routledge, New York, USA.

WHO-IARC. 1991. IARC Monographs on the Evaluation of Carcinogenic Risk to Humans: Caffeine. International Agency for Research on Cancer - World Health Organization 51.

CHAPTER 4

Chemistry and Crystal Structures of Complexes of Caffeine and Tea Catechin

TAKASHI ISHIZU*[1] AND HIROYUKI TSUTSUMI[2]

[1] Professor of Organic and Bio-organic Chemistry, Faculty of Pharmacy and Pharmaceutical Sciences, Fukuyama University, 1 Sanzo, Gakuen-cho, Fukuyama Hiroshima, 729-0292, Japan; [2] Research Assistant of Organic and Bio-organic Chemistry, Faculty of Pharmacy and Pharmaceutical Sciences, Fukuyama University, 1 Sanzo, Gakuen-cho, Fukuyama Hiroshima, 729-0292, Japan
*E-mail: ishizu@fupharm.fukuyama-u.ac.jp

4.1 Introduction

Tea (*Camellia sinensis*, Camelliaceae) has been consumed throughout the world since ancient times to maintain and improve health. It is well known that tea protects against lifestyle-related diseases such as cancer, high blood pressure, diabetes, obesity, and arteriosclerosis (Kuroda and Hara 2004). Tea leaves contain many characteristic constituents, caffeine, catechins, theanine, vitamins, saponin, *etc*.

Caffeine (2009), which was isolated from coffee in 1820 by a German chemist, Friedlieb Ferdinand Runge, is a bitter and white crystalline xanthine alkaloid that is a psychoactive stimulant. Caffeine is a naturally occurring stimulant found in the leaves, seeds, or fruit of over sixty plants, where it acts as a natural pesticide that paralyzes and kills certain insects feeding on the

Food and Nutritional Components in Focus No. 2
Caffeine: Chemistry, Analysis, Function and Effects
Edited by Victor R Preedy

plants (Nathanson 1984). In humans, caffeine acts as a central nervous system (CNS) stimulant, temporarily warding off drowsiness and restoring alertness. Then caffeine is best known for its stimulant, or "wake-up," effect. Caffeine mildly stimulates the nervous and cardiovascular systems. It affects the brain and results in elevated mood, decreased fatigue, and increased attentiveness, so a person can think more clearly and work harder.

And the catechins, which are also found in tea leaves as one of the major ingredients, are a group of polyphenols that show various physiologically modulating effects such as anti-carcinogenic (Ahmad *et al* 2000; Lambert and Yang 2003), anti-metastatic (Sazuka *et al* 1997; Maeda-Yamamoto *et al* 1999), and anti-oxidative (Kimura *et al* 2002; Hashimoto *et al* 2003). The major eight catechins in green tea are mainly classified into two categories by the existence of a galloyl group on the oxygen atom at the C3 position, non-galloylated and galloylated catechins (Fig. 4.1) (Hayashi and Ujihara 2008). Generally, galloylated catechins show higher activities than non-galloylated catechins (Hara and Watanabe 1989; Mimura *et al* 1994; Tezuka *et al* 1997; Okabe *et al* 1997).

Interestingly, it is known that caffeine forms complexes with catechins in black tea (Horman and Viani 1972; Martin *et al* 1986; Gaffney *et al* 1986). Such complexes are thought to be unique stereochemical structures with interesting intermolecular interactions between caffeine and catechins. Also, it is very important to elucidate the stereochemical structures and the intermolecular interactions to explain the character and function of the complexes of caffeine and catechins.

Thus, many researchers have been investigating the structure of the complexes of caffeine and catechins. Maruyama *et al* noted that some galloylated catechins have a strong affinity for caffeine, and assumed that these catechins bound caffeine molecules in the space formed from B and B' rings on the basis of ^{1}H NMR chemical shift changes in galloylated catechins (Maruyama *et al* 1991). Haslam *et al* reported that in non-galloylated catechins such as (+)-catechin (CA) and (−)-epicatechin (EC), the A and C rings provided a general site for caffeine association, but in galloylated catechins such as (+)-catechin-3-*O*-gallate (Cg) and (−)-epigallocatechin-3-*O*-gallate (EGCg), the galloyl ester becomes the preferred site for complexation (Cai *et al* 1990). Furthermore, Hayashi *et al* reported that an investigation of the ^{1}H NMR chemical shift change and NOESY spectra in catechins and caffeine solution showed the participation of A rings of catechins in complexation, as well as B or B' rings (Hayashi *et al* 2004). These structural studies of the complexes of caffeine and catechins were performed in solution using NMR techniques, but their complete structures were still unclear and the detailed interactions between caffeine and catechins have not been elucidated sufficiently.

We have prepared crystals of complexes of caffeine and various tea catechins, and investigated their stereochemical structures and intermolecular interaction by X-ray crystallographic analysis (Ishizu *et al* 2009a; 2009b). First

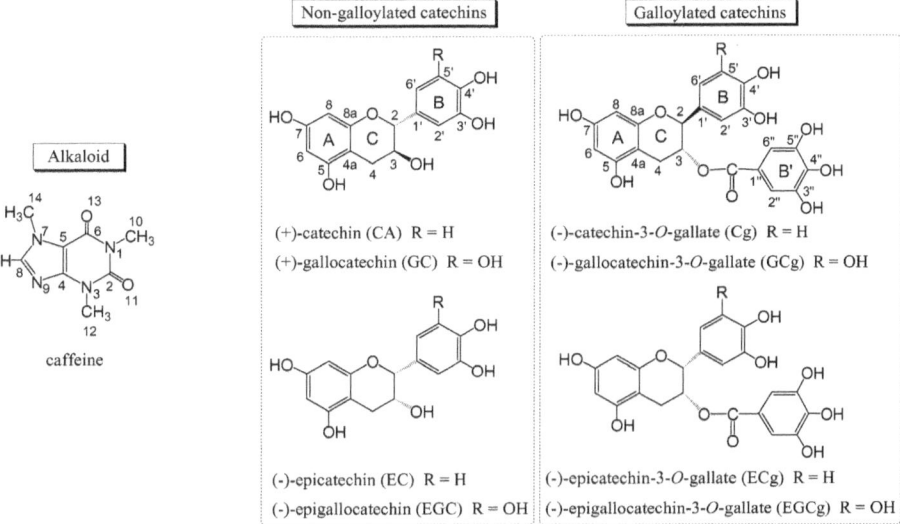

Figure 4.1   Structures of caffeine and tea catechins. Major catechins in green tea are mainly classified into two categories by the existence of a galloyl group on the oxygen atom at the C3 position, non-galloylated and galloylated catechins.

the stereochemical structure of complexes of caffeine and CA and EC of non-galloylated catechins were investigated. Crystals of the complexes of caffeine and two non-galloylated catechins, CA and EC, were prepared, and the stereochemical structure and intermolecular interactions were investigated and compared with those of the complexes of caffeine and CA, and caffeine and EC.

Next, we focused on CA of non-galloylated catechins and Cg of galloylated catechins. The structural difference between CA and Cg is the presence or absence of a galloyl group on the oxygen atom at the C3 position, and all others are the same. X-Ray crystallographic analysis of complexes of caffeine and CA, Cg were performed to determine the stereochemical structures and to elucidate the detailed non-covalent interaction between caffeine and CA, Cg moieties. Furthermore, we investigated the difference of non-covalent bonds in the stereochemical structures having a galloyl group or not.

4.2 Stereochemical Structures of Complexes of Caffeine and Catechins

4.2.1 Complex of Caffeine and CA

A solution containing equimolecular amounts of caffeine and CA in water was lyophilized to give a pale orange powder, which was recrystallized from methanol to afford colorless needles. The crystal structure was determined to

(a)   (b)

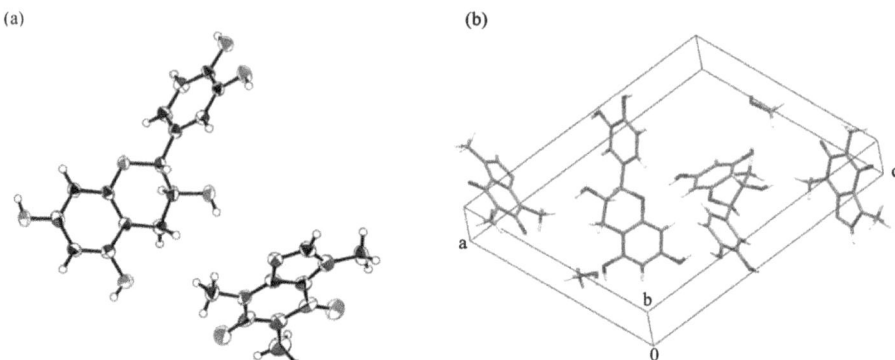

Figure 4.2   Crystal structure of 1 : 1 complex of caffeine and CA. (a) ORTEP
drawing (b) One unit cell.

be a 1 : 1 complex of caffeine and CA by X-ray crystallographic analysis
(Tsutsumi *et al* 2011). One unit cell contained 2 caffeine, 2 CA, and 2 methanol
molecules as crystal solvent (Fig. 4.2(b)).

The crystal structure of CA alone was also determined by X-ray crystallographic analysis to compare with the CA moiety of the 1 : 1 complex of
caffeine and CA. One unit cell contained two CA molecules and four methanol
molecules as crystal solvent (Fig. 4.3(b)).

Torsion angles of caffeine moieties of all complexes of caffeine and CA, EC,
Cg are shown in Table 4.1, indicating that every caffeine moiety has an almost
plain and rigid xanthine skeleton. On the other hand, catechin molecules have
conformational flexibility, including orientation of the linkage between B, B'
and C rings owing to puckering of the pyran C ring. The torsion angles of H2-
C2-C3-H3 and C1'-C2-C3-O in the CA moiety of the 1 : 1 complex of caffeine
and CA were 169° and 48.3(7)°, and those in CA alone were 170° and 53.3(6)°,
respectively. These results indicated that both B ring and hydroxyl group of
CA moiety of the 1 : 1 complex and CA alone were in equatorial positions with
respect to the C ring of the CA molecule.

(a)   (b)

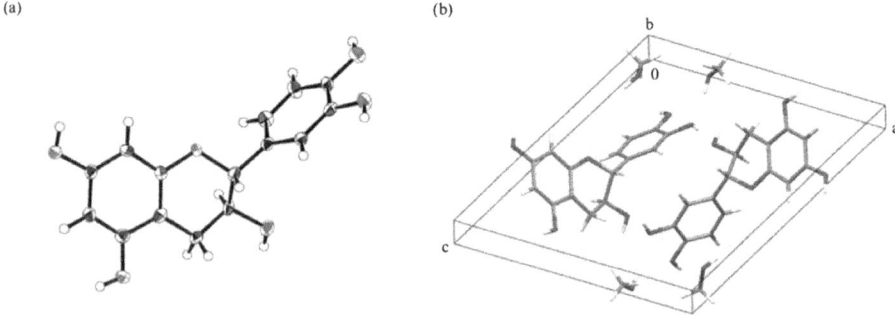

Figure 4.3   Crystal structure of CA alone. (a) ORTEP drawing (b) One unit cell.

Table 4.1 Torsion angles of caffeine moiety in complexes of caffeine and catechins.

	1:1 complex of caffeine and CA	*1:1 complex of caffeine and EC*	*2:1 complex of caffeine and Cg*	*2:1:1 complex of caffeine and CA, EC*
Torsion angle of caffeine A				
∠N7-C5-C6-O13	1.3(11)°	1.3(15)°	11(5)°	2(3)°
∠O13-C6-N1-C10	4.2(9)°	0.6(7)°	6(4)°	11(2)°
∠O11-C2-N3-C12	4.9(9)°	2.3(12)°	3(4)°	4(2)°
∠C14-N7-C8-N9	176.0(5)°	178.7(7)°	180(2)°	169(2)°
∠C10-N1-C2-O11	3.8(9)°	1.8(12)°	1(3)°	4(2)°
∠C12-N3-C4-N9	5.9(9)°	1.0(12)°	4(3)°	5(3)°
Torsion angle of caffeine B				
∠N7-C5-C6-O13	–	–	3(3)°	1(2)°
∠O13-C6-N1-C10	–	–	0.5(7)°	5(2)°
∠O11-C2-N3-C12	–	–	2(3)°	2(3)°
∠C14-N7-C8-N9	–	–	175.8(13)°	178(2)°
∠C10-N1-C2-O11	–	–	1(3)°	4(2)°
∠C12-N3-C4-N9	–	–	4(3)°	9(2)°

Numbers in parentheses indicate standard deviations.

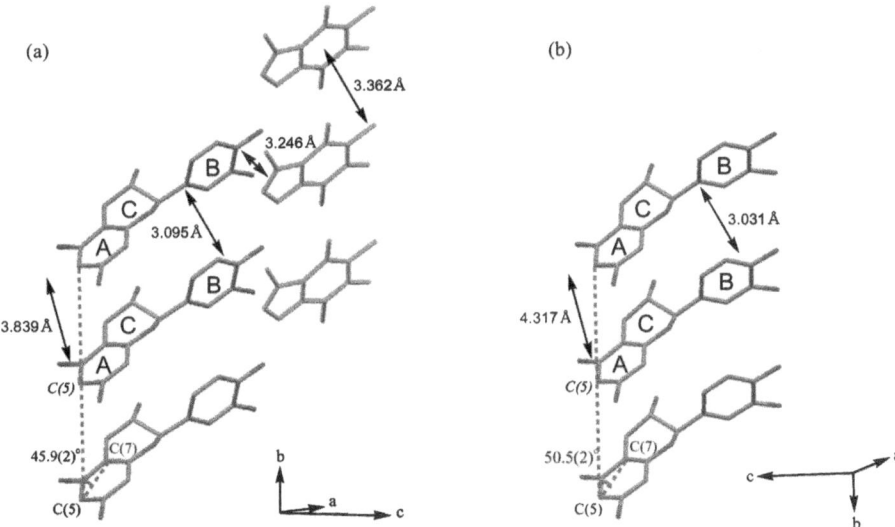

Figure 4.4 Layer structures of 1 : 1 complex of caffeine and CA, and CA alone. Crystal solvent and hydrogen atoms are omitted for clarity. Both black arrows indicate the distances between planes. (a) 1 : 1 complex of caffeine and CA (b) CA alone.

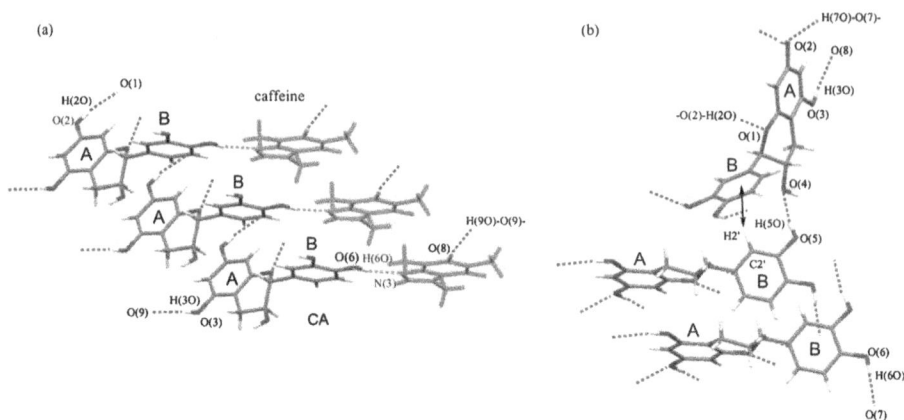

Figure 4.5 Intermolecular interactions of the 1 : 1 complex of caffeine and CA, and CA alone. (a) 1 : 1 complex of caffeine and CA. Black and gray dotted-lines indicate O–H...O and O–H...N hydrogen bonds, respectively. (b) CA alone. Black arrow indicates CH–π interaction.

Table 4.2 Hydrogen bonds in 1 : 1 complex of caffeine and CA.

D-H	A	D...A	D-H	H...A	∠D-H...A
OH(2O)[a]	O(1)	2.823(6)	0.83	2.00	172.69
OH(3O)	O(9)	2.672(7)	0.84	1.84	167.73
OH(6O)	N(3)	2.770(7)	0.82	1.98	161.96
OH(9O)	O(8)	2.727(6)	0.82	1.93	163.65

[a]Atom label in ORTEP drawing. D, H and A indicate hydrogen-bonding donor, hydrogen atom and hydrogen-bonding acceptor, respectively. A unit of length is Å.

Table 4.3 Hydrogen bonds in CA alone.

D-H	A	D...A	D-H	H...A	∠D-H...A
OH(2O)[a]	O(1)	2.801(5)	0.84	1.97	169.74
OH(3O)	O(8)	2.762(7)	0.84	1.93	171.84
OH(5O)	O(4)	2.684(7)	0.84	1.90	153.77
OH(6O)	O(7)	2.685(7)	0.84	1.87	162.53
OH(7O)	O(2)	2.856(8)	0.84	2.04	163.82

[a]Atom label in ORTEP drawing. D, H and A indicate hydrogen-bonding donor, hydrogen atom and hydrogen-bonding acceptor, respectively. A unit of length is Å.

The layer structures of the 1 : 1 complex of caffeine and CA and CA alone were investigated. As shown in Fig. 4.4(a), units of the 1 : 1 complex piled up parallel to the b-axis, and the angle of C(7)-C(5)-$C(5)$ was 45.9(2)°. There is caffeine and the B ring of CA in the same plane, and the five-membered ring of caffeine and the B ring mutually shifted and accumulated. And in the layer structure of CA alone, CA faced the same direction and piled up parallel to the b-axis (Fig. 4.4(b)); the angle of C(7)-C(5)-$C(5)$ was 50.5(2)°. Herein, the CA moieties of the 1:1 complex resemble CAs in the layer structure.

In the layer structure of the 1 : 1 complex of caffeine and CA, three O–H...O and one O–H...N intermolecular hydrogen bonds were formed between CAs, caffeine and CA, caffeine and methanol, and five O–H...O intermolecular hydrogen bonds between CAs, CA and methanol were observed in the crystal structure of CA alone (Fig. 4.5, Tables 4.2 and 4.3).

4.2.2 Complex of Caffeine and EC

A solution containing equimolecular amounts of caffeine and EC in water was lyophilized to give a colorless powder. The powder was recrystallized from methanol to afford colorless block crystals. The single crystal was determined to be a 1 : 1 complex of caffeine and EC by X-ray crystallographic analysis, and one unit cell contained four units of the 1 : 1 complex of caffeine and EC and eight water molecules as crystal solvent (Fig. 4.6(b)).

The torsion angles of H2-C2-C3-H3 and C1'-C2-C3-O in the EC moiety of the 1 : 1 complex of caffeine and EC were 60.23° and 58.4(7)°, respectively. These results indicated that the B ring and 3-OH group of EC were in equatorial and axial positions with respect to the C ring of the EC molecule, respectively. In the layer structure shown in Fig. 4.7, units of the 1 : 1 complex of caffeine and EC stacked in parallel in the same direction as the b-axis. The six-membered ring of caffeine and the A ring of EC appear in turn along the b-axis, and the six-membered rings of caffeines were located in almost the middle

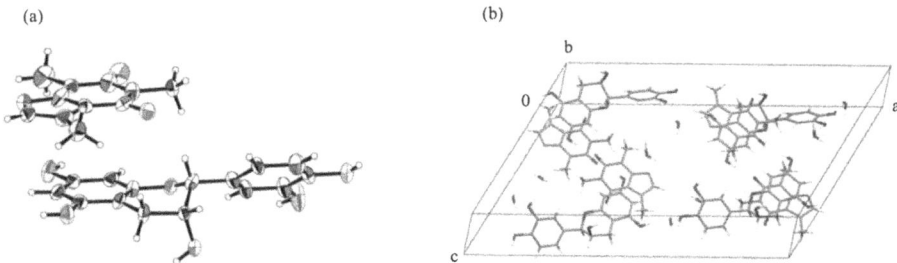

(a)   (b)

Figure 4.6 Crystal structure of the 1 : 1 complex of caffeine and EC. (a) ORTEP drawing with thermal ellipsoids at 30% probability level, (b) One unit cell.

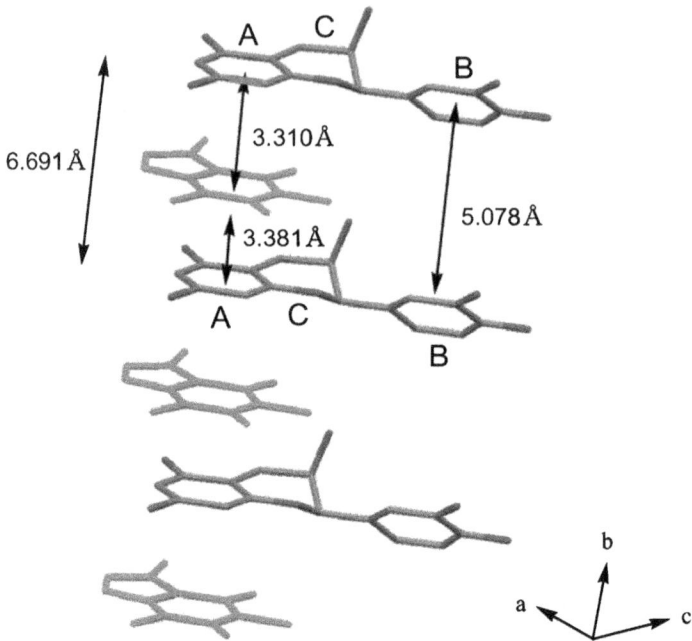

Figure 4.7   Layer of the 1 : 1 complex of caffeine and EC. Crystal solvent and hydrogen atoms are omitted for clarity. Black arrows indicate the distances between planes.

of the A rings of ECs. The A rings of both upper and lower ECs faced the six-membered ring of the caffeine (Fig. 4.7).

Non-covalent interactions are weaker forces than a covalent bond, but play a very important role in forming complexes; therefore non-covalent interactions among caffeine and catechin moieties in the complexes were investigated. In the 1 : 1 complex of caffeine and EC, face-to-face π–π stacking interaction formed between a six-membered ring of caffeine and the A ring of EC (Fig. 4.8). Two O–H...O intermolecular hydrogen bonds were also observed (Table 4.4).

Table 4.4   Hydrogen bonds in 1 : 1 complex of caffeine and EC.

D-H	A	D...A	D-H	H...A	∠D-H...A
OH(2O)[a]	O(7)	2.721(6)	0.83	1.89	177.95
OH(6O)	O(4)	2.719(7)	0.83	1.91	166.47

[a]Atom label in ORTEP drawing. D, H and A indicate hydrogen-bonding donor, hydrogen atom and hydrogen-bonding acceptor, respectively. A unit of length is Å.

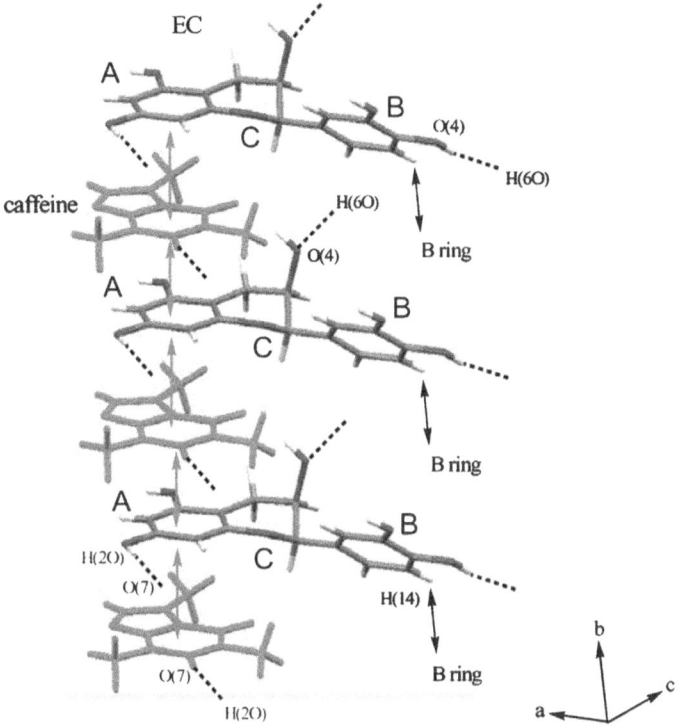

Figure 4.8 Intermolecular interactions in the 1 : 1 complex of caffeine and EC. Black dotted-lines and gray and black arrows indicate O-H...O hydrogen bonds and face-to-face π-π and CH-π interactions, respectively.

4.2.3 Complex of caffeine and CA, EC

Solutions containing caffeine and CA, EC of molar ratios 2 : 1 : 1, 4 : 1 : 1, and 4 : 2 : 1 in water gave sticky substances of molar ratios 2 : 1 : 1, 4 : 1 : 1, and 4 : 2 : 1 based on measurement of the integral volume of ^{1}H NMR signals, respectively, while a solution containing caffeine and CA, EC of molar ratio 1 : 1 : 1 in water did not give such a sticky substance. In the three sticky substances, only the sticky substance with molar ratio 2 : 1 : 1 gave colorless needles successfully. The single crystal structure was determined to be a 2 : 1 : 1 complex of caffeine and CA, EC by X-ray crystallographic analysis. One unit cell contained two units of the 2 : 1 : 1 complex and twelve water molecules as crystal solvent (Fig. 4.9(b)).

The torsion angles of H2-C2-C3-H3 and C1'-C2-C3-O in the CA and EC moieties of the 2 : 1 : 1 complex of caffeine and CA, EC were 169°and 48.3(7)°, 60°and 58.4(8)°, respectively. These results indicated that the B ring and hydroxyl group of the CA moiety of the 2 : 1 : 1 complex were both in equatorial positions with respect to the C ring of the CA molecule, while those

(a)   (b)

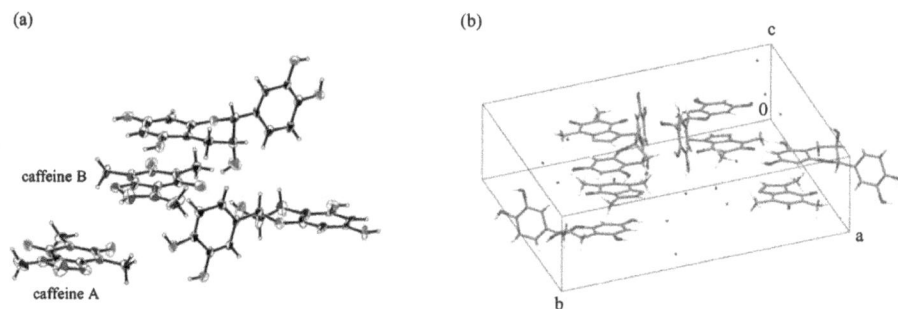

Figure 4.9 Crystal structure of the 2 : 1 : 1 complex of caffeine and CA, EC. (a)
ORTEP drawing with thermal ellipsoids at 30% probability level, (b) One
unit cell.

of the EC moiety of the 2 : 1 : 1 complex were in equatorial and axial positions
with respect to the C ring of the EC molecule, respectively. The crystal
structure of the 2 : 1 : 1 complex of caffeine and CA, EC had two layers; one
layer in which caffeine and CA formed alternate lines (Layer A) and the other
layer in which caffeine and EC formed alternate lines (Layer B) (Fig. 4.10).

Offset π–π interactions were formed between caffeine and CA in Layer A,
and between caffeine and EC in Layer B. In addition, CH–π interactions were
formed between the methyl group of N7 of caffeine and the B ring of CA, and
the methyl group of N7 of caffeine and the B ring of EC, and the CH–π
interactions connected Layers A and B. One O–H...O intermolecular hydrogen

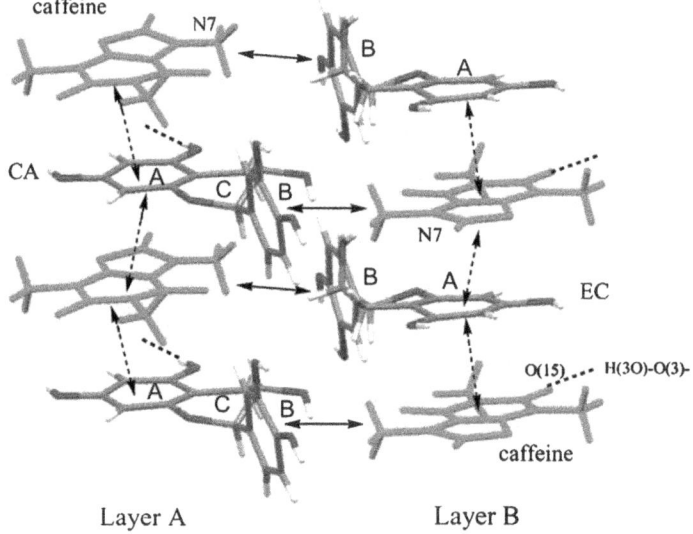

Figure 4.10 Intermolecular interactions in the 2 : 1 : 1 complex of caffeine and CA,
EC. Black dotted-lines and black arrows indicate O–H...O hydrogen
bonds and offset π–π and CH–π interaction, respectively.

Table 4.5   Hydrogen bonds in 2 : 1 : 1 complex of caffeine and CA, EC.

D-H	A	D...A	D-H	H...A	∠D-H...A
OH(3O)[a]	O(15)	2.67(2)	0.85	1.83	167.16

[a]Atom label in ORTEP drawing. D, H and A indicate hydrogen-bonding donor, hydrogen atom and hydrogen-bonding acceptor, respectively. A unit of length is Å.

bond between caffeine and CA was observed in the 2 : 1 : 1 complex (Table 4.5). Layer A was different from the layer structure of the 1 : 1 complex of caffeine and CA, which was formed by four intermolecular hydrogen bonds (Fig. 4.5(a)), while Layer B was very similar to the layer structure of the 1 : 1 complex of caffeine and EC, which was formed by face-to-face π–π interaction between the six-membered ring of caffeine and the A ring of EC, and two intermolecular hydrogen bonds (Fig. 4.8) (Ishizu *et al* 2010).

4.2.4   Complex of Caffeine and Cg

A suspension containing equimolecular amounts of caffeine and Cg in water gave a sticky substance, which afforded a colorless powder. The powder was recrystallized from water to give colorless plates of the complex of caffeine and Cg. The crystal structure was determined to be a 2 : 1 complex of caffeine and Cg by X-ray crystallographic analysis. In a unit of the 2 : 1 complex of caffeine and Cg, one caffeine molecule was located below the A ring of a Cg molecule, and the other was located above the B' ring of a Cg molecule. One unit cell contained one Cg molecule and two caffeine molecules (Fig. 4.11(b)).

The torsion angles of H2-C2-C3-H3 and C1'-C2-C3-O in the Cg moiety of the 2 : 1 complex of caffeine and Cg were 169° and 47.6(16)°, respectively. This

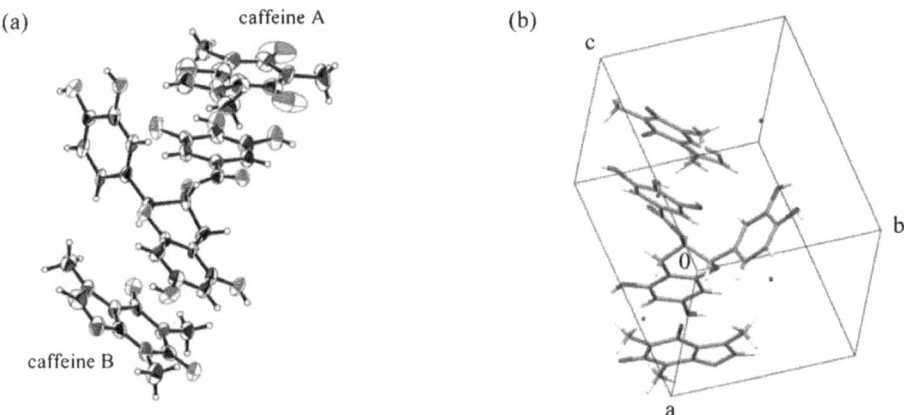

(a)   caffeine A

caffeine B

(b)

Figure 4.11   Crystal structure of the 2 : 1 complex of caffeine and Cg. (a) ORTEP drawing, (b) One unit cell.

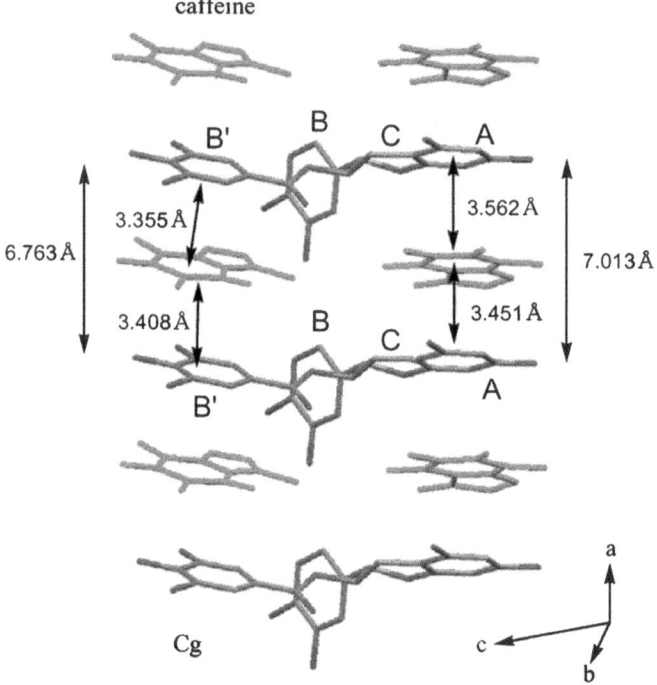

Figure 4.12 Layer of the 2 : 1 complex of caffeine and Cg. Crystal solvent and hydrogen atoms are omitted for clarity. Black arrows indicate the distance between planes.

indicated that both B and B' rings of Cg were both in equatorial positions with respect to the C ring of the Cg molecule. The layer structure of 2 : 1 complex of caffeine and Cg was investigated. In the layer structure of the 2 : 1 complex of caffeine and Cg, units of the 2 : 1 complex piled up parallel to the a-axis (Fig. 4.12). The distances between A rings of Cgs, and B' rings of Cgs were 7.013Å and 6.763Å, respectively. Two caffeine molecules were located in almost the middle of the A rings of Cgs and B' rings of Cgs. The A and B' rings of the upper Cg and the B' rings of the lower Cg slightly shifted to the six-membered rings of caffeine, and the A rings of the lower Cg were face to face with six-membered rings of caffeine.

Intermolecular interactions between caffeine and Cg moieties to form the 2 : 1 complex were investigated. The two caffeine molecules were sandwiched between the A and B' rings of Cgs by offset π–π interactions (black dotted arrows in Fig. 4.13), except between six-membered rings of caffeine and the A ring of Cg marked by black arrow in Fig. 4.13. The molecular interaction of the black arrow was face-to-face π–π interaction. As shown in Fig. 4.13 and Table 4.6, six O–H...O hydrogen bonds were observed in the 2 : 1 complex of caffeine and Cg.

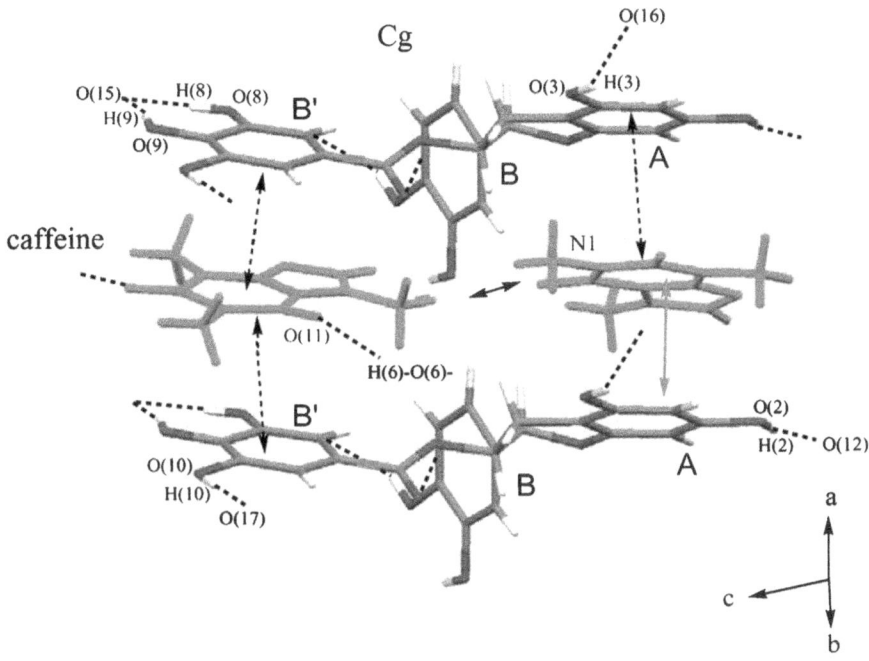

Figure 4.13   Intermolecular interactions in the 2 : 1 complex of caffeine and Cg. Black dotted-lines and gray, black dotted and black arrows indicate O–H...O hydrogen bonds and face-to-face and offset π–π interaction and CH–π interaction, respectively.

Table 4.6   Hydrogen bonds in 2 : 1 complex of caffeine and Cg.

D-H	A	D...A	D-H	H...A	∠D-H...A
OH(2O)^a	O(12)	2.84(6)	0.83	2.02	169.05
OH(3O)	O(16)	2.71(2)	0.83	1.96	150.20
OH(6O)	O(11)	2.65(2)	0.83	1.88	154.16
OH(8O)	O(15)	2.68(1)	0.84	1.91	152.91
OH(9O)	O(15)	3.02(7)	0.83	2.20	167.93
OH(10O)	O(17)	2.68(2)	0.82	1.85	178.36

^{a}Atom label in ORTEP drawing. D, H and A indicate hydrogen-bonding donor, hydrogen atom and hydrogen-bonding acceptor, respectively. A unit of length is Å.

4.3   Conclusion

Crystals of complexes of caffeine and tea catechins, CA, EC, Cg were prepared and X-ray crystallographic analysis of the complexes were performed to determine the stereochemical structures and to elucidate the detailed non-covalent interaction. Caffeine formed a 1 : 1 complex with CA by intermolecular hydrogen bonds, and a 1 : 1 complex with EC by face-to-face π–π interaction and intermolecular hydrogen bonds. A solution of caffeine and two kinds of non-galloylated catechin, CA and EC (molar ratio 2 : 1 : 1) in water afforded a 2 : 1 : 1 complex, the crystal structure of which had two layers, one layer in which caffeine and CA formed alternate lines by offset π–π interaction and the other layer in which caffeine and EC formed alternate lines by offset π–π interaction. The two layers were connected by CH–π interactions and intermolecular hydrogen bonds.

The structural difference between CA and Cg is the presence of a galloyl group on the oxygen atom at the C3 position or its absence, and all others are the same; however, quite different intermolecular interactions were observed between the 2 : 1 complex of caffeine and Cg and the 1 : 1 complex of caffeine and CA. In the 2 : 1 complex of caffeine and Cg, face-to-face, offset π–π interactions and intermolecular hydrogen bonds were observed, whereas in the 1 : 1 complex of caffeine and CA, only intermolecular hydrogen bonds were observed. Face-to-face π–π interactions and offset π–π interactions could not be found in the crystalline state of the 1 : 1 complex of caffeine and CA, unlike the 2 : 1 complex of caffeine and Cg.

Summary Points

- A solution containing equimolecular amounts of caffeine and CA in water afforded colorless needles of 1 : 1 complex of caffeine and CA, which was formed by intermolecular hydrogen bonds.
- Torsion angles of caffeine moieties of all complexes of caffeine and CA, EC, Cg are shown in Table 4.1, indicating that every caffeine moiety has an almost plain and rigid xanthine skeleton.
- Both B ring and hydroxyl group of CA moiety were in equatorial positions with respect to the C ring of the CA molecule.
- A solution containing equimolecular amounts of caffeine and EC in water afforded colorless block crystals of 1 : 1 complex of caffeine and EC, which was formed by face-to-face π–π interaction and intermolecular hydrogen bonds. The face-to-face π–π interaction formed between a six-membered ring of caffeine and the A ring of EC.
- The B ring and 3-OH group of EC moiety of the 1 : 1 complex of caffeine and EC were in equatorial and axial positions with respect to the C ring of the EC molecule, respectively.
- A solution of caffeine and two kinds of non-galloylated catechin and CA, EC (molar ratio 2 : 1 : 1) in water afforded a 2 : 1 : 1 complex. Both B ring and hydroxyl group of CA moiety of the 2 : 1 : 1 complex of caffeine and CA, EC were in equatorial positions with respect to the C ring of the CA molecule. The B ring and 3-OH group of EC moiety of the 2 : 1 : 1 complex were in equatorial and axial positions with respect to the C ring of the EC molecule, respectively.
- The crystal structure of the 2 : 1 : 1 complex of caffeine and CA, EC had two layers, one layer in which caffeine and CA formed alternate lines by offset

π–π interactions and another layer in which caffeine and EC formed alternate lines by offset π–π interactions. The two layers of the 2 : 1 : 1 complex of caffeine and CA, EC were connected by CH–π interactions and intermolecular hydrogen bonds.

- A suspension containing equimolecular amounts of caffeine and Cg in water afforded colorless plates of 2 : 1 complex of caffeine and Cg. In a unit of the 2 : 1 complex of caffeine and Cg, one caffeine molecule was located below the A ring of a Cg molecule, and the other was located above the B' ring of a Cg molecule.
- The B and B' rings of Cg moieties of the 2 : 1 complex were both in equatorial positions with respect to the C rings of Cg molecule.
- The two caffeine molecules were sandwiched between the A and B' rings of Cgs by π–π interactions. And also six O–H...O hydrogen bonds were observed in the 2 : 1 complex of caffeine and Cg.

Key Facts of Caffeine and Catechins

- It is well known that tea protects against lifestyle-related diseases, such as cancer, high blood pressure, diabetes, obesity, and arteriosclerosis, and many studies on the ingredients of tea have been performed.
- Caffeine and catechins are included in the leaves and buds of the tea plant as major ingredients, and they have various bioactivities. Interestingly, it is known that catechins form complexes with caffeine, especially in black tea. Such complexation is not only an interesting chemical phenomenon, but also may show interesting unique biological activities.
- Thus, many researchers have been investigating the soluble structure of the complexes of caffeine and catechins by NMR and calculation techniques. Unfortunately, these studies have not fully determined the structures of the complexes of caffeine and catechins.
- We have prepared crystals of complexes of caffeine and various tea catechins, and investigated their stereochemical structures and intermolecular interaction by X-ray crystallographic analysis.

Key Facts of Complex of Caffeine and Non-galloylated Catechins

- A solution containing equimolecular amounts of caffeine and CA in water afforded colorless needles of 1 : 1 complex of caffeine and CA, which was formed by intermolecular hydrogen bonds.
- A solution containing equimolecular amounts of caffeine and EC in water afforded colorless block crystals of 1 : 1 complex of caffeine and EC, which was formed by face-to-face π–π interaction and intermolecular hydrogen bonds.

- A solution of caffeine and two kinds of non-galloylated catechin, CA and EC (molar ratio 2 : 1 : 1) in water afforded a 2 : 1 : 1 complex. The crystal structure of the 2 : 1 : 1 complex of caffeine and CA, EC had two layers, one layer (Layer A) in which caffeine and CA formed alternate lines by offset π–π interactions and another layer (Layer B) in which caffeine and EC formed alternate lines by offset π–π interactions. The two layers of the 2 : 1 : 1 complex of caffeine and CA, EC were connected by CH–π interactions and intermolecular hydrogen bonds.
- Layer A was different from the layer structure of the 1 : 1 complex of caffeine and CA, which was formed by four intermolecular hydrogen bonds, while Layer B was very similar to the layer structure of the 1 : 1 complex of caffeine and EC, which was formed by face-to-face π–π interaction between the six-membered ring of caffeine and the A ring of EC, and two intermolecular hydrogen bonds.

Key Facts of Complex of Caffeine and Galloylated Catechins

- A suspension containing equimolecular amounts of caffeine and Cg in water afforded colorless plates of 2 : 1 complex of caffeine and Cg. In a unit of the 2 : 1 complex of caffeine and Cg, one caffeine molecule was located below the A ring of a Cg molecule, and the other was located above the B' ring of a Cg molecule.
- The two caffeine molecules were sandwiched between the A and B' rings of Cgs by π–π interactions. And also six O–H...O hydrogen bonds were observed in the 2:1 complex of caffeine and Cg.
- The structural difference between CA and Cg is the presence of a galloyl group on the oxygen atom at the C3 position or its absence, and all others are the same; however, quite different intermolecular interactions were observed between the 2 : 1 complex of caffeine and Cg and the 1 : 1 complex of caffeine and CA. In the 2 : 1 complex of caffeine and Cg, face-to-face, offset π–π interactions and intermolecular hydrogen bonds were observed, whereas in the 1 : 1 complex of CA and caffeine, only intermolecular hydrogen bonds were observed.

Definition of Words and Terms

CA : One of major eight tea catechins, non-galloylated 2,3-*trans*-catechin.
EC: One of major eight tea catechins, non-galloylated 2,3-*cis*-catechin.
Cg: One of major eight tea catechins, galloylated 2,3-*trans*-catechin.
Caffeine: Xanthine alkaloid, a central nervous system-stimulating agent.
Complex: In this article, compound in which plural molecules are connected by a non-covalent bond.

X-Ray crystallographic analysis: Analysis for structural determination based on the elastic scattering of X-rays from the electron clouds of the individual atoms.

Face-to-face π–π interaction: Interaction which occurs when two aromatic rings face each other.

Offset π-π interaction: Interaction which occurs when two aromatic rings shift and face each other.

CH–π interaction: Interaction which occurs when alkyl group and aromatic ring face each other.

Intermolecular hydrogen bond: Hydrogen bonding to be formed between molecules.

List of Abbreviations

CA	(+)-Catechin
Cg	(–)-Catechin-3-*O*-gallate
EC	(–)-Epicatechin
EGCg	(–)-Epigallocatechin-3-*O*-gallate
NMR	Nuclear Magnetic Resonance
NOE	Nuclear Overhauser Effect

References

Ahmad, N., Cheng, P. and Mukhtar, H., 2000. Cell cycle dysregulation by green tea polyphenol epigallocatechin-3-gallate. Biochem. Biophys. Res. Commun. 275: 328–334.

Cai, Y., Gaffney, S. H., Lilley, T. H., Magnolato, D., Martin, R., Spencer, C. M., Haslam, E., 1990. Polyphenol interactions. Part 4. Model studies with caffeine and cyclodextrins. Chem. Soc. Perkin Trans. 2: 2197–2209.

Collins English Dictionary (10th ed.) HarperCollins Publishers, 2009. caffeine or caffein. Available at: http://dictionary.reference.com/browse/caffein. Retrieved 2010-11-08.

Gaffney, S. H., Martin, R., Lilley, T. H., Haslam, E. and Magnolato, D., 1986. The association of polyphenols with caffeine and α- and β-cyclodextrin in aqueous media. Chem. Commun. 2: 107–109.

Hara, Y. and Watanabe, M., 1989. Antibacterial activity of tea polyphenols against clostridium botulinum. Nippon Shokuhin Kogyo Gakkaish. 36: 951–955.

Hashimoto, F., Ono, M., Masuoka, C., Ito, Y., Sakata, Y., Shimizu, K., Nonaka, G., Nishioka, I. and Nohara, T., 2003. Evaluation of the anti-oxidative effect (in vitro) of tea polyphenols. Biosci. Biotechnol. Biochem. 67: 396–401.

Hayashi, N. and Ujihara, T., 2008. A water-soluble acyclic phane receptor recognizing 3-trans-gallate-type catechins. J. Org. Chem. 73: 4848–4854.

Hayashi, N., Ujihara, T. and Kohata, K., 2004. Binding energy of tea catechin/ caffeine complexes in water evaluated by titration experiments with ^{1}H-NMR. Biosci.Biotechnol. Biochem. 68: 2512–2518.

Horman, I. and Viani, R., 1972. The nature and conformation of the caffeine-chlorogenate complex of coffee. J. Food Sci. 37: 925–927.

Ishizu, T., Tsutsumi, H., Sato, T., Yamamoto, H. and Shiro, M., 2009. Crystal structure of complex of gallocatechin gallate and caffeine. Chem. Lett. 38: 230–231.

Ishizu, T., Tsutsumi, H. and Sato, T., 2009. Interaction between gallocatechin gallate and caffeine in crystal structure of 1 : 2 and 2 : 2 complexes. Tetrahedron Lett. 50: 4121–4124.

Ishizu, T., Sato, T., Tsutsumi, H. and Yamamoto, H., 2010. Stereochemical structure determination of caffeine complexes with galloylated and non-galloylated catechins. Chem. Lett. 39: 607–609.

Tsutsumi, H., Kinoshita, Y., Sato, T. and Ishizu, T., 2011. Configurational studies of complexes of various tea catechins and caffeine in crystal state. Chem. Pharm. Bull. 59:1008–1015.

Kimura, M., Umegaki, K., Kasuya, Y., Sugisawa, A. and Higuchi, M., 2002. The relation between single/double or repeated tea catechin ingestions and plasma antioxidant activity in humans. Eur. J. Clin. Nutr. 56: 1186–1193.

Kuroda, Y. and Hara, Y., 2004. Health effects of tea and its catechins. Kluwer Academic / Plenum Publishers, New York, 11–60 pp.

Lambert, J. D. and Yang, C. S., 2003. Cancer chemopreventive activity and bioavailability of tea and tea polyphenols. Mutat. Res. 523–524: 201–208.

Maeda-Yamamoto, M., Kawahara, H., Tahara, N., Tsuji, K., Hara, Y. and Isemura, M., Effects of tea polyphenols on the invasion and matrix metalloproteinases [join to rest of reference below]
activities of human fibrosarcoma HT1080 cells. J. Agric. Food. Chem. 47: 2354.

Martin, R., Lilley, T. H., Falshaw C. P., Haslam, E., Begley, M. J. and Magnolato, D., 1986. The caffeine-potassium chlorogenate molecular complex. Phytochemistry. 26: 279.

Maruyama, N., Suzuki, Y., Sakata, K., Yagi, A. and Ina, K., 1991. NMR spectroscopic and computer graphics studies on the creaming down of tea. Processing International Symposium Tea Science. 145–149.

Mimura, S., Watanabe, J., Tomita, T., Sano, M. and Tomita, I., 1994. The inhibitory effects of tea polyphenols (flavan-3-ol derivatives) on Cu^{2+} mediated oxidative modification of low density lipoprotein. Biol. Pharm. Bull. 17: 1567–1572.

Okabe, S., Suganuma, M., Hayashi, M., Sueoka, E., Komori, A. and Fujiki, H., 1997. Mechanisms of growth inhibition of human lung cancer cell line, PC-9, by tea polyphenols. Jpn. J. Cancer Res. 88: 639–643.

Nathanson, J. A. 1984. Caffeine and related methylxanthines: possible naturally occurring pesticides. Science. 226: 184–187.

Sazuka, M., Imazawa, H., Shoji, Y., Mita, T., Hara, Y. and Isemura, M., 1997. Inhibition of collagenases from mouse lung carcinoma cells by green tea catechins and black tea theaflavins. Biosci. Biotechnol. Biochem. 61: 1504–1506.

Tezuka, M., Suzuki, H., Suzuki, Y., Hara, Y. and Okada, S., 1997. Inactivation effect of tea leaf catechins on human type-A influenza virus. Jpn. J. Toxicol. Environ. Health. 43: 311–315.

CHAPTER 5
Synthesis of Labeled Caffeine

FRÉDÉRIC BALSSA

Laboratoire des Courses Hippiques, 15 rue de Paradis, ZAC Pompidou,
91370 Verrières le Buisson, France
E-mail: f.balssa@lchfrance.fr

5.1 Introduction

This chapter presents an overview of strategies and methods used for the
preparation of labeled caffeine. Due to its specific physico-chemical properties,
the use of labeled caffeine constitutes a powerful analytical tool with
applications in various fields of biology, pharmacology, medicine and
botanics.

Stable isotope labeled caffeine exhibits a mass difference with native caffeine
which can easily be evidenced by mass spectrometry. If the introduced isotope
is unstable, the native molecule or its metabolites become radioactive and can
therefore be detected by appropriate β-radiation spectrometry.

The numerous applications of labeled caffeine made its synthesis of
particular interest. Since each specific application requires a proper choice of
the isotope nature as well as the number and the location of labeling sites,
chemists have developed various strategies for selective isotope introduction
into the caffeine molecule.

5.2 Strategies for Caffeine Labeling

Three synthetic strategies for the synthesis of labeled caffeine are now well
established: catalyzed exchange of hydrogen for deuterium or tritium, total

Food and Nutritional Components in Focus No. 2
Caffeine: Chemistry, Analysis, Function and Effects
Edited by Victor R Preedy
© The Royal Society of Chemistry 2012
Published by the Royal Society of Chemistry, www.rsc.org

synthesis of caffeine from labeled building blocks and partial synthesis by alkylation of an appropriate xanthine derivative with a labeled methyl group.

5.2.1 Synthesis of Labeled Caffeine by Hydrogen Exchange

Under specific conditions, caffeine undergoes hydrogen exchange: one or several hydrogen atoms in the molecule are exchanged for deuterium or tritium. A labile hydrogen atom (H-8, on the imidazole ring) is easily exchanged for tritium from [^{3}H$_2$]O. The exchange of non-labile hydrogen atoms with hydrogen isotopes from deuterated or tritiated water, or with deuterium or tritium gas requires a metal catalyst.

The regioselectivity of the labeled atom introduction and the isotope enrichment ratio depend on the procedure used and especially on the nature of the metal catalyst.

This labeling procedure sounds attractive because caffeine is the starting material, and so long multistep synthesis is avoided. Quantitative yields are usually high and isotope composition is satisfactory.

5.2.2 Synthesis of Labeled Caffeine through Total Synthesis

This method is used when the molecule is to be labeled at the xanthine ring skeleton. In this case, the introduced isotopes are not subject to exchange in aqueous medium (as for H-8 hydrogen) or to metabolic elimination (such as methyl groups on nitrogen atoms).

Starting from the appropriate labeled building block, this method allows the regioselective labeling of any atom on the xanthine ring.

The synthetic procedure is based on the historical Traube synthesis (Traube 1900). Fortunately, caffeine can be prepared with basic starting compounds such as urea, formic acid or formamide labeled with various isotopes and which are commercially available.

Tedious multistep synthesis is the major drawback of this procedure, especially when the isotopes are introduced at the earlier step, leading to some extensive loss of labeled material.

5.2.3 Synthesis of Labeled Caffeine by Partial Synthesis with a Xanthine Derivative

Labeled caffeine is readily prepared from xanthine or its mono- or dimethyl-derivatives by alkylation of the unsubstituted nitrogen atoms with an appropriate labeled methylating agent. Thus, unsubstituted nitrogen atoms of the xanthine derivative are methylated to afford caffeine as the sole product. The number and position of labeled methyl groups depend on the initial xanthine derivative.

Methyl iodide labeled with various isotopes of carbon or hydrogen is one of the most convenient methylating agents. Furthermore, xanthine, methyl-xanthine and dimethylxanthine are commercially available and their preparation is well documented. Consequently, direct methylation of the appropriate xanthine derivative with labeled methyl iodide is obviously the most direct procedure for the preparation of labeled caffeine. So far this method, which allows for the synthesis of labeled caffeine in a single step, has been the most investigated.

The methylation of unsubstituted nitrogen atoms is carried out in a polar solvent in the presence of a base at room temperature or with moderate heating.

The isotopic composition of the labeled caffeine depends only on the isotopic purity of the labeled methylating agent.

The alkylation rate with labeled methylating agents of unsubstituted nitrogen in the xanthine ring has been poorly investigated. Nevertheless, it has been reported that N-3 is more reactive than N-1 (Balssa and Bonnaire 2007) and that the methylation rate of N-7 is superior to N-1 (Pierce *et al* 1984).

5.3 Syntheses of Labeled Caffeine

5.3.1 Synthesis of $[^{15}N_2]$Caffeine

$[^{15}N_2]$caffeine was prepared from $[^{15}N_2]$urea by total synthesis (Kenani *et al* 1995). $[^{15}N_2]$urea (1) was converted into $[1,3-^{15}N_2]$-5,6-diaminouracil (4) in three steps. The imidazole ring was then closed by the action of formamide to give $[1,3-^{15}N_2]$xanthine (5).

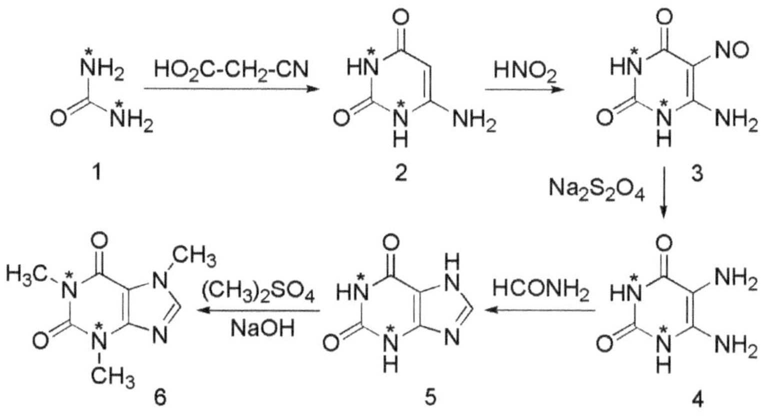

Scheme 5.1   Synthesis of $[1,3-^{15}N_2]$caffeine. $[1,3-^{15}N_2]$Caffeine was prepared from $[^{15}N_2]$urea by total synthesis in five steps. $[1,3-^{15}N_2]$caffeine (6) was obtained by methylation of the intermediate $[1,3-^{15}N_2]$xanthine (5) with dimethylsulfate. Labeled atoms are indicated by an asterisk.

[1,3-^{15}N$_2$]caffeine **(6)** was obtained by methylation of [1,3-^{15}N$_2$]xanthine **(5)** with DMS in aqueous alkaline solution (Scheme 5.1). The authors reported an overall yield of 85% after crystallization of [1,3-^{15}N$_2$]caffeine **(6)** in absolute ethanol.

5.3.2 Syntheses of [^{11}C]Caffeine

Three isotopomers of [^{11}C]caffeine were prepared by partial synthesis from dimethylxanthine. 3,7-dimethylxanthine (theobromine) **(7)**, 1,7-dimethylxanthine (paraxanthine) **(8)** and 1,3-dimethylxanthine (theophylline) **(9)** were submitted to the action of [^{11}C]H$_3$I in presence of base to give respectively [1-^{11}CH$_3$]caffeine **(10)** (Comar *et al* 1975; Funaki *et al* 1992; Maziere *et al* 1974), [3-^{11}CH$_3$]caffeine **(11)** (Funaki *et al* 1992) and [7-^{11}CH$_3$]caffeine **(12)** (Denutte *et al* 1982; Funaki *et al* 1992; Saji *et al* 1978) (Scheme 5.2).

Reported solvent–base systems were: methanol–sodium carbonate (Maziere *et al* 1974), DMF–sodium hydroxide (Funaki *et al* 1992) and dimethylsulfoxide–sodium hydride (Denutte *et al* 1982).

The latter system is of particular interest because the base is not used to scavenge HI produced during the reaction, but to generate an N-7 anion by hydrogen abstraction prior to the [^{11}C]H$_3$I addition. Since deprotonated theophylline is more reactive towards [^{11}C]H$_3$I than theophylline, this enhancement of reactivity allows shortened reaction time. This improvement is very useful, especially when a short half-life isotope is to be introduced in the caffeine molecule (^{11}C half-life: 20.4 min). This allows to attain high specific radioactivity and thereby high detection sensitivity.

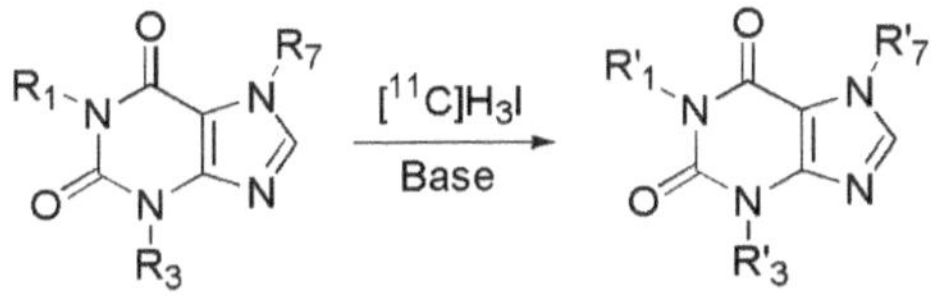

7: 3,7-dimethylxanthine
R$_1$=H, R$_3$=R$_7$=CH$_3$

10: [1-^{11}CH$_3$]Caffeine
R'$_1$=[^{11}C]H$_3$, R'$_3$=R'$_7$=CH$_3$

8: 1,7-dimethylxanthine
R$_3$=H, R$_1$=R$_7$=CH$_3$

11: [3-^{11}CH$_3$]Caffeine
R'$_3$=[^{11}C]H$_3$, R'$_1$=R'$_7$=CH$_3$

9: 1,3-dimethylxanthine
R$_7$=H, R$_1$=R$_3$=CH$_3$

12: [7-^{11}CH$_3$]Caffeine
R'$_7$=[^{11}C]H$_3$, R'$_1$=R'$_3$=CH$_3$

Scheme 5.2 Syntheses of [^{11}C]caffeine from dimethylxanthines. Three isotopomers of [^{11}C]caffeine were prepared by partial synthesis from dimethylxanthine. 3,7-Dimethylxanthine (theobromine) **(7)**, 1,7-dimethylxanthine (paraxanthine) **(8)** and 1,3-dimethylxanthine (theophylline) **(9)** were submitted to the action of [^{11}C]H$_3$I in presence of base to give respectively [1-^{11}CH$_3$]caffeine **(10)**, [3-^{11}CH$_3$]caffeine **(11)** and [7-^{11}CH$_3$]caffeine **(12)**.

The purification of the [¹¹C]caffeine was performed by HPLC (Denutte *et al* 1982) or semi preparative HPLC (Funaki *et al* 1992).

5.3.3 Syntheses of [¹³C]Caffeine

[¹³C]caffeine was prepared by partial synthesis, starting from xanthine and 3-methylxanthine (Scheme 5.3).

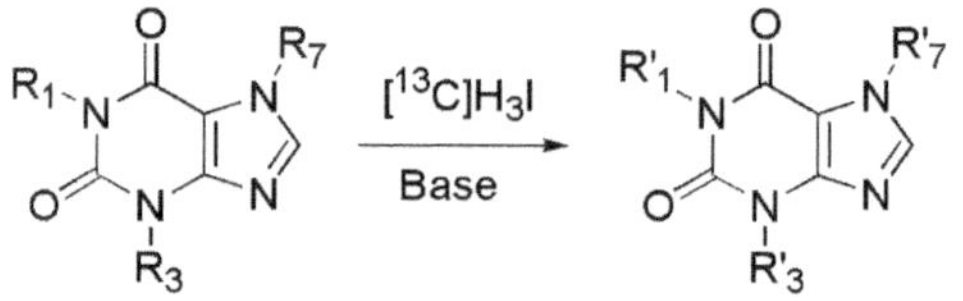

13: xanthine
$R_1=R_3=R_7=H$

15: 3-methylxanthine
$R_3=CH_3, R_1=R_7=H$

14: [1,3,7-¹³CH₃]Caffeine
$R'_1=R'_3=R'_7=[^{13}C]H_3$

16: [1,7-¹³CH₃]Caffeine
$R'_1=R'_7=[^{13}C]H_3, R'_3=CH_3$

Scheme 5.3 Syntheses of [¹³C]caffeine from xanthine and 3-methylxanthine. [¹³C]caffeine was prepared by partial synthesis, starting from xanthine and 3-methylxanthine, by the action of [¹³C]H₃I in aqueous sodium hydroxide or methanolic potassium hydroxide solution.

Xanthine (**13**) was methylated by the action of [¹³C]H₃I in aqueous sodium hydroxide to afford [1,3,7-¹³CH₃]caffeine (**14**). The labeled caffeine was purified by column chromatography (Arnaud *et al* 1980).

In the same way, [1,7-¹³CH₃]caffeine (**16**) was prepared from 3-methyl-xanthine (**15**) by the action of [¹³C]H₃I in a methanolic potassium hydroxide solution. After 24 h stirring at room temperature and liquid–liquid extraction, pure [1,7-¹³CH₃]caffeine (**16**) was obtained with 82% yield (Pierce *et al* 1984).

5.3.4 Syntheses of [¹⁴C]Caffeine

Synthesis of [¹⁴C]caffeine was extensively investigated since it has been the most widely used in metabolic studies.

Numerous strategies were set up, both by partial and total synthesis.

5.3.4.1 *Syntheses of [¹⁴C]Caffeine by Partial Synthesis*

Two strategies were investigated for this purpose: the methylation of a labeled xanthine derivative with unlabeled methylating agent and the methylation of an unlabeled xanthine derivative with a labeled methylating agent.

[2-¹⁴C]xanthine (**17**) was methylated with DMS in aqueous sodium hydroxide to give [2-¹⁴C]caffeine (**18**) with high chemical purity (Heftmann

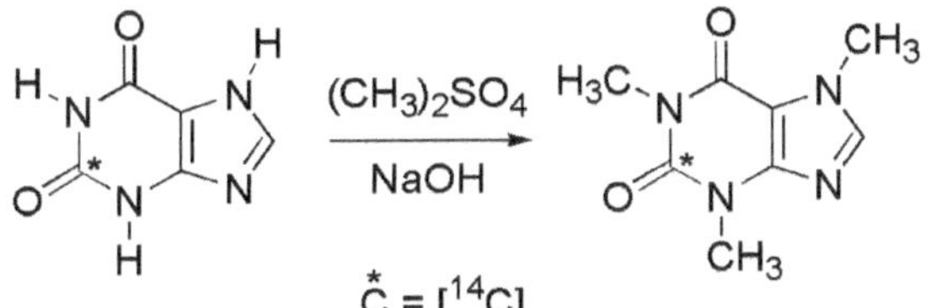

Scheme 5.4   Synthesis of [2-^{14}C]caffeine from [2-^{14}C]xanthine. [2-^{14}C]caffeine (18) was prepared with high chemical purity by methylation of [2-^{14}C]xanthine (17) with dimethylsulfate in aqueous sodium hydroxide solution.

1971; Suzuki and Waller 1984) (Scheme 5.4). The labeled caffeine was purified by repeated paper chromatography and TLC (Suzuki and Waller 1984).

[1-^{14}CH$_3$]caffeine (19) was prepared by the action of a labeled methylating agent on 3,7-dimethylxanthine (7) (Carr *et al* 2010; Kalberer 1964). In the same way, [7-^{14}CH$_3$]caffeine (20) was prepared from 1,3-dimethylxanthine (9) by N-7 methylation with [^{14}C]H$_3$I (Kalberer 1964) (Scheme 5.5).

Labeled methyl iodide is one of the most convenient reagents for the preparation of labeled compounds, but it is toxic and volatile. Furthermore, the shelf life of [^{14}C]H$_3$I is limited due to its autoradiolyse and its shipping by air is restricted by safety regulations. Therefore new ^{14}C labeled methylating agents were developed. The synthesis of [^{14}C-methyl]-methyl nosylate (Figure 5.1) offered a valuable alternative to the use of [^{14}C]H$_3$I (Carr *et al* 2010).

[1-^{14}CH$_3$]Caffeine (19) was prepared by the action of [^{14}C-methyl]-methyl nosylate on 3,7-dimethylxanthine (7) in the presence of sodium hydrogencarbonate in DMF–ACN mixture (Carr *et al* 2010).

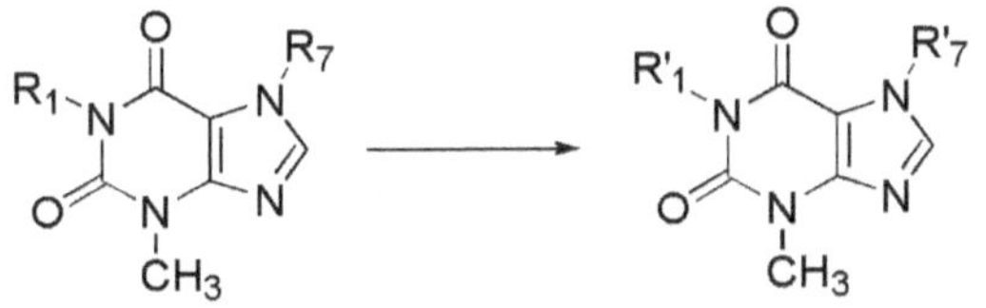

7: 3,7-dimethylxanthine
R$_1$=H, R$_7$=CH$_3$

19: [1-^{14}CH$_3$]Caffeine
R'$_1$=[^{14}C]H$_3$, R'$_7$=CH$_3$

9: 1,3-dimethylxanthine
R$_1$=CH$_3$, R$_7$=H

20: [7-^{14}CH$_3$]Caffeine
R'$_1$=CH$_3$, R'$_7$=[^{14}C]H$_3$

Scheme 5.5   Syntheses of [^{14}C]caffeine from theobromine and theophylline. [1-^{14}CH$_3$]Caffeine (19) and [7-^{14}CH$_3$]caffeine (20) were prepared by the action of a labeled methylating agent respectively on 3,7-dimethylxanthine (7) and on 1,3-dimethylxanthine (9).

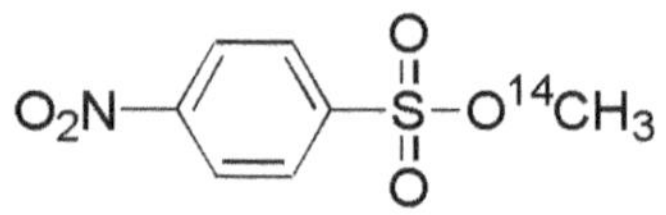

Figure 5.1   Chemical structure of [^{14}C-methyl]-methyl nosylate. [^{14}C-methyl]-methyl nosylate was a valuable alternative to the use of hazardous [^{14}C]H$_3$I for the methylation of a xanthine derivative in the synthetic procedure leading to [^{14}C]caffeine.

5.3.4.2   Syntheses of [^{14}C]Caffeine by Total Synthesis

Similarly to [1,3-^{15}N$_2$]caffeine, [2-^{14}C]caffeine **(25)** was prepared from [^{14}C]urea by total synthesis based on the Traube procedure (Ayrey and Yeomans 1976; Kalberer 1964). [2-^{14}C]urea **(21)** was converted into [2-^{14}C]-5,6-diaminouracil **(24)** in three steps and the imidazole ring was then closed by the action of formamide to give [2-^{14}C]xanthine **(17)**.

[2-^{14}C]caffeine **(25)** was obtained by methylation of [2-^{14}C]xanthine **(17)** with DMS in aqueous alkaline solution (Scheme 5.6).

Caffeine was also labeled on carbon C-8. The first published method consists of a three step procedure starting from 5,6-diaminouracil **(26)**. The imidazole ring was synthesized by the subsequent action of [^{14}C]formic acid and sodium hydroxide to give [8-^{14}C]xanthine **(27)**. [8-^{14}C]caffeine (28) was obtained by methylation of [8-^{14}C]xanthine **(27)** with DMS in aqueous sodium hydroxide (Kalberer 1964) (Scheme 5.7).

The second procedure (Scheme 5.8) differs from the previous one (Scheme 5.7) in the methylation step. Indeed, the 6-[^{14}C]formyl-5,6-diaminouracil **(29)** was dimethylated prior to the imidazole ring closure. This procedure allows the preparation of the intermediate [8-^{14}C]-1,3-dimethylxanthine ([8-^{14}C]theophylline)

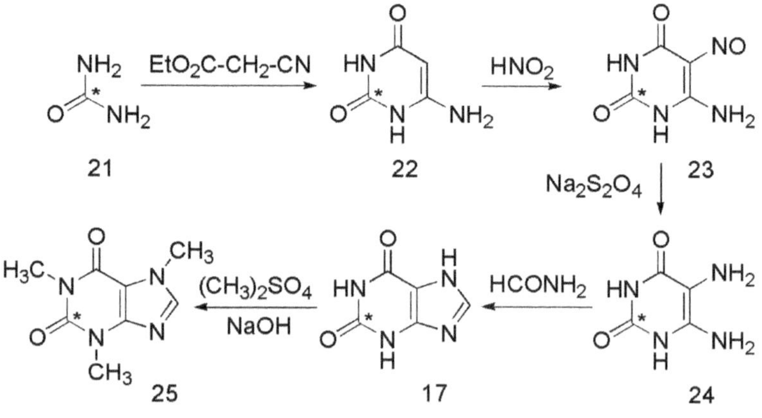

Scheme 5.6   Synthesis of [2-^{14}C]caffeine. [2-^{14}C]caffeine **(25)** was prepared from [^{14}C]urea **(21)** by total synthesis in five steps.

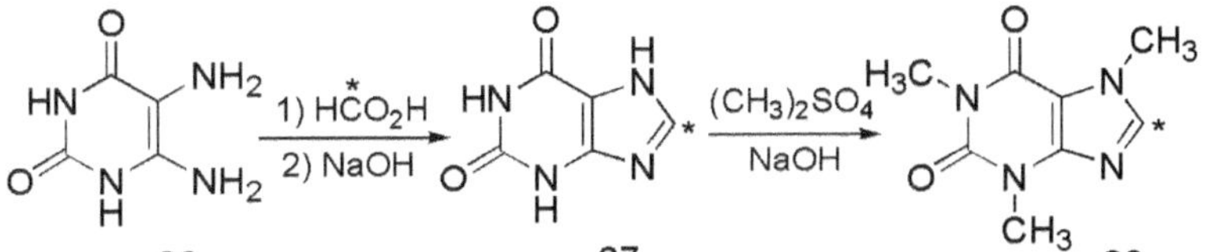

Scheme 5.7 Synthesis of [8-^{14}C]caffeine. (Kalberer 1964). [8-^{14}C]caffeine (**28**) was obtained in three steps from 5,6-diaminouracil (**26**) by the action of [^{14}C]formic acid.

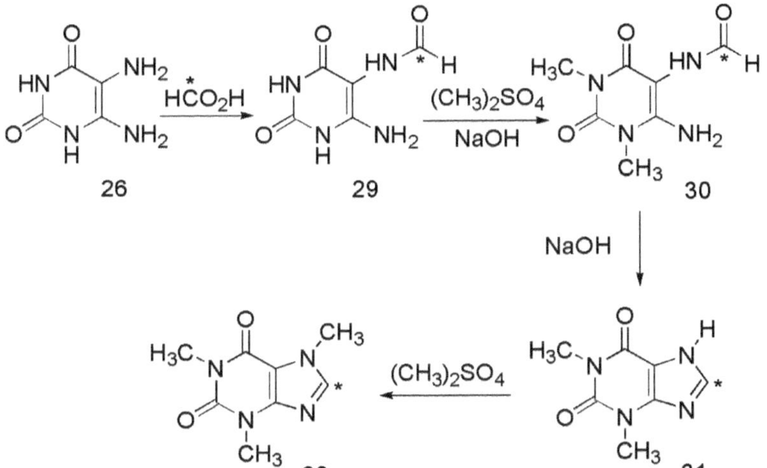

Scheme 5.8 Synthesis of [8-^{14}C]caffeine. (Ayrey and Yeomans 1976). [8-^{14}C]-1,3-dimethylxanthine ([8-^{14}C]theophylline) (**31**) was obtained in three steps from 5,6-diaminouracil (**29**). [8-^{14}C]caffeine (**28**) was prepared by methylation of [8-^{14}C]-1,3-dimethylxanthine (**31**).

(**31**) which gives the [8-^{14}C]caffeine (**28**) after N-7 methylation with DMS (Ayrey and Yeomans 1976).

5.3.5 Syntheses of [^{2}H]Caffeine

Deuterium labeled caffeine has been most investigated since deuterated building blocks or deuterium sources such as C[^{2}H$_3$]I or [^{2}H$_2$]O are commercially available. Furthermore, deuterium is a stable isotope, so its use is not submitted to safety restrictions common to radioactive isotopes (such as ^{11}C, ^{14}C and ^{3}H).

Deuterium incorporation into caffeine has been performed by two methods: hydrogen–deuterium catalytic exchange and partial synthesis by xanthine derivative methylation.

5.3.5.1 Syntheses of [²H]Caffeine by Hydrogen–Deuterium Exchange

Structural investigations on the carbon–hydrogen–metal bond led to the discovery of an iridium complex of caffeine allowing regioselective hydrogen–deuterium exchange (Crabtree *et al* 1985).

[Ir(cod)(PPh$_3$)$_2$]$^+$ (32) was treated successively with caffeine and hydrogen to give the complex (33) which was exposed to deuterium gas at 1 atm pressure and room temperature. Under these conditions, the caffeine complex undergoes a regioselective hydrogen–deuterium exchange at the N-3 methyl group. The labeled caffeine was recovered by displacement of the metal complex to give highly enriched [3-CD$_3$]caffeine (35): the exchange was complete in 8 h (Scheme 5.9).

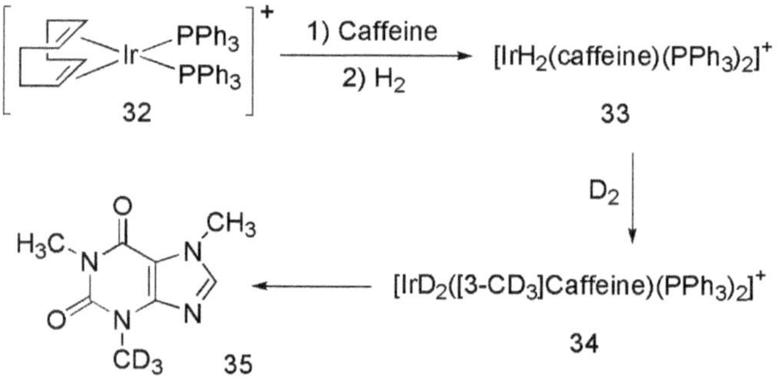

Scheme 5.9 Iridium catalyzed synthesis of [3-CD$_3$]caffeine. Under 1 atm deuterium gas pressure, an iridium-caffeine complex (33) undergoes a regioselective hydrogen–deuterium exchange at the N-3 methyl group. The labeled caffeine was recovered by removal of the metal complex to give highly enriched [3-CD$_3$]caffeine (35): exchange was complete in 8 h.

Despite a great interest in fundamental carbon–hydrogen bond activation, this method is not used as a preparative procedure because [3-CD$_3$]caffeine (35) can be more easily prepared by partial synthesis from 1,7-dimethylxanthine (8).

Recently, a procedure for multi-deuterium incorporation was developed. Heating caffeine (36) at 160–180 °C in deuterated water in the presence of palladium on carbon catalyst gave multi-deuterated caffeine (37) with an excellent chemical yield (85 to 98%) and high deuterium incorporation. Deuterated water is the source of deuterium, and hydrogen is introduced at the beginning of the reaction only for activation of the metal catalyst (Scheme 5.10) (Modultwa *et al* 2010).

Deuterium incorporation depends mainly on the reaction temperature (there is no deuterium incorporation on the methyl at N-1 and N-7 at 90 °C). Above 160 °C, all hydrogen atoms are exchanged, however the incorporation ratio depends on the steric hindrance at the exchanging atoms. Indeed, maximum

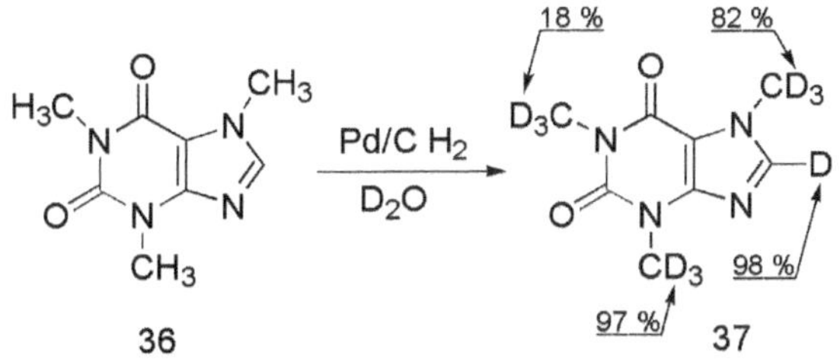

Scheme 5.10 Synthesis of multi-deuterated caffeine. Heating caffeine (**36**) at 160–180 °C in deuterated water in the presence of palladium on carbon catalyst gave multi-deuterated caffeine (**37**) with excellent chemical yield (85 to 98%) and high deuterium incorporation. Labeled caffeine was recovered by distillation of the solvent after removal of the catalyst by filtration.

deuterium incorporation at the N-3 and N-7 methyl groups was respectively 97% and 82%, while the deuterium incorporation at N-1 methyl was only 18%, after two successive hydrogen–deuterium exchange reactions.

Labeled caffeine was recovered by distillation of the solvent after removal of the catalyst by filtration. This procedure is especially valuable for a rapid, easy and cost effective synthesis of deuterium labeled caffeine.

5.3.5.2   *Syntheses of [²H]Caffeine by Partial Synthesis*

Caffeine labeling by trideuteromethylation of unsubstituted nitrogen atoms of a xanthine derivative was extensively investigated. All possible isotopomers have been prepared using this method (Scheme 5.11).

Reported yields ranged from 10 to 83.5%. However, most of these procedures were used for small scale synthesis in which chemical and isotopic purity were the major requirements, rather than the chemical yield. Highest yield was obtained in optimized large scale synthesis (Balssa and Bonnaire 2007).

Syntheses of [CD₃]caffeine

The three isotopomers of [CD₃]caffeine were made by starting from appropriate dimethylxanthine by the action of CD₃I in the presence of base.

[1-CD₃]caffeine (**38**) was prepared by methylation of 3,7-dimethylxanthine (**7**) in an acetone–water mixture in the presence of sodium hydroxide (Cherrah *et al* 1987; Falconnet *et al* 1986; Horning *et al* 1979). This compound was purified by preparative TLC (Falconnet *et al* 1986), silica column chromatography (Cherrah *et al* 1987) and liquid–liquid extraction followed by recrystallization from ethanol (Horning *et al* 1979) or from water (Telo and Vieira 1997).

[3-CD₃]caffeine (**35**) was prepared by methylation of 1,7-dimethylxanthine (**8**) in an acetone–water mixture in the presence of sodium hydroxide (Cherrah *et al* 1987; Falconnet *et al* 1986; Horning *et al* 1979) or in DMF in the presence of potassium carbonate (Regal *et al* 2005). This compound was purified by

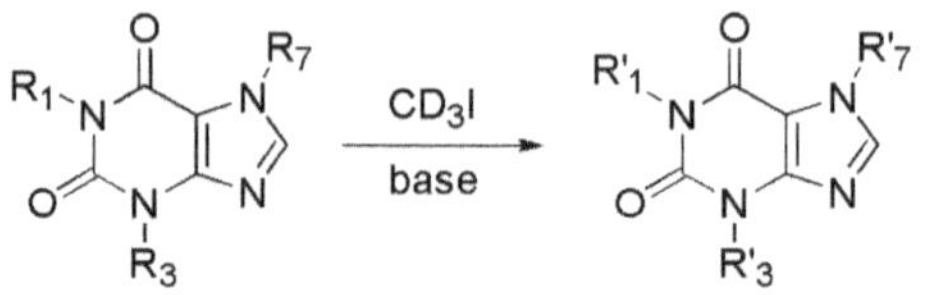

7: 3,7-dimethylxanthine R_1=H, R_3=R_7=CH_3	38: [1-CD_3]Caffeine R'_1=CD_3, R'_3=R'_7=CH_3
8: 1,7-dimethylxanthine R_3=H, R_1=R_7=CH_3	35: [3-CD_3]Caffeine R'_3=CD_3, R'_1=R'_7=CH_3
9: 1,3-dimethylxanthine R_7=H, R_1=R_3=CH_3	39: [7-CD_3]Caffeine R'_7=CD_3, R'_1=R'_3=CH_3
40: 7-methylxanthine R_1=R_3=H, R_7=CH_3	43: [1,3-CD_3]Caffeine R'_1=R'_3=CD_3, R'_7=CH_3
41: 3-methylxanthine R_1=R_7=H, R_3=CH_3	44: [1,7-CD_3]Caffeine R'_1=R'_7=CD_3, R'_3=CH_3
42: 1-methylxanthine R_3=R_7=H, R_1=CH_3	45: [3,7-CD_3]Caffeine R'_3=R'_7=CD_3, R'_1=CH_3
13: xanthine R_1=R_3=R_7=H	46: [1,3,7-CD_3]Caffeine R'_1=R'_3=R'_7=CD_3

Scheme 5.11   Syntheses of [CD_3]caffeine from xanthine derivatives. Caffeine labeling of unsubstituted nitrogen from a xanthine derivative by the action of CD_3I in the presence of base allowed for the preparation of all possible isotopomers of [CD_3]caffeine.

preparative TLC (Falconnet *et al* 1986), silica column chromatography (Cherrah *et al* 1987), liquid–liquid extraction followed by recrystallization from ethanol (Horning *et al* 1979) and preparative HPLC (Regal *et al* 2005).

[7-CD_3]caffeine **(39)** was prepared by methylation of 1,3-dimethylxanthine **(9)** in an acetone–water mixture in the presence of sodium hydroxide (Cherrah *et al* 1987; Falconnet *et al* 1986; Horning *et al* 1979), or in DMF in the presence of potassium carbonate (Tserng 1983), or in DMF in the presence of sodium hydroxide (Hill *et al* 1988), or finally in methanol in the presence of sodium methylate (Balssa and Bonnaire 2007). This compound was purified by preparative TLC (Falconnet *et al* 1986), silica column chromatography (Cherrah *et al* 1987), liquid–liquid extraction followed by recrystallization from ethanol (Horning *et al* 1979), liquid–liquid extraction (Tserng 1983), precipita-tion (Hill *et al* 1988), recrystallization from water (Telo and Vieira 1997) or recrystallization from chloroform–diethylether (Balssa and Bonnaire 2007).

Syntheses of [(CD_3)$_2$]caffeine

[1,3-(CD_3)$_2$]caffeine **(43)**, [1,7-(CD_3)$_2$]caffeine **(44)** and [3,7-(CD_3)$_2$]caffeine **(45)** were made from respectively 7-methylxanthine **(40)** (Falconnet *et al* 1986), 3-methylxanthine **(41)** (Cherrah *et al* 1987; Falconnet *et al* 1986) and 1-methylxanthine **(42)** (Cherrah *et al* 1987; Falconnet *et al* 1986; Regal *et al* 2005). These isotopomers were prepared by the action of CD_3I in an acetone–

water mixture in the presence of sodium hydroxide (Cherrah *et al* 1987; Falconnet *et al* 1986), or in DMF in the presence of potassium carbonate (Regal *et al* 2005). They were purified by preparative TLC (Falconnet *et al* 1986), silica column chromatography (Cherrah *et al* 1987) and preparative HPLC (Regal *et al* 2005).

Syntheses of [1,3,7-(CD$_3$)$_3$]caffeine

[1,3,7-(CD$_3$)$_3$]caffeine (46) was prepared from xanthine (13) by the action of CD$_3$I in an acetone–water mixture in the presence of sodium hydroxide (Cherrah *et al* 1987; Falconnet *et al* 1986; Horning *et al* 1979). The purification of [1,3,7-(CD$_3$)$_3$]caffeine was achieved by preparative TLC (Falconnet *et al* 1986), silica column chromatography (Cherrah *et al* 1987) and repeated recrystallization from ethanol (Horning *et al* 1979).

5.3.6 Syntheses of [^{3}H]Caffeine

Tritium labeled caffeine was prepared by catalyzed hydrogen–tritium exchange. This reaction occurs regioselectively at the H-8 position. Tritium sources were tritiated water, a tritiated water–tritiated acetic acid mixture or tritium gas. Catalysts such as alkali or palladium and platinum have been used (Scheme 5.12).

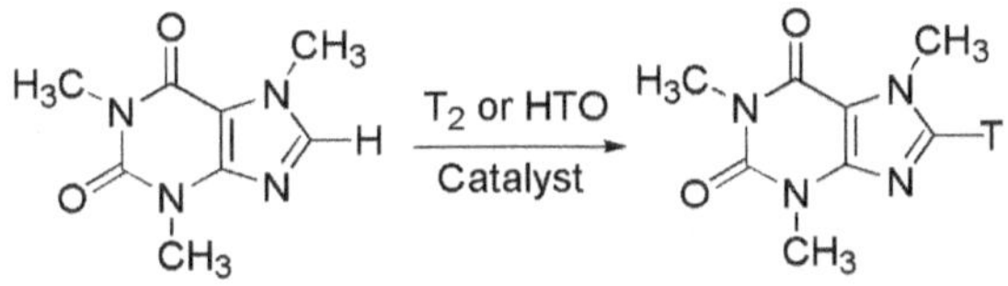

Scheme 5.12 Synthesis of [8-T]caffeine by hydrogen–tritium exchange. H-8 tritiated caffeine was prepared by catalyzed hydrogen–tritium exchange with tritiated water, a tritiated water–tritiated acetic acid mixture or tritium gas.

[8-T]caffeine (47) was prepared by heating caffeine (36) in tritiated water and alkaline pH was found to improve the tritium incorporation rate (Jelinska and Sobkowski 1977).

This hydrogen–tritium exchange reaction was performed in tritiated water–tritiated acetic acid mixture in the presence of platinum black catalyst (Neiman 1981).

The H-8 substitution as in [8-T]caffeine was also performed by the action of tritium gas in the presence of palladium on barium sulfate (47) (Bloxsidge *et al* 1981).

5.4 Characterization of Labeled Caffeine

Labeled caffeine is usually characterized by mass spectrometry (MS) and nuclear magnetic resonance (NMR).

5.4.1 Characterization by Nuclear Magnetic Resonance

Detailed ^1H, ^2H and ^{13}C NMR data for seven kinds of [CD$_3$]caffeine have been published (Falconnet *et al* 1986). Additional data have been reported for ^1H NMR of [7-CD$_3$]caffeine (Balssa and Bonnaire 2007), ^1H NMR of [1,3-^{15}N$_2$]caffeine (Kenani *et al* 1995) and ^{13}C and ^{15}N NMR of caffeine (Stikowski 1995).

5.4.2 Characterization by Mass Spectrometry

Mass spectrometry has been widely used for the characterization of labeled caffeine. Detailed data for [1-CD$_3$]caffeine (Houghton 1982), [3-CD$_3$]caffeine (Houghton 1982), [7-CD$_3$]caffeine (Balssa and Bonnaire 2007; Hill *et al* 1988; Houghton 1982; Tserng 1983), [1,3-^{15}N$_2$]caffeine (Kenani *et al* 1995) and [1,7-^{13}CH$_3$]caffeine (Pierce *et al* 1984) have been published.

Summary Points

This chapter focuses on labeled caffeine synthesis.

Syntheses of caffeine labeled with ^2H, ^3H, ^{11}C, ^{13}C, ^{14}C and ^{15}N have been described.

Labeled caffeine can be prepared by base or metal catalyzed hydrogen–deuterium or hydrogen–tritium exchange.

Labeled caffeine can be prepared by partial synthesis such as methylation of xanthine derivatives (*e.g.* xanthine, methylxanthine and dimethylxanthine).

Labeled caffeine can be prepared by total synthesis from common labeled building blocks such as labeled urea, formic acid or formamide.

Labeled caffeine can be easily characterized by nuclear magnetic resonance or mass spectrometry.

Key Facts of Labeled Caffeine

Labeled caffeine constitutes a potent and irreplaceable tool in analytical chemistry applied to biology, pharmacology, medicine, botanics …

Labeled caffeine is the native caffeine in which one or more atoms have been substituted by one or more isotopes.

The introduced isotope in the caffeine molecule may be stable or radioactive.

Labeled caffeine is detected in a biological matrix by mass spectrometry. Radioactively labeled caffeine is detected by an appropriate radiation detector.

Preparation of caffeine labeled with ^2H, ^3H, ^{11}C, ^{13}C, ^{14}C and ^{15}N are described.

Labeled caffeine can be prepared by hydrogen–deuterium or hydrogen–tritium exchange, by partial synthesis from xanthine derivative or total synthesis from labeled building blocks.

Labeled caffeine can be characterized by mass spectrometry and by nuclear magnetic resonance.

Definitions of Words and Terms

^{11}C: Radioactive isotope of carbon (half-life: 20.3 minutes)

^{13}C: Stable isotope of carbon (natural abundance: 1.10%).

^{14}C: Radioactive isotope of carbon (half-life: 5700 years).

^{2}H *(deuterium: D)*, Stable isotope of hydrogen (natural abundance: 0.015%). Water in which hydrogen is substituted by deuterium is called "heavy water".

^{3}H *(tritium: T)*, Radioactive isotope of hydrogen (half-life: 12.3 years).

^{15}N: Stable isotope of nitrogen (natural abundance: 0.366%).

High-performance liquid chromatography (HPLC): is an analytical separation technique connected with a detector (*e.g.* UV spectroscopy, fluorometry, mass spectrometry). Preparative scale HPLC is used to purify milligram to gram quantities.

Thin layer chromatography (TLC): is a technique used to separate a mixture of compounds on a thin layer of stationary phase coated on a support (glass plates, aluminium or plastic foils). TLC is convenient to follow up the course of a chemical reaction. Preparative scale TLC is used to purify up to about 100 milligrams of substance.

Isotope: Two atoms of an element are named "isotopes" when their nucleus contains the same number of protons but a different number of neutrons.

Labeled caffeine: Native caffeine in which one or more atoms have been substituted by one or more isotopes.

Paraxanthine: 1,7-dimethylxanthine, major metabolite of caffeine in man.

Theobromine: 3,7-dimethylxanthine, the principal alkaloid of cacao bean.

Theophylline: 1,3-dimethylxanthine, alkaloid found in tea.

Xanthine: is the heterocyclic nucleus of the caffeine molecule which can be viewed as a fusion of a uracil heterocycle (six membered ring) with an imidazole heterocycle (five membered ring).

Abbreviations

ACN	Acetonitrile
DMF	Dimethylformamide
DMS	Dimethylsulfate
HPLC	High-performance liquid chromatography
TLC	Thin layer chromatography

References

Arnaud, M. J., Thelin-Doemer, A., Ravussin, E. and Acheson, K. J., 1980. Study of the demethylation of [1,3,7-Me-13C]-caffeine in man using respiratory exchange measurements. Biomedical Mass Spectrometry. 7: 521–524.

Ayrey, G. and Yeaomans, M. A., 1976. Carbon-14 ring labelling of some methylxanthines of pharmaceutical importance. Journal of Labelled Compounds and Radiopharmaceuticals. 12: 323–345.

Balssa, F. and Bonnaire, Y., 2007. Easy preparative scale syntheses of labelled xanthines: caffeine, theophylline and theobromine. Journal of Labelled Compounds and Radiopharmaceuticals. 50: 33–41.

Bloxsidge, J. P., Elvidge, J. A., Gower, M., Jones, J. R., Evans, E. A., Kitcher, J. P. and Warrell, D. C., 1981. Tritium nuclear magnetic resonance spectroscopy. Part 13. Tritium labelled neurochemicals. Journal of Labelled Compounds and Radiopharmaceuticals. 18: 1141–1165.

Carr, S. A., Lacy, L. P., Pounds, S. and Filer, C. N., 2010. The synthesis, characterization and chemistry of [methyl-^{14}C] methyl nosylate. Journal of Labelled Compounds and Radiopharmaceuticals. Special issue: 10th international symposium on the synthesis and applications of isotopes and isotopically labelled compounds. 53: 406–489.

Cherrah, Y., Falconnet, J. B., Desage, M., Brazier, J. L., Zini, R. and Tillement, J. P., 1987. Study of deuterium isotope effects on protein binding by gas chromatography/mass spectrometry. Caffeine and deuterated isotopomers. Biomedical and Environmental Mass Spectrometry. 14: 653–657.

Comar, D., Maziere, M., Marazano, C. and Raynaud, C., 1975. Carbon-11 labelled psychoactive drugs. In: Proceedings of the 22nd annual meeting. Journal of Nuclear Medicine. 16: 521.

Crabtree, R. H., Holt, E. M., Lavin, M. and Morehouse, S. M., 1985. Inter- vs. intramolecular C–H activation: a C–H–Ir bridge in [IrH$_2$(8-methylquinoline)L$_2$]BF$_4$ and a C–H + M → C–M–H reaction trajectory. Inorganic Chemistry. 24: 1986–1992.

Denutte, H. R., Vandewalle, T., Cattoir, H. J., Vandecasteele, C., Jonckheere, J. A., Slegers, C., Gelijkens, C. F. and De Leenheer, A. P., 1982. The preparation of ^{11}C-labelled caffeine. Journal of Labelled Compounds and Radiopharmaceuticals. 19: 735–743.

Falconnet, J. B., Brazier, J. L. and Desage, M., 1986. Synthesis of seven deuteromethyl-caffeine analogues. Observation of deuterium isotope effects on CMR analysis. Journal of Labelled Compounds and Radiopharmaceuticals. 23: 267–276.

Funaki, Y., Iwata, R. and Ido, T., 1992. Preparation of [^{11}C]-caffeines from [^{11}C]-methyl iodide. CYRIC annual report 1992, chapter III. 2. Tohoku University, Japan. Available at: http://www.cyric.tohoku.ac.jp/english/report/repo1992/92b02.pdf. Accessed 11 April 2011.

Heftmann, E., 1971. Synthesis of caffeine-2-^{14}C. Journal of Labelled Compounds and Radiopharmaceuticals. 7: 463–465.

Hill, D. W., McSharry, B. T. and Trzupek L. S., 1988. Quantitative analysis by isotope dilution using mass spectroscopy. Journal of Chemical Education. 65: 907–910.

Horning, M. G., Nowlin, J., Thenot, J.-P. and Bouwsma, O. J., 1979. Effects of deuterium substitution on the rate of caffeine metabolism. In: Stable isotopes: Proceedings of the third international conference. Academic Press, Inc., New York, USA, pp 379–384.

Houghton, E., 1982. Deuteromethylation of dimethylxanthines: a gas chromatographic mass spectrometric method for confirmatory analysis in horse urine extracts. Biomedical Mass Spectrometry. 9: 103–107.

Jelinska, M. and Sobkowski, J., 1977. The kinetics of tritium-hydrogen exchange between xanthine, theophylline, caffeine and water. Tetrahedron. 33: 803–806.

Kalberer, P. 1964. Untersuchengen zum Abbau des Kaffeins in der Blättern von Coffea arabica. Berichte der Schweizerischen Botanischen Gesellschaft. 74: 62–107. Available at: http://retro.seals.ch/digbib/view? rid=bhl-001:1964:74::64&id=browse&id2=browse1&id3= Accessed 11 April 2011.

Kenani, A., Bernier, J.-L. and Henichart J.-P., 1995. Synthesis and E. I. M. S. fragmentation analysis of [1,3-^{15}N$_2$]-xanthine and [1,3-^{15}N$_2$]-caffeine. Journal of Labelled Compounds and Radiopharmaceuticals. 36: 187–192.

Maziere, M., Maranzano, C. and Comar, D., 1974. Colloque de médecine nucléaire de langue française (I)-58. In: G. Meyniel, (ed.) Clermont-Ferrand, France.

Modutlwa, N., Maegawa, T., Monguchi, Y. and Sajiki, H., 2010. Synthesis of deuterium-labelled drugs by hydrogen-deuterium (H-D) exchange using heterogeneous catalysis. Journal of Labelled Compounds and Radiopharmaceuticals. Special issue: Metal catalyzed exchange. 53: 686–692.

Neiman, L.A. 1981. The introduction of tritium into biologically active compounds with the aid of exchange methods. Russian Chemical Reviews. 50: 109–123.

Pierce Jr, W. M., Schlager, J. J., Madden, R. J. and Hurst, H. E., 1984. A simple, rapid synthesis of caffeine-1,7-^{13}CH$_3$. Journal of Labelled Compounds and Radiopharmaceuticals. 21: 187–192.

Regal, K. A., Kunze, K. L., Peter, R. M. and Nelson, S. D., 2005. Oxidation of caffeine by CPY1A2: isotope effects and metabolic switching. Drug Metabolism and Disposition. 33: 1837–1844.

Saji, H., Ido, T., Iwata, R., Suzuku, K., Tamate, K., Yoshikawa, K. and Kasida, Y., 1978. Caffeine-^{11}C, ephedrine-^{11}C, and methylephedrine-^{11}C: synthesis and distribution in man. Radioisotopes. 27: 451–455.

Sitkowski, J., Stefaniak, L., Nicol, L., Martin, M. L., Martin, J. G. and Webb, G. A., 1995. Complete assignments of ^{1}H, ^{13}C and ^{15}N NMR spectra of caffeine. Spectrochimica Acta. 51A: 839–841.

Suzuki, T. and Waller, G. R., 1984. Biosynthesis and biodegradation of caffeine, theobromine and theophylline in *coffea arabica* L. fruits. Journal of Agricultural and Food Chemistry. 32: 845–848.

Telo, J. P. and Vieira, A. J. S. C., 1997. Mechanism of free radical oxidation of caffeine in aqueous solution. Journal of the Chemical Society, Perkin Transaction 2. 1755–1757.

Traube, W. 1900. Der synthetische aufbau der harnsäure, des xanthins, theobromins, theophyllins und caffeïns aus der cyanessigsäure. Berichte der Deutschen Chemischen Gesellschaft. 33: 3035–3056.

Tserng, K.-Y. 1983. Gas chromatographic-mass spectrometric quantification of theophylline and its metabolites in biological fluids. Journal of Pharmaceutical Sciences. 72: 526–529.

CHAPTER 6

Diffusion of Caffeine in Different Aqueous Media at Physiological Temperature

ANA C. F. RIBEIRO*[1], VICTOR M. M. LOBO[1],
CECÍLIA I. A. V. SANTOS[2] AND MIGUEL A. ESTESO[2]

[1] Department of Chemistry, University of Coimbra, 3004 - 535 Coimbra,
Portugal; [2] Departamento de Química Física, Facultad de Farmacia,
Universidad de Alcalá, 28871 Alcalá de Henares, Spain
*E-mail: anacfrib@ci.uc.pt

6.1 Introduction

It is known that the solubility of a drug is one of the main factors for its effectiveness. In fact, its absorption from the gastrointestinal tract to pass into the circulatory system by passive diffusion depends on its solubility in water. Moreover, the passage of a drug through a membrane, and its availability in the cellular environment, also depends on the existence of a proper balance between water and lipid solubility. In summary, to be effective, most of the drugs must be administered in a water soluble dosage form (Thomas 2007; Zhou 2008).

Since water is both the most abundant component in the human body (up to 75%) and, moreover, the most common solvent used to carry out a chemical reaction, the importance of knowing the physicochemical properties of a drug in aqueous solution is easily inferred. However, if the pharmaceutical form of the drug in question is not soluble enough, functional excipients such as

Food and Nutritional Components in Focus No. 2
Caffeine: Chemistry, Analysis, Function and Effects
Edited by Victor R Preedy
© The Royal Society of Chemistry 2012
Published by the Royal Society of Chemistry, www.rsc.org

cyclodextrins[†] (CDs) can be added. Due to their truncated cone structure, CDs have the capacity to interact with the drug molecules, through a solubilization mechanism based on the establishment of non-covalent binding sites and complex formation (which may or may not be of inclusion) (Brewster and Loftsson 2007) and to ascertain a commonly called controlled-release system for delivery of drugs. The development of those systems continues to be an active area of research to which our group is paying attention (Ribeiro *et al* 2009a; 2009b; 2010).

Natural and chemically modified CDs have been extensively used to improve several drug properties such as solubility, dissolution rate, stability and bioavailability (Albers and Müller 1995; Fromming and Szejtli 1994; Stella and Rajewski 1997; Uekama *et al* 1994; Uekama and Irie 1996). In an oral drug delivery system, hydrophilic and ionizable CDs can serve as immediate and delay release formulations, while hydrophobic CDs can retard the release rate of water soluble drugs. So, the use of a given CD molecule depends on the drug molecule and its target. Among them, we are particularly interested in the most popular β-cyclodextrin (β-CD) and hydroxyl-β-cyclodextrin (HP-β-CD) in aqueous solutions.

Concerning drugs, particular relevance has been given to caffeine (1,3,7-trimethylxanthine), an alkaloid that naturally occurs in some beverages, such as coffee, tea and "cola" drinks (Murty and Manonmani 2009). This drug has a great significance in many fields, such as biological, biomedical, medical and pharmaceutical applications (*e.g.*, it is widely used as a stimulant drug acting on the central nervous system (CNS), on muscles, including the cardiac muscle, and on kidneys). In particular we are concerned about systems containing caffeine and cyclodextrins, (that is, β-CD and HP-β-CD) in aqueous solutions, especially at physiological temperature for *in vivo* pharmaceutical applications (like analgesics, diet aids, and cold remedies), and in a low concentration range, corresponding to the therapeutic dosage. Despite the tremendous importance of these systems, and the considerable work already done, the diffusion behavior is still poorly understood. Due to the fact that this information is essential for the design of those systems, we have been obtaining multicomponent chemical mutual diffusion (interdiffusion) coefficients for solutions of caffeine and the above mentioned cyclodextrins at physiological temperature.

6.2   Theoretical Aspects

6.2.1   Concepts of Diffusion

Mutual diffusion coefficient, D, in a binary system, may be defined in terms of the concentration gradient (without convection or migration) by a phenomenological relationship (Robinson and Stokes 1959)

$$J = -D\frac{\partial c}{\partial x} \tag{1}$$

known as Ficks first law, where J represents the flow of matter across a suitable chosen reference plane per area unit and per time unit, in a one-dimensional system (along to the x axis), and c is the concentration of solute in moles per volume unit at the point considered; Equation (1) may be used to determine D.[‡] The diffusion coefficient may also be obtained by considering Fick's second law,

$$\frac{\partial c}{\partial t} = \frac{\partial}{\partial x}\left(D\frac{\partial c}{\partial x}\right) \tag{2}$$

In general, the available methods are categorized into two groups: steady- and unsteady-state methods, according to Equations (1) and (2). In most of the processes, diffusion is a three-dimensional phenomenon. However, the majority of the experimental methods used, analyze the diffusion phenomenon by restricting it to a one-dimensional process, because it is much easier to manage the necessary mathematical treatment in one-dimension (being afterwards generalized to a three-dimensional space).

The resolution of Equation (2) for a one-dimensional process is much easier if we consider D as a constant. This approximation is applicable only when there are small differences of concentration, which is the case in the Taylor technique (Robinson and Stokes 1959; Tyrrel and Harris 1984). In these conditions, it is legitimate to consider that our determinations of differential diffusion coefficients are parameters with a well-defined thermodynamic meaning (Robinson and Stokes 1959; Tyrrel and Harris 1984).

Diffusion in a ternary solution is described by the extension of the Fick diffusion Equations (eqn (3) and eqn (4)),

$$-(J_1) = (D_{11})_v\frac{\partial c_1}{\partial x} + (D_{12})_v\frac{\partial c_2}{\partial x} \tag{3}$$

$$-(J_2) = (D_{21})_v\frac{\partial c_1}{\partial x} + (D_{22})_v\frac{\partial c_2}{\partial x} \tag{4}$$

where J_1, J_2, $\partial c_1/\partial x$ and $\partial c_2/\partial x$ are the molar fluxes and the gradients in concentration of solute 1 and solute 2, respectively. Main diffusion coefficients, D_{11} and D_{22}, give the flux of each solute produced by its own concentration gradient. Cross diffusion coefficients, D_{12} and D_{21}, give the coupled flux of each solute driven by a concentration gradient in the other solute. A positive D_{ik} cross-coefficient ($i \neq k$) indicates a co-current coupled transport of solute i from regions of higher concentration of solute k to regions of lower concentration of solute k. However, a negative D_{ik} coefficient indicates a counter-current coupled transport of solute i from regions of lower to higher concentration of solute k.

This phenomenon is an irreversible process. In fact, the gradient of chemical potential in the real solution is treated as the true virtual force producing diffusion. However, in most cases, that force can be quantified by the gradient of the concentration at constant temperature. Thus, we may consider the following approaches to describe the isothermal diffusion: the thermodynamics of irreversible processes and Ficks laws (Robinson and Stokes 1959).

6.2.2   Taylor Dispersion Technique

In the Taylor dispersion technique, a small amount of a given solution is injected into a laminar carrier stream of solvent, or of solution at a different concentration, to flow throughout a long capillary tube (Barthel *et al* 1996; Callendar and Leaist 2006; Tyrrell and Harris 1984). The length of the Teflon®️ dispersion tube used in the present study was measured directly by stretching the tube in a large hall and using two high quality theodolites and appropriate mirrors to accurately focus on the tube ends. This technique gave a tube length of 3.2799 (± 0.0001) $\times$ 10^4 mm, in agreement with less-precise control measurements using a good-quality measuring tape. The radius of the tube, 0.5570 ($\pm$ 0.00003) mm, was calculated from the tube volume obtained by accurately weighing (resolution 0.1 mg) the tube when empty and when filled with distilled water of known density.

In a pattern run, a sample of 0.063 mL of the solution under study ($c_j \pm \Delta c$) is injected into the laminar carrier stream (c_j) through a 6-port Teflon®️ valve (Rheodyne, model 5020). The flow rate is kept constant (0.17 mL min^{-1}) with the assistance of a metering pump (Gilson model Minipuls 3) which allows retention times of about 1.1×10^4 s. Both the dispersion tube and the injection valve are placed into an air thermostat bath to keep the temperature constant at 303.15 K ($\pm$ 0.01 K).

The dispersion of the injected samples is monitored by using a differential refractometer (Waters model 2410) at the outlet of the dispersion tube. Voltage values as a function of the elapsed time, $V(t)$, are measured at accurate 5 s intervals by using a digital voltmeter (Agilent 34401 A) provided with an IEEE interface. Binary diffusion coefficients are calculated from the dispersion equation,

$$V(t) = V_0 + V_1 t + V_{max}(t_R/t)^{1/2} \exp[12D(t t_R)^2/r^2 t] \qquad (5)$$

being the additional fitting parameters: t_R, the mean sample retention time; V_{max}, the peak height; V_0, the baseline voltage; and V_1, the baseline slope.

Extensions of the Taylor technique have been used to determine ternary mutual diffusion coefficients (D_{ik}) for multicomponent solutions. These D_{ik} coefficients, defined by eqn (3) and eqn (4), are evaluated from the fitting of two or more replicate pairs of peaks for each carrier-stream, to the ternary dispersion equation,

$$V(t) = V_o + V_1 t + V_{\max}(t_R/t)^{1/2}\left[w_1 \exp\left(-\frac{12D_1(t-t_R)^2}{r^2 t}\right) + (1-W_1)\exp\left(-\frac{12D_2(t-t_R)^2}{r^2 t}\right)\right] \quad (6)$$

Two pairs of refractive–index profiles, D_1 and D_2, are the eigenvalues of the matrix of the ternary D_{ik} coefficients.

In these experiments, small volumes of the sample solutions, of compositions $(\bar{c}_1 \pm \overline{\Delta c_1})$ and $(\bar{c}_2 \pm \overline{\Delta c_2})$ are injected into the carrier solutions of compositions $\bar{c}_1$ and $\bar{c}_2$, respectively, at time t = 0.

6.3   Experimental Aspects

The Taylor dispersion method allowed us to study the diffusion behavior of several electrolytes and non-electrolytes (Ribeiro *et al* 2009a; 2009b) in aqueous solutions, as well as some multicomponent chemical systems under different conditions. Our interest is the characterization of some binary, ternary (and in some cases quaternary) systems, involving electrolytes in different media (*e.g.*, β-CD–caffeine–water and HP-β-CD–caffeine–water (Ribeiro *et al* 2006; 2007)). Moreover, inasmuch as it is important to approach the physiological conditions for a suitable characterization of the controlled-release systems, our studies were carried out at 310.15 K.

6.3.1   Diffusion of Aqueous Caffeine

Table 6.1 shows average diffusion coefficient values, D, measured by using the Taylor dispersion technique for aqueous solutions of caffeine at 310.15 K, and concentrations from (0.001 to 0.010) mol dm^{-3}. For each carrier solution, a minimum of 4 profiles were generated by injecting samples that were more concentrated or less concentrated than the carrier solution (uncertainties of 1–2 %).

6.3.2   Diffusion of Caffeine in the Presence of β-CD

Average diffusion coefficient values, D_{11}, D_{12}, D_{21} and D_{22}, for the ternary system β-CD + caffeine + water at 310.15 K, are summarized in Table 6.2.

Main coefficients D_{11} and D_{22} give the molar flux of both the β-CD (1) and caffeine (2) components driven by their own concentration gradient.

As it can be observed, in general these coefficients are lower than the binary diffusion coefficients of aqueous β-CD and caffeine (deviations <3%) measured previously by the same technique (Table 6.1). The addition of caffeine produces relatively small changes in D_{11} for β-CD (at the most 3%), while added β-CD gets to major changes in D_{22} for caffeine (8% for the upper concentration). These facts lead us to deduce that some caffeine molecules could be present in solution as eventual inclusion complexes of β-CD, as a

Table 6.1   Mutual diffusion coefficients for aqueous caffeine solutions and the respective standard deviations, $D \pm S_D$, at different concentrations, c, and 310.15 K (unpublished and own published data).

c (mol dm⁻³)	$D \pm S_D$ (10^{-9} m² s⁻¹)
0.001	1.053 ± 0.012
0.002	1.031 ± 0.009
0.004	0.995 ± 0.008
0.005	0.980 ± 0.009^a
0.008	0.952 ± 0.005
0.010	0.944 ± 0.003^a

[a]Taylor binary D value for aqueous caffeine at 310.15 K (Ribeiro *et al* 2007)

result of the binding interactions between their molecules and the interior surface of the truncated cone of β-CD molecules. As a consequence, they would have less mobility and, therefore, they can be responsible by the relatively large decrease in D_{22}. Accordingly, this effect is less accentuated when the influence of caffeine on the transport of β-CD is considered. One acceptable explanation may be attempted by considering the similarity of the mobilities of both the free β-CD species and the eventual aggregates of β-CD and caffeine.

By analysing, at finite concentrations, the values of the cross-coefficients D_{12} and D_{21} (that should be zero at infinitesimal concentration, within the experimental error), we are able to understand the influence of the macromolecular solutes on the diffusion of the solution components. In our

Table 6.2   Ternary diffusion coefficients for aqueous β-CD (1) + caffeine (2) solutions and their respective standard deviations, S_D, at 310.15 K (unpublished plus referenced data).

$c_1{}^a$	$c_2{}^a$	D_{11}	D_{12}	D_{21}	D_{22}	$D_{12}/D_{22}{}^d$	$D_{21}/D_1{}^e$
0.000	0.002				1.031		
0.000	0.005				0.980^c		
0.000	0.010				0.944^c		
0.002	0.000	0.443^b					
0.005	0.000	0.441^b					
0.010	0.000	0.436^b					
0.002	0.002	0.443 ± 0.007	-0.017 ± 0.012	0.031 ± 0.008	1.009 ± 0.016	-0.017	0.070
0.005	0.005	0.441 ± 0.007	0.006 ± 0.035	0.051 ± 0.033	0.958 ± 0.011	0.007	0.116
0.010	0.002	0.429 ± 0.008	-0.005 ± 0.021	0.012 ± 0.017	0.999 ± 0.022	-0.005	0.028
0.010	0.005	0.428 ± 0.013	-0.008 ± 0.024	0.037 ± 0.020	0.937 ± 0.015	-0.009	0.085
0.010	0.010	0.424 ± 0.008	0.020 ± 0.011	0.054 ± 0.031	0.868 ± 0.012	0.029	0.127

[a]c_i in units of (mol dm⁻³), D_{ik} in units of (10^{-9} m² s⁻¹); [b]Taylor binary D value for aqueous β-CD (Barthel *et al* 1996); [c]Taylor binary D value for aqueous caffeine at 310.15 K (Ribeiro *et al* 2007); [d]D_{12}/D_{22} give the number of moles of β-CD transported per mole of caffeine; [e]D_{21}/D_{11} give the number of moles of caffeine co-transported per mole of β-CD

case, it is easy to see that the gradient in concentration of β-CD produces a co-current coupled flow of caffeine (positive values for D_{21}). Having in mind that D_{21}/D_{11} gives the number of moles of caffeine co-transported per mole of β-CD driven by the concentration gradient of the last one, we can observe, at the compositions studied, that a mole of diffusing β-CD co-transports at most 0.13 mol of caffeine, increasing this co-transport with the increase of the caffeine concentration, but decreasing it with the increase of the β-CD concentration.

However, by looking at the D_{12} values in Table 6.2, it is not possible to assume any kind of transport of β-CD occurring as a consequence of the caffeine concentration gradient (small positive and–or negative values similar to the experimental errors). Thus, the almost negligible D_{12}/D_{22} values found, at the same compositions, reflect that a caffeine flux is unable to produce a coupled flux of β-CD.

6.3.3 Diffusion of Caffeine in the Presence of HP-β-CD

Average diffusion coefficient values, D_{11}, D_{12}, D_{21} and D_{22}, for the ternary system HP-β-CD + caffeine + water at 310.15 K, are summarized in Table 6.3.

For the case of caffeine diffusion in the presence of a hydrophilic derivative of β-CD, we can easily appreciate that the above mentioned effects are more accentuated. In the concentration ranges studied, there is no appreciable influence on the main coefficient D_{11} for HP-β-CD either of the presence of caffeine or the concentration gradients of both species. On the contrary, D_{22} values are much lower than the binary diffusion coefficient previously measured (deviation between 3% and 15%) and even lower (between 3% and 12%) than in the presence of β-CD. In this last case, for a given HP-β-CD concentration the decrease is less pronounced with the increase of caffeine. A possible explanation for these facts can be given on the basis that some caffeine molecules and HP-β-CD molecules may exist in solution as eventual inclusion complexes 1 : 1, as a result of the binding interactions between caffeine molecules and the interior surface of the truncated cone of HP-β-CD molecules and that this interaction occurs more extensively than in the case of the β-CD molecules.

Also, by looking at the cross-coefficients D_{12} and D_{21}, we can observe no negligible increase in their values, within the precision of this method, in relation to those for the (β-CD + caffeine) system. In this case, both solutes generate coupled flows to each other. At the compositions range used, a mole of diffusing HP-β-CD co-transports at most 0.28 mol of caffeine and, although in small quantity, a mole of diffusing caffeine can co-transport up to 0.09 mol of HP-β-CD.

6.4 Conclusions

The behavior of the diffusion of caffeine in aqueous solutions changes in the presence of each carbohydrate at 310.15 K. In fact, at the compositions used, a

Table 6.3   Ternary diffusion coefficients for aqueous HP-β-CD (1) + caffeine (2) solutions and their respective standard deviations, S_D, at 310.15 K (unpublished plus referenced data).

$c_1{}^a$	$c_2{}^a$	D_{11}	D_{12}	D_{21}	D_{22}	$D_{12}/D_{22}{}^d$	$D_{21}/D_{11}{}^e$
0.000	0.002				1.031		
0.000	0.005				0.980[c]		
0.000	0.010				0.944[c]		
0.002	0.000	0.408[b]					
0.005	0.000	0.407[b]					
0.010	0.000	0.406[b]					
0.002	0.002	0.407 ± 0.015	−0.038 ± 0.047	0.070 ± 0.061	0.997 ± 0.029	−0.038	0.172
0.005	0.005	0.406 ± 0.006	0.057 ± 0.035	0.080 ± 0.033	0.883 ± 0.013	0.065	0.197
0.010	0.002	0.403 ± 0.009	0.077 ± 0.040	0.031 ± 0.033	0.875 ± 0.011	0.088	0.077
0.010	0.005	0.403 ± 0.004	0.061 ± 0.023	0.033 ± 0.013	0.852 ± 0.018	0.072	0.081
0.010	0.010	0.401 ± 0.013	0.024 ± 0.014	0.113 ± 0.031	0.827 ± 0.012	0.029	0.282

[a]c_i in units of (mol dm^{-3}), D_{ik} in units of (10^{-9} m^2 s^{-1}) ; [b]Taylor binary D value for aqueous HP-β-CD (Ribeiro *et al* 2006); [c]Taylor binary D value for aqueous caffeine at 310.15 K (Ribeiro *et al* 2007); [d]D_{12}/D_{22} give the number of moles of HP-β-CD transported per mole of caffeine; [e]D_{21}/D_{11} give the number of moles of caffeine co-transported per mole of HP-β-CD

mole of diffusing HP-β-CD co-transports at most 0.28 mol of caffeine, while a mole of diffusing β-CD is only able to co-transport 0.13 mol of caffeine. Thus, we can consider that the HP-β-CD is more favorable to the transport of that drug.

The diffusion coefficients determined for aqueous solutions of β-CD and HP-β-CD provide transport data necessary to model the diffusion in pharmaceutical and engineering applications.

Summary Points

Caffeine is a weakly soluble drug, whose prominent pharmacologic action is to stimulate the central nervous system but it is also used in a wide range of pharmaceutical preparations.

Its solubility, as with other physical properties of this drug, can be improved by applying one carrier, *e.g.* a cyclodextrin as an excipient, and establishing a drug release system.

Depending on the goal to achieve, it is important to choose the adequate cyclodextrin molecule but also, in order to adequately characterize the controlled-release system, it is important to obtain information on its behavior at physiological temperature.

Diffusion of caffeine at physiological temperature, in the presence of both β-CD and HP-β-CD, changes with respect to the aqueous system.

The changes in the diffusion coefficient values for caffeine, accomplished by the coupled transport of caffeine by the cyclodextrin molecules, allow us to deduce the existence of interactions between these solutes. Moreover this interaction appears to be stronger for the hydroxypropylated derivative.

Key Facts of the Cyclodextrins

Cyclodextrins (CDs) are cyclic oligosaccharides (Li and Purdy 1992; Uekama *et al* 1998) where the primary and secondary groups of the glucose unities are turned to the exterior ends of the molecule, making them hydrophilic as a whole. However, the interior surface of their truncated cone structure (*i.e.*, the cavity), normally considered as the site of the guest molecules, is largely hydrophobic. These hydrophobic cavities provide a favorable host potential to form inclusion complexes with a large variety of both organic and inorganic compounds in different solvents (water included).

Key Facts of the Diffusion

It is necessary to make a distinction between two distinct processes: self-diffusion (also named as intradiffusion, tracer diffusion, single ion diffusion, or ionic diffusion), characterized through D^*, and mutual diffusion (also known as interdiffusion, concentration diffusion, or salt diffusion), distinguished by means of D (Mills and Lobo, 1989). Methods such as those based on NMR, polarography, and capillary-tube techniques with radioactive isotopes, give self-diffusion coefficients, D^*, but not mutual diffusion ones, D. However, for bulk substance transport, the appropriate parameter to be considered is the mutual diffusion coefficient, D. Theoretical relationships derived between both self-diffusion and mutual diffusion coefficients, D^* and D, respectively, have had limited success for the estimation of D values (as well as theoretical expressions for the calculation of D) and consequently experimental mutual diffusion coefficients are absolutely necessary.

Definitions of Words and Terms

Bioavailability: The degree to which a drug is absorbed or becomes available at the site of physiological activity after administration.

Co-current coupled transport: Coupled diffusion of a solute, generated by the concentration gradient of another solute also present in the solution, from areas of high concentration to areas of low concentration of the latter.

Counter-current coupled transport: Coupled diffusion of a solute, generated by the concentration gradient of another solute also present in the solution, from areas of low concentration to areas of high concentration of the latter.

Cyclodextrins: Cyclic oligosaccharides composed of 6 or more D-(+) glucopyranoside units linked by α-1,4-glucosidic bonds, forming a ring. Native cyclodextrins are called α–cyclodextrin (six glucose units), β–cyclodextrin (seven glucose units) and γ–cyclodextrin (eight glucose units).

Controlled drug release system: The drug delivery occurs when a system or matrix (vehicle) is methodically combined with a drug or other active agent, so that the active agent is released from the material in a predetermined form. The release of the active agent may be constant or cyclic over a long period of time, or can be triggered by the environment or other external events.

Excipient: Inert substance included in a pharmaceutical preparation to give suitable consistency or form to the drug or act as a carrier.

Inclusion complex: A complex in which one component (the host) features a cavity in which molecular entities of a second chemical component (the guest) are positioned without establishing covalent bonds (usually the interactions are only due to van der Waals forces).

Passive diffusion: Diffusion of a solute that occurs by moving down the concentration gradient from areas of high concentration to areas of low concentration of this solute.

Mutual diffusion coefficient: Parameter D in Ficks law diffusion (eqn (1)) that quantifies the diffusion of electrolytes or non-electrolytes as a whole, as a consequence of a concentration gradient.

Self-diffusion coefficient: Parameter D^* concerning the phenomenon of intra-diffusion of one of the species of a solution (see footnote[‡]).

Abbreviations

CNS	Central Nervous System
CD(s)	Cyclodextrin(s)
β-CD	β-Cyclodextrin
HP-β-CD	Hydroxypropyl-β-cyclodextrin
D	Diffusion Coefficient

Acknowledgements

C.I.A.V.S. is grateful for SFRH/BD/45669/2008 from "Fundação para a Ciência e Tecnologia". Financial support from FCT (FEDER)-PTDC/AAC-CLI/098308/2008 is gratefully acknowledged. M.A.E. is grateful to the University of Alcalá (Spain) for the financial assistance (Mobility Grants for Researchers Program).

References

Albers, E. and Müller, B.W., 1995. Cyclodextrin derivatives in pharmaceutics, CRC Critical Reviews in Therapeutic Drug Carrier Systems. 12: 311–337.

Barthel, J., Gores, H. J., Lohr, C. M. and Seidl, J. J., 1996. Taylor dispersion measurements at low electrolyte concentrations. 1. Tetraalkylammonium perchlorate aqueous solutions. Journal of Solution Chemistry. 25: 921–935.

Brewster, M. E. and Loftsson, T., 2007. Cyclodextrins as pharmaceutical solubilizers. Advanced Drug Delivery Reviews. 59: 645–666.

Callendar, R. and Leist, D. G., 2006. Diffusion coefficients for binary, ternary, and polydisperse solutions from peak-width analysis of Taylor dispersion profiles. Journal of Solution Chemistry. 35: 353–379.

Fromming, K. H. and Szejtli, J. (ed.), 1994. Cyclodextrins in Pharmacy. Kluwer, Dordrecht, pp. 19–32.

Li, S. and Purdy, W. C., 1992. Cyclodextrins and their applications in analytical chemistry. Chemical Reviews. 92: 1457–1470.

Lobo, V. M. M. 1990. Handbook of Electrolyte Solutions. Elsevier, Amsterdam, 1168 pp.

Mills, R. and Lobo, V. M. M. 1989. Self-diffusion in electrolyte solutions. A critical examination of data compiled from the literature. Elsevier Science Pub., Amsterdam, 346 pp.

Pushpa S., Murthy and Manonmani, H. K., 2009. Physico-chemical, antioxidant and antimicrobial properties of Indian monsooned coffee. European Food Research and Technology. 229: 645–650.

Ribeiro, A. C. F., Leist, D. G., Lobo, V. M. M., Esteso, M. A., Valente, A. J. M., Santos, C. I. A. V., Cabral, A. M. T. D. P. V. and Veiga, F. J. B., 2006. Binary mutual diffusion coefficients of aqueous solutions of ß-cyclodextrin at temperatures from 298.15 K to 312.15 K. Journal of Chemical & Engineering Data. 51: 1368–1371.

Ribeiro, A. C. F., Valente, A. J. M., Santos, C. I. A. V., Prazeres, P. M. R. A., Lobo, V. M. M.; Burrows, H. D., Esteso, M. A., Cabral, A. M. T. D. P. V. and Veiga, F. J. B., 2007. Binary mutual diffusion coefficients of aqueous solutions of α-cyclodextrin, 2-hydroxypropyl-α-cyclodextrin and 2-hydroxypropyl-ß-cyclodextrin at temperatures from 298.15 to 312.15 K. Journal of Chemical & Engineering Data. 52: 586–590.

Ribeiro, A. C. F., Santos, C. I. A. V., Lobo, V. M. M., Cabral, A. M. T. D. P. V., Veiga, F. J. B. and Esteso, M. A., 2009. Diffusion coefficients of the ternary system β-cyclodextrin + caffeine + water at 298.15 K. Journal of Chemical & Engineering Data. 54: 115–117.

Ribeiro, A. C. F., Santos, C. I. A. V., Lobo, V. M. M., Cabral, A. M. T. D. P. V., Veiga, F. J. B. and Esteso, M. A., 2009. Interactions between 2-hydroxypropyl-β-cyclodextrin and caffeine as seen by diffusion at 25 °C. Journal of Chemical Thermodynamics. 41: 1324–1328.

Ribeiro, A. C. F., Barros, M. C. F., Lobo, V. M. M., Quintanilha, G. and Esteso, M. A., 2010. Diffusion coefficients of the ternary system calcium chloride -caffeine-water at 25 °C and 37 °C. Journal of Chemical & Engineering Data. 55: 897–900.

Ribeiro, A. C. F., Santos, C. I. A. V., Lobo, V. M. M. and Esteso, M. A., 2010. Quaternary diffusion coefficients of β-cyclodextrin + KCl + caffeine + water at 298.15 K using a Taylor dispersion method. Journal of Chemical & Engineering Data. 55: 2610–2612.

Robinson, R. A and R. H. Stokes., 1959. Electrolyte Solutions. 2nd Edition, Butterworths, London, 559 pp.

Stella, V. J. and Rajewski, R. A., 1997. Cyclodextrins: their future in drug formulation and delivery. Pharmaceutical Research. 14: 556–567.

Thomas, G. 2007. Medicinal Chemistry: an introduction, 2nd Edition, Wiley-Interscience, pp. 44–46.

Tyrrell, H. J. V. and Harris, K. R., 1984. Diffusion in Liquids. 2nd Edition, Butterworths, London, 448 pp.

Uekama, K., Hirayama, F. and Irie, T., 1994. Application of cyclodextrins. Drug absorption enhancement: concepts, possibilities, limitations and trends. In: Boer, A. G. (ed.) Drug Targeting Delivery. Vol 3. Harwood Academic, Swizerland, pp. 411–456.

Uekama, K. and Irie, T., 1996. Pharmaceutical use of cyclodextrins in various drug formulations. In: Szejtli, J. and Osa, T., (ed.) Comprehensive Supramolecular Chemistry. Vol 3. Cyclodextrins. Elsevier Science Ltd., New York, pp. 451–482.

Uekama. K, Hirayama, F. and Irie, T., 1998. Cyclodextrin drug carrier systems. Chemical Reviews. 98: 2045–2076.

Zhou, H. 2008. Water insoluble drug formulation. CRC Press, chp. 5, pp. 91–100.

Analysis

CHAPTER 7

Analysis of Caffeine by Liquid Chromatography-Mass Spectrometry

DANIEL PERRONE[1] AND ADRIANA FARAH*[1,2]

[1] Laboratório de Bioquímica Nutricional e de Alimentos, Departamento de Bioquímica, Instituto de Química, Universidade Federal do Rio de Janeiro, Brazil; [2] Laboratório de Química e Bioatividade de Alimentos & Núcleo de Pesquisa em Café Prof. Luiz Carlos Trugo, Instituto de Nutrição, Universidade Federal do Rio de Janeiro, Brazil
*E-mail: afarah@nutricao.ufrj.br

7.1 Introduction

Caffeine, 1,3,7-trimethyl-xanthine, is a purine-based xanthine alkaloid (Figure 7.1). Caffeine was first isolated from coffee in 1820 by the German chemist Friedlieb Ferdinand Runge and in 1821 by a team of French chemists, Robiquet, Pelletier and Caventou. Pelletier first conceived the word "caféine" from the French word *café* (coffee) and this term became the English word "caffeine". Caffeine has been the subject of extensive research for its wide occurrence in nature and long history of use. Over 60 plant species contain caffeine and it has been suggested that its consumption dates back from the Stone Age (Barone and Roberts 1996). Nowadays, caffeine is one of the most widely consumed substances in the world and probably the psychoactive substance with the largest consumption among humans. Caffeine has become

Food and Nutritional Components in Focus No. 2
Caffeine: Chemistry, Analysis, Function and Effects
Edited by Victor R Preedy
© The Royal Society of Chemistry 2012
Published by the Royal Society of Chemistry, www.rsc.org

Figure 7.1   structure of caffeine: $C_8H_{10}N_4O_2$. MW: 194.19.

an almost universal component of the diet, either as a natural constituent or as a common additive in many different food products (Begas *et al* 2007).

7.1.1   Caffeine Sources

Caffeine is found in varying quantities in the seeds, leaves and fruits of some plants and its botanical functions are thought to be both antifungal and insecticidal (Nathanson *et al* 1984; Arora *et al* 1997; Hollingsworth *et al* 2002; Araque *et al* 2007). Among the main food sources of caffeine, some are internationally recognized, such as coffee (*Coffea* sp.), tea (*Camellia sinensis*) and cocoa (*Theobroma cacao*), but others are locally consumed, *e.g.*, mate (*Ilex paraguariensis*) and guaraná (*Paullinia cupana*) in South America (Clifford and Ramirez-Martinez 1990; Pelozo *et al* 2008), Yaupon Holly (*Ilex vomitoria*) in the United States and Mexico (Power and Chestnut 1919; Palumbo *et al* 2007), and Miang (pickled tea) in Thailand (Phromrukachat *et al* 2010). Caffeine is also a common ingredient of soft drinks, such as cola, originally prepared from kola nuts, and energy drinks, where it is found in association with other stimulating substances (Barone and Roberts 1996; Aranda and Morlock 2006). In addition to food sources, caffeine is also found in a wide variety of prescription and over-the-counter pharmaceutical products (Benowitz 1990; Barone and Roberts 1996).

Caffeine content in different beverages suffers from wide variation, as a consequence of differences in the methods of preparation (methods of brewing coffee such as, percolation, *espresso*, French press, *etc.*, and of preparing tea infusions) (Bell *et al* 1996; Astill *et al* 2001), production source (*e.g.*, Arabica or Robusta coffee; black or green tea) (Casal *et al* 2000; Astill *et al* 2001) and analytical methods (Watson 2003).

Worldwide, it is estimated that at least 80% of the adult population consume caffeine on a daily basis, reaching up to 90% in North America (Frary *et al* 2005). Studies suggest a mean daily caffeine intake for US consumers of 3 mg kg^{-1}. Among children younger than 18 years of age who are consumers of caffeine-containing foods, the mean daily caffeine intake is about 1 mg kg^{-1}. Both adults and children in Denmark and UK have higher levels of caffeine intake, of 7 and 4 mg kg^{-1}, respectively (Barone and Roberts 1996). The major caffeine vehicle also changes according to the country. In the UK, India and China, most adults consume tea, whereas in France, Italy, Germany and

Brazil, coffee drinking is prevalent. In the USA, it is estimated that only half of the adult population consumes coffee, but more soft-drinks beverages are consumed than in other countries around the world (Watson 2003; Frary *et al* 2005).

7.1.2 Pharmacokinetics and Physiological Effects of Caffeine

Caffeine is rapidly absorbed in the gastrointestinal tract; in approximately 20 min, about 90% of the caffeine contained in a cup of coffee is cleared from the stomach, although some absorption may occur. The time for caffeine to reach its peak plasma concentrations widely varies, ranging from 15 min to 2 hours, depending on the presence of food in the intestine or when large amounts of caffeine are consumed (*e.g.*, after taking caffeine-containing pharmaceuticals) (Arnaud 1999).

After absorption, due to its lipophilic nature, caffeine can pass through biological membranes, including the blood-brain barrier. The plasma half-life of caffeine is between 3 and 7 h in normal adults. Whereas increasing age does not affect caffeine distribution, premature and full-term neonates show a higher plasma half-life, mainly due to the lower activity of caffeine metabolizing enzymes and other elimination pathways. No major differences are found among men and women, although the menstrual cycle and pregnancy significantly affect caffeine distribution. Caffeine half-life in adult smokers is half of that observed for non-smokers (Arnaud 1999).

Caffeine is metabolized in the cytochrome P450 (CYP) enzyme system, a large and diverse group of enzymes responsible for, among other things, the metabolism of xenobiotic substances. The metabolism of caffeine includes multiple and separate pathways with demethylation to dimethyl- and monomethylxanthines; ring oxidation with the formation of trimethyl-, dimethyl- and monomethyl-urates; and ring opening yielding substituted diaminouracils (Figure 7.2). The major pathway is demethylation at N3 position of caffeine to produce paraxanthine (1,7-dimethyl-xanthine, 17X), which accounts for up to 80% of caffeine metabolites. Paraxanthine can be further demethylated by CYP1A2 at N7 position to produce 1-methyl-xanthine (1X), hydroxylated by CYP2A6 and CYP1A2 to produce 1,7-dimethyl-uric acid (17U), or it can undergo ring opening by *N*-acetyl-transferase 2 to produce 5-acetylamino-6-formylamino-3-methyluracil (AFMU). Another 22 metabolites have been described (Arnaud 1999), including theophylline (1,3-dimethyl-xanthine), theobromine (3,7-dimethyl-xanthine), and both uric acid and uracil derivatives (*e.g.*, 1-methyl-uric acid, 1U and 5-acetylamino-6-amino-3-methyluracil, AAMU, respectively).

Caffeine mechanism of action occurs through its binding to adenosine receptors, which are widely distributed throughout the body, and competitive antagonism of adenosine. Caffeine is usually described as a stimulant of the central nervous system, causing the increase of the synthesis and release of a myriad of neurotransmitters, such as catecholamines, dopamine, adrenaline,

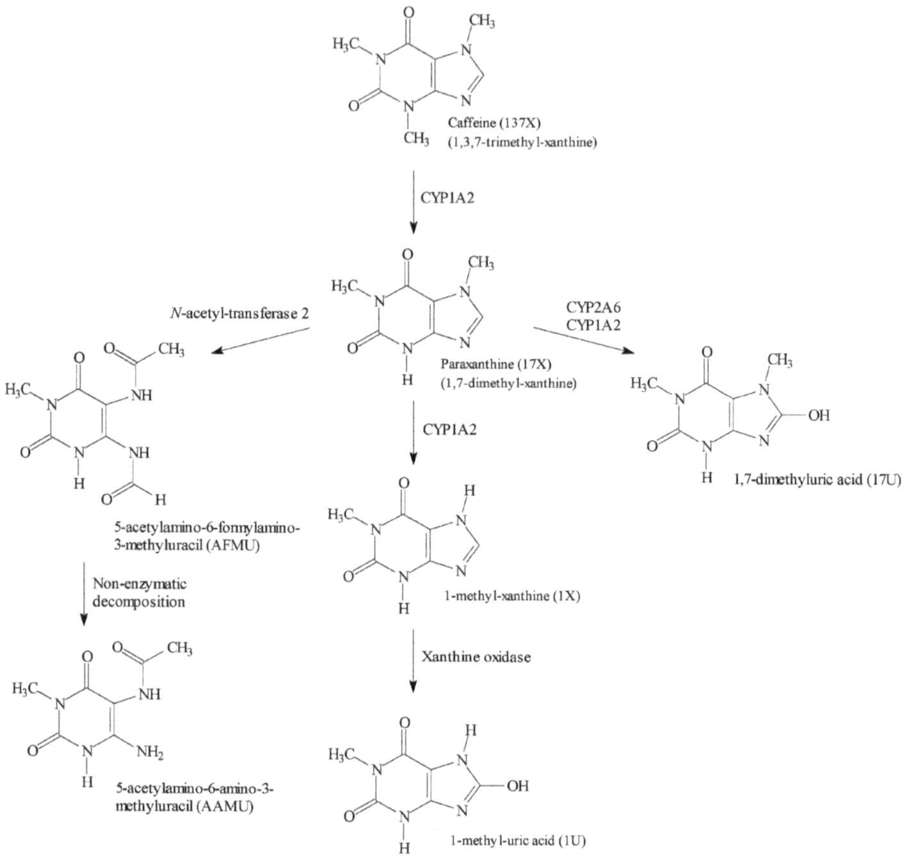

Figure 7.2   Hepatic bioconversion of caffeine in its major metabolites.

serotonin and acetylcholine and, therefore, improving vigilance, alertness, wellbeing, motivation, clarity of mind, mental concentration and sense of energy. Caffeine also affects the rate and force of contraction of the heart through the direct stimulus of the myocardial tissue. Although high caffeine doses (over 250 mg) may cause a moderate raise in the blood pressure, this effect is not observed in habitual caffeine consumers. Caffeine induces a mild diuresis owing to an increase in renal blood flow, an increased glomerular filtration rate and a decrease in tubular reabsorption of sodium ions (Arnaud 1999).

7.2   Analysis of Caffeine and Related Compounds by LC-MS

In this chapter, we addressed caffeine analysis by LC-MS techniques in food, water and biological matrices. In addition to caffeine, we discussed LC-MS analysis of other xanthine alkaloids, such as theobromine and theophylline, as well as caffeine metabolites. Among biological matrices, we divided analytical methods in four sub-sections, according to the field of interest, as follows: caffeine analysis in commonly investigated biological fluids (urine, plasma, saliva); caffeine analysis in unusually investigated biological fluids (breast milk) and tissues; caffeine metabolites; other fields, including metabolomics, doping control and pharmacokinetic studies.

Caffeine analysis in major food sources, such as coffee and tea, may be routinely performed by colorimetric methods or HPLC with UV detection. In biological fluids, major caffeine metabolites, such as paraxanthine and 1-methyl-uric acid, may also be analyzed using HPLC-UV. However, this technique is not appropriate if one desires to investigate caffeine and other xanthine alkaloids in foods that are not major sources of these compounds or minor caffeine metabolites in biological fluids and tissues. The analysis of caffeine in aqueous media with environmental implications cannot also be performed by HPLC-UV or colorimetric methods. In these cases, analytes are present from low to trace levels (ppb to low ppt) in the samples and–or may be subjected to matrix effects, which could cause qualitative and quantitative analytical errors.

LC-MS is the method of choice for determining caffeine at low levels and–or in very complex matrices. The three main advantages of LC-MS over conventional HPLC methods can be represented by 3 "S", Sensitivity, Selectivity and Speed. However, due to their high cost, LC-MS techniques are not economically justifiable for the analysis of caffeine in major sources of caffeine, such as coffee.

Sensitivity relates to the potential of mass spectrometry (MS) to determine analytes even when present in trace levels (low ppt) in the sample. MS sensitivity allows the analysis of trace levels of caffeine in food and water, as well as minor metabolites in biological fluids and tissues. Selectivity relates to the ability of MS detectors to discriminate a specific analyte among a wide diversity of compounds present in the sample. Finally, speed is related to the possibility of rapid chromatographic runs, due to the ability of MS for analyzing different analytes even if they are not chromatographically separated. Both selectivity and speed are MS features that allow the simultaneous and fast analysis of caffeine together with many additional compounds of interest, such as other food constituents, pharmaceuticals, metabolites and biological molecules.

7.2.1 Food

Despite coffee being the major food source of caffeine, LC-MS methods have been largely used for the analysis of this alkaloid in tea samples, usually in association with other components, such as catechins. Caffeine analysis in coffee is usually done by simpler and more straightforward methods, such as colorimetric and HPLC-UV, as high amounts of caffeine are usually found (with the exception of decaffeinated coffee) and low interference from the food matrix is expected. LC-MS methods are only worthwhile when other components besides caffeine are to be simultaneously analyzed in coffee.

The apparently first two works that used LC-MS for the analysis of caffeine in food samples were published in 2004. Del Rio *et al* (2004) described the application of high-performance liquid chromatography-*tandem* mass spectrometry (HPLC-MS^n) methods for the rapid and routine analysis of caffeine, theobromine and more than 30 phenolic compounds in green and black tea. Infusions were injected directly onto a reversed phase C12 column and two different mobile phase gradients were employed to separate out first the catechin-derived compounds and second the flavonols and theaflavins. Mobile phase contained 1% aqueous formic acid to aid the formation of both positive and negative ions in the electrospray probe. Caffeine and theobromine were identified on the basis of their retention time, absorbance spectrum and pseudo-molecular ions [M–H]$^-$ (*m/z*) 195 and 181. In addition to these data, MS fragmentation pattern was also used to identify phenolic compounds, such as (+)-catechin, (–)-epicatechin, theaflavin and their various gallate derivatives, quercetin and kaempferol mono-, di-, and triglycosides, quinic acid esters of gallic acid and hydroxycinnamates.

Zhu *et al* (2004) established a methodology for analyzing multiple bioactive compounds in tea and tea extracts, including caffeine, by LC-DAD-ESI-MS (liquid chromatography-diode array detector-electrospray ionization-mass spectrometry). Caffeine and other purine alkaloids (theobromine and theophylline) were ionized in the positive mode. Caffeine identification was based on the molecular weight of its pseudomolecular ion, [M+H]$^+$, *m/z* 195. Since caffeine coeluted with chlorogenic acid, quantification with DAD data was not suitable. However, measurement of these two compounds could be completed using single ion monitoring (SIM) mode of ESI-MS. The proposed method provided more qualitative and quantitative information compared with general HPLC in the analysis of multi-components in tea, and complex extraction or sample pretreatment was unnecessary.

The differentiation of tea products and cultivars according to their mass spectral fingerprints, which included caffeine, was reported by LC-MS methodologies. To differentiate green, oolong and jasmine tea, Chen *et al* (2007) developed a novel method based on ambient mass spectrometry, a set of related techniques which are increasingly used for the direct detection of trace amounts of components present in complex matrices with minimal sample pretreatment. The desorption electrospray ionization (DESI) technique (Tákats *et al* 2004) uses a high pressure (*ca.* 200 psi) sheath gas and a methanol–water solution (1 : 1, *v/v*) which is electrosprayed directly onto samples to maintain a stable signal. The direct analysis in real time (DART) technique (Cody *et al* 2005) is operated without spraying methanol but with inert gases such as helium and was recently found to be less sensitive than DESI (Williams *et al* 2006), especially for compounds absorbed chemically by solid surfaces. Even though DESI and DART techniques could be employed to

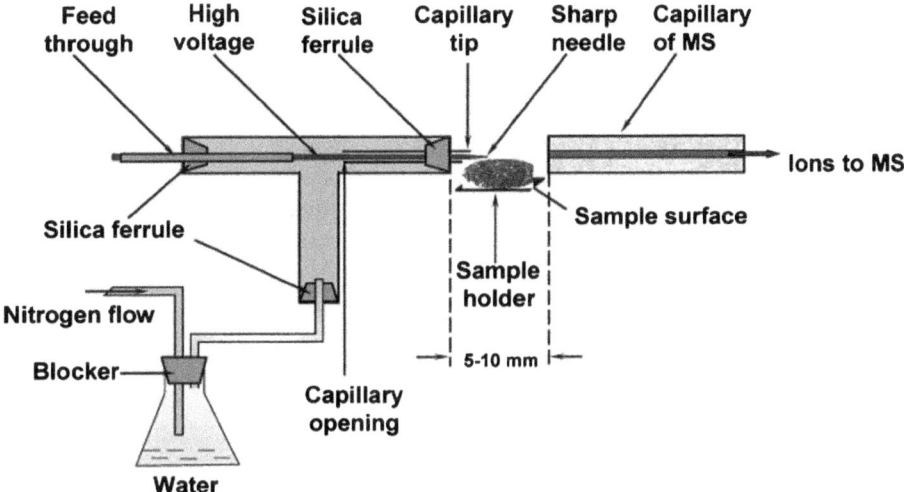

Figure 7.3 Schematic diagram of the DAPCI source used by Chen *et al* (2007) for direct analysis.

analyze tea infusions, it would require an extra step to make the tea brew. On the other hand, desorption chemical ionization (DAPCI) has been demonstrated for direct surface analysis with enhanced sensitivity, thus enabling the analysis directly on tea leaves (Figure 7.3). The DAPCI-MS and mass-mass spectrometry (MS-MS) spectra of tea products were recorded in a linear ion trap mass spectrometer (LTQ-MS). Caffeine was detected from tea samples as protonated molecules at *m/z* 195. In the collision induced dissociation (CID) experiments, caffeine ions generated two major fragments of *m/z* 138 and 110 by the loss of CH_3NCO and CO, successively. This fragmentation pattern was in agreement with previous observations in the literature (Ford *et al* 2005; Venzie *et al* 2007) and with the data obtained using the standard compound, thus confirming the assignment of caffeine.

Lin *et al* (2008) applied a LC-DAD-ESI-MS method to establish the profile of phenolic compounds and purine alkaloids of 41 green and 25 fermented teas. These samples could be organized in five distinct groups according to their profiles; (1) high-grade teas, consisting of the younger buds and leaves harvested in the early leaf-growing stage; (2) green teas; (3) partially fermented oolong, tikuanyin and puerh teas; (4) fully fermented black teas; (5) highly overfermented tou and chitsebenng black teas.

Wang *et al* (2008) used an HPLC-DAD-ESI-MS-MS method to analyze the polyphenols and purine alkaloids in 22 different tea cultivars, from three distinct origins (Fujian Province, Guangdong Province and Taiwan). Their method, which was adapted from that of del Rio *et al* (2004), entailed ultrasonic extraction in acetonitrile–water (1 : 1, v/v) and chromatographic separation on an octadecylsilane (ODS) column with linear gradient elution by the mobile phase consisting of acetonitrile and 1% formic acid aqueous solution. Caffeine was identified by comparison of retention time and MS spectra of standard compound, and quantified by DAD. Differentiation according to the origin of the tea was established, but mainly due to differences in polyphenols and not caffeine contents among samples.

The use of other ionization techniques besides ESI and ambient MS for the analysis of caffeine in food samples, in particular tea, was demonstrated by Castro *et al* (2010). They presented an LC-MS method which used particle beam glow discharge (PB-GD) electron ionization (EI) for the determination of caffeine and catechins in proposed green tea standard reference materials.

While ESI-MS can provide molecular weight information of polar compounds without extensive fragmentation, in many cases the addition of MS-MS methods are necessary for the complete species-specific identification. The PB-GD interface (Figure 7.4) provides a means of introducing LC effluents into low pressure sources (*i.e.,* electron impact and chemical ionization) while maintaining chromatographic characteristics such as retention and elution quality (Davis *et al* 2003). Other advantages of the PB interface are its mechanical–operational simplicity, compatibility with a wide range of solvent polarities and flow rates, as well as its high efficiency in solvent removal. As such, the PB interface is well suited for the introduction of solution-phase analytes into GD sources. The coupling of the particle beam interface to a glow discharge mass spectrometry (PB-GD-MS) ion source

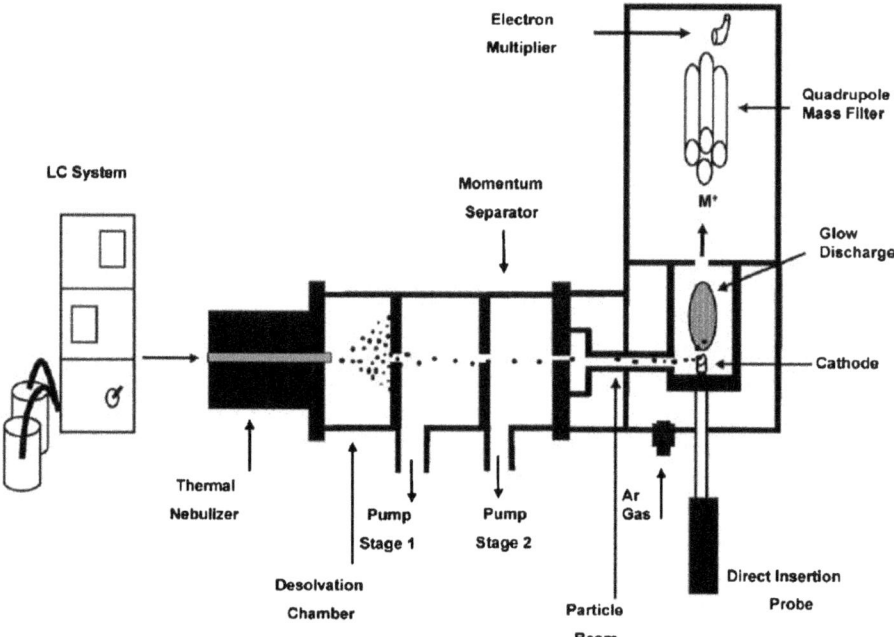

Figure 7.4   Diagrammatic representation of the basic components of the particle-beam glow-discharge-mass spectrometry (PB-GD-MS) apparatus (Davis *et al* 2003).

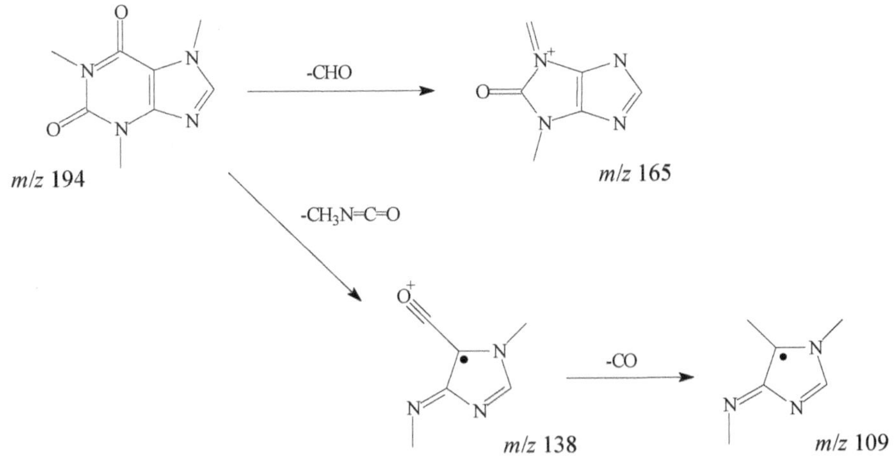

Figure 7.5 Fragmentation patterns of caffeine in PB-GD-MS proposed by Castro *et al* (2010)

produces a high level of information, specifically, the mass spectra of a wide variety of compounds are very EI-like in nature, yielding molecular ions and easily identified fragmentation patterns.

Using PB-GD-MS, the mass spectra they obtained were clear and showed easy-to-interpret fragmentation patterns. Mass spectrum of caffeine showed a base peak corresponding to the molecular ion $[M]^{+\bullet}$ of *m/z* 194 (the fact that the characteristic fragment for catechins had the same mass was mere coincidence) with characteristic fragment peaks of *m/z* 165, 138, and 109 (Figure 7.5). The absolute limit of detection for caffeine was determined to be on the nanogram level. Quantification of caffeine was carried out using the standard addition and internal standard methods, with the latter providing appreciable improvements in precision and recovery.

The selectivity and speed inherent of MS suit the employment of this technique for the analysis of caffeine in food samples in combination with other components of interest. Perrone *et al* (2008) reported the development and validation of a rapid LC-MS method for the simultaneous quantification of caffeine, trigonelline, nicotinic acid and sucrose in coffee. The method involved extraction with hot water, clarification with basic lead acetate and membrane filtration. The ESI source was operated in the negative mode to generate sucrose ions and in the positive mode to generate caffeine, trigonelline and nicotinic acid ions. Suppression of response due to matrix effect was observed for all the analytes. This effect, which has been extensively reported, especially when analyzing complex matrices, such as biofluids, environmental and food samples (Matuszewski *et al* 2003), should be of major concern during the development of LC-MS methods "to ensure that precision, selectivity, and sensitivity of the proposed method would not be compromised" (FDA, Food and Department of Health, 2001). Caffeine limits of detection and quantification were 11.9 and 39.6 ng mL^{-1}, respectively, approximately 5 times lower than those reported for HPLC with UV detection (Casal *et al* 1998) and similar to other LC-MS methods (Zhu *et al* 2004). The method applicability was evaluated by the analysis of different types of coffee available in the market (green and commercial samples of regular and decaffeinated roasted and instant coffees, roasted to different degrees). The method showed to be suitable, thus presenting as a fast and reliable alternative method for routine coffee analysis.

7.2.2   Water

It is well known that the source, presence, and destiny of pharmaceutical compounds in the aquatic environment are of concern. These compounds may be excreted unmetabolized or partially metabolized by humans, resulting in their eventual passage into the environment. Monitoring the infiltration of domestic wastewater and untreated sewage into aquatic ecosystems has been a crucial challenge for environmental agencies whose efforts are devoted to protect the delicate balance between biological systems at risk and the inevitable urban colonization, especially of coastal areas.

To unequivocally distinguish the contributions of different sources to the overall contamination found in specific ecosystems, molecular tracers, which are unique chemical compounds, can be used. Since caffeine is ranked number one drug worldwide, it may be used for this purpose, providing a potential way of tracing human-derived wastewater intrusions in coastal ecosystems by analyzing the concentrations of caffeine in surface and ground water.

HPLC-UV methods are inadequate for the analysis of caffeine in water since they show inadequate detection limits (in the order of 0.05 ppb) to properly determine the trace amounts of caffeine likely to be found in environmental samples (in the order of low ppt). In this sense, due to its high sensitivity, LC-MS is the method of choice for the analysis of trace levels of caffeine in surface waters.

Gardinali and Zhao (2002) developed a new LC-MS method using atmospheric pressure chemical ionization (APCI) to determine trace amounts of caffeine in surface water. The method allowed the determination of caffeine at levels as low as 4.0 ng L^{-1} (ppt) in both salt and freshwater. Sample preparation entailed membrane filtration, addition of deuterated atrazine as internal standard, liquid/liquid extraction with methylene chloride, evaporation of the organic solvent and reconstitution in mobile phase. Chromatographic separation was achieved on a C18 column using gradient elution with water and methanol. Quantitative determination of caffeine in the final extracts was carried out by APCI in positive mode under SIM (ions with *m/z* 195 and 196).

More recently, LC-MS-MS, which shows higher selectivity and sensitivity than LC-MS, has been employed for the analysis of trace levels of caffeine in water samples. However, matrix effects pose a great challenge for the quantitative analysis of environmental samples by LC-MS-MS as signal suppression or enhancement can compromise the accuracy of analytical results. Zhao and Metcalfe (2008) approached this subject by determining the effects of sample matrix on the analysis of six neutral pharmaceuticals, including caffeine, in samples of municipal wastewater using LC-APCI-MS-MS. They also evaluated whether isotope-labeled internal standards could be used to compensate for matrix effects. Wastewater samples were filtered, applied onto solid-phase extraction (SPE) cartridges and the eluates were concentrated. Samples were then spiked with isotope-labeled internal standards, including caffeine-$^{13}C_3$. The analytes were separated on a C18 column using a mobile phase consisting of a gradient of 10 mM ammonium acetate and acetonitrile. MS determination was performed using a quadrupole-linear trap instrument equipped with an APCI source. Multiple reaction monitoring (MRM) in the positive ion mode was used for quantification (*m/z* transition 195 → 138, in the case of caffeine). Isotope-labeled compounds corrected for overestimates that occurred as a result of interferences from the sample matrix.

Conley *et al* (2008) also used LC-MS-MS for the analysis of caffeine, together with another 12 different pharmaceuticals and one metabolite in surface water at low ng L^{-1} levels. Pharmaceuticals were chosen based on the estimated usage rate inferred from the number of prescriptions dispensed per year in the US, and commonly consumed non-prescription drugs. Sample preparation conditions were similar to those described by Zhao and Metcalfe (2008). Separation of compounds were performed on a bridged-ethyl-siloxane–silica hybrid C18 ultra-performance liquid chromatography (UPLC) column using a step-wise binary elution gradient consisting of 0.1% aqueous formic acid and acetonitrile. For detection, a triple quadrupole MS fitted with an ESI source operated in the positive mode was employed.

Very recently, Wang *et al* (2011) developed and validated a comprehensive LC-MS-MS method for the simultaneous analysis of caffeine and another 15 pharmaceutical compounds, including antibiotics, hormones, analgesics, stimulants, antiepileptics, and X-ray contrast media, in two different water matrices. A quadrupole-linear trap MS equipped with an ESI source was employed in both positive and negative modes for the detection of analytes and their quantification was performed by MRM.

A slightly different approach was addressed by Huerta-Fontela *et al* (2007), who used UPLC-MS-MS to analyze simultaneously 15 stimulatory drugs in water. In addition to caffeine and nicotine, which are non-controlled drugs, they studied controlled drugs such as cocaine, amphetamine-related compounds, lysergic acid, ketamine, and fentanyl. Metabolites of these stimulatory drugs were also investigated. Using a bridged-ethyl-siloxane–silica hybrid C18 UPLC column with gradient elution (0.1% aqueous formic acid and acetonitrile), chromatographic separation was achieved in less than 4.5 min. Detection was performed by a triple quadrupole MS equipped with an ESI source working in positive mode. Quantification and confirmation MRM *m/z* transitions for caffeine were 195 → 138 and 195 → 110, respectively.

7.2.3   Biological Matrices

7.2.3.1   *Urine, Plasma and Saliva*

The first reports of the use of LC-MS for the analysis of caffeine in biological samples date back to the end of the 1980's. Setchell *et al* (1987) reported the development of an LC-MS method employing thermospray ionization for the analysis of caffeine in human serum and saliva. In the following year, Sakairi and Kambara (1988) published a paper on the first use of atmospheric pressure ionization (API) for the analysis of drugs, including caffeine in human serum.

In 1995, Hieda *et al* (1995) established a sensitive and reliable analytical procedure for the detection of theophylline, theobromine and caffeine in human plasma and urine by gradient capillary HPLC-frit-fast atom bombardment (FAB) MS. FAB is a relatively soft ionization technique and produces primarily pseudomolecular ions. The material to be analyzed is mixed with a non-volatile chemical protection environment called a matrix and is bombarded under vacuum with a high energy beam of electrons (4 to 10 keV). The nature of its ionization products places FAB close to ESI.

In the background of assessing the exposure of restaurant and hotel workers to environmental tobacco smoke, Tuomi *et al* (1999) reported a method for the simultaneous analysis of nicotine and its metabolites, together with caffeine, in urine by HPLC-MS[n]. Identification and quantification were performed using ESI on a quadrupole ion trap mass analyzer. Caffeine detection limit of the whole procedure was 5 μg L[-1], with a 92% recovery.

Many analytical methods were reported over the following years, usually including improvements in aspects such as sample preparation, selectivity, sensitivity and speed.

Recently, Arinobu *et al* (2009) established an automated on-line method for ultra-fast determination (2 min) of caffeine and theophylline with backflush column-switching and with a monolithic separation column. Serum sample

Step 1.
Elimination of large molecules and trapping of the compounds

Step 2.
Backflush elution, chromatographic separation and mass spectrometric detection of the compounds

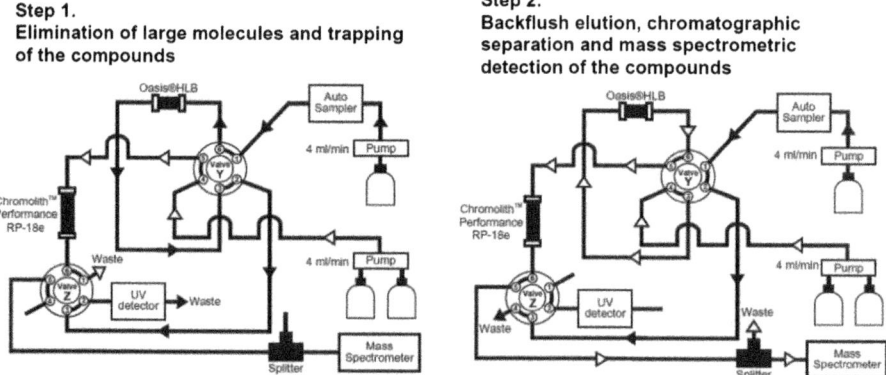

Figure 7.6   Diagrams of a backflush column-switching system employed by Arinobu *et al* (2009).

diluted with ultrapure water was injected and transferred to an Oasis HLB cartridge used as a precolumn for extraction. After switching the valves, the analytes trapped in the precolumn were eluted in the backflush mode and separated with a Chromolith Performance RP-18e column (C18-bonded monolithic silica) (Figure 7.6). Detection was performed by APCI-MS.

The development of ultra-fast methods for the analysis of caffeine in human plasma culminated with the publication of the fastest MS determination technique to date using conventional HPLC instrumentation by Ptolemy *et al* (2010). This group described a single ultra-centrifugation based sample pretreatment combined with a designed LC-MS-MS protocol which provides selective quantification of caffeine and theobromine in human saliva, plasma and urine samples. Their protocol, in addition to avoiding extensive sample preparation (*e.g.*, desalting, protein removal and–or preconcentration), permitted elution of both analytes within 1.3 min of the applied gradient. After inoculation with $^{13}C_3$ isotopically labeled caffeine in order to improve accuracy and precision, samples were centrifuged and the supernatant was filtered through a 10 kDa weight cut-off membrane. The filtrate containing low molecular weight compounds was diluted 20-fold prior to injection in the LC-MS system, which effectively minimized the deleterious contributions of the biological matrices to quantitation. Positive-mode ESI and a triple quadruple MS-MS instrument operated in MRM mode (195.1 $\rightarrow$ 137.9 transition) were used for detection.

LC-MS methods for the analysis of caffeine in plasma of specific populations, such as premature infants (gestational age less than 37 weeks) have also been reported. This patient population is considered to be vulnerable due to their immaturity at birth. Caffeine and other drugs are widely used in the pharmacotherapy of premature and term neonates. As pharmacokinetic studies in this population are hampered by limitations in the number and volume of plasma samples, Zhang *et al* (2008) developed an LC-MS-MS assay for the simultaneous determination of caffeine, acetaminophen, phenytoin, ranitidine, and theophylline in small volume human plasma specimens for pharmacokinetic evaluations in neonates. Sample preparation was performed by protein precipitation with methanol after addition of phenacetin (internal standard) to a 50 µL-aliquot of plasma. After chromatographic separation on a C18 column using gradient elution, analytes were detected using a triple quadrupole MS that was operated in positive ion mode with ESI.

7.2.3.2 Breast Milk and Tissues

Besides urine, saliva and plasma, caffeine analysis by LC-MS in other biological fluids and tissues has been reported. Pellegrini *et al* (2007) described a procedure based on HPLC-MS^n for the simultaneous analysis of caffeine, arecoline and nicotine and its metabolites in breast milk. The major reason for drug investigation in human milk is to calculate excretion of certain compounds in this fluid and, consequently, the approximate dose ingested by breastfeeding infants. Caffeine is known to be consumed by pregnant and lactating mothers in large quantities in beverages and in prescription and non-prescription drugs despite the suggestion that its consumption should be reduced during pregnancy. As mentioned before, caffeine elimination in the newborn is slow due to an immaturity in the hepatic metabolic pathway, implying that the breast-fed newborns with a significant daily intake of caffeine *via* maternal breast milk could potentially accumulate toxic concentrations of this substance with consequent adverse reactions. Their methodology entailed liquid/liquid extraction with chloroform–isopropanol (95 : 5, *v/v*) and chromato-graphy on a C8 reversed-phase column using a gradient of 50 mM ammonium formate, pH 5.0, and acetonitrile as a mobile phase at a flow rate of 0.5 mL min^{-1}. ESI was employed and the triple quadrupole MS was operated in the MRM mode, in which the protonated molecule [M+H]$^+$ was selected in the first quadrupole and the collision energy was adjusted to optimize the signal for the most abundant product ions. Using MRM, caffeine protonated ion at *m/z* 195 showed a major product ion at *m/z* 138, corresponding to loss of C_2H_3NO. This ion then lost CO to produce an ion at *m/z* 110. A secondary fragmentation of caffeine pseudomolecular ion yielded an ion at *m/z* 180, corresponding to the loss of CH_3. The concerted loss of three CH_3 groups produced an ion at *m/z* 150 (Figure 7.7).

Nowadays, pharmacokinetic studies of minor food constituents tend to go further than analyzing their metabolites in biological fluids, investigating their behavior in all the tissues of the body. However, the employment of LC-MS

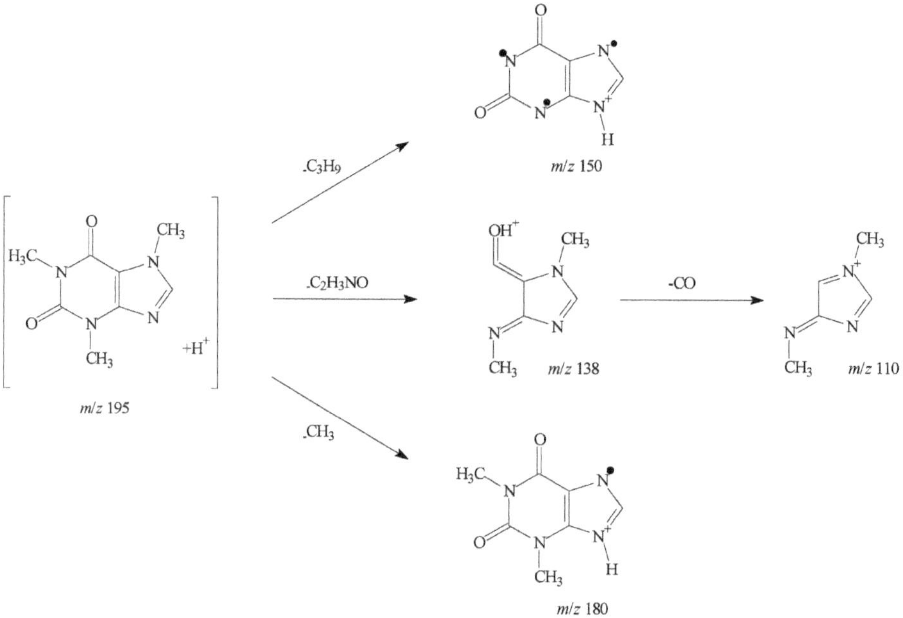

Figure 7.7   Fragmentation pathway of caffeine proposed by Pellegrini *et al* (2007).

methodologies for the analysis of caffeine in biological tissues is very recent. Serra *et al* (2011) developed and validated a rapid, selective and sensitive method to determine caffeine, theobromine, procyanidins and anthocyanins in different rat tissues, such as liver, brain, the aorta vein and adipose tissue. Their method entailed the extraction of homogenized tissues by off-line liquid/ solid extraction followed by solid-phase extraction. Extracts were then analyzed by UPLC-ESI-MS^n, using a triple quadrupole as the analyzer. Two MRM transitions were studied, selecting the most sensitive transition for quantification and a second one for confirmation purposes (195 $\rightarrow$ 138 and 195 $\rightarrow$ 110 for caffeine, respectively).

Prenatal exposure to caffeine and tobacco has been discouraged by the healthcare profession for decades. However, caffeine has been the most prevalent xenobiotic consumed by pregnant mothers, with excessive maternal consumption being related to a number of health issues such as spontaneous abortion and intrauterine growth retardation. The consumption of nicotine, the addictive ingredient in tobacco, has been correlated with a number of severe health issues. The determination of the prevalence of prenatal caffeine and cotinine (the primary metabolite of nicotine) exposure may be done by its measurement in urine, meconium or breast milk. However, such biological fluids are difficult to collect and quantities are limited. In this sense, Jones *et al* (2011) validated an LC-MS^n method for the simultaneous detection of caffeine and cotinine in umbilical cord tissue. Samples were simply homogenized with acetonitrile, centrifuged and the supernatants were evaporated and reconstituted in the LC mobile phase. The detector for the system was a triple quadrupole fitted with an ESI source in the positive mode. Both analytes were quantified using MRM and the corresponding stable isotope internal standards (cotinine-d_3 and caffeine-$^{13}C_3$). The same quantification and confirmation MRM transitions for caffeine (195 $\rightarrow$ 138 and 195 $\rightarrow$ 110, respectively) studied by Serra *et al* (2011) were used in this work.

7.2.3.3 Caffeine Metabolites

Because of the large human consumption of caffeine, its metabolism and the concentrations of caffeine metabolites in plasma and other biological fluids and of end-products recovered in urine may be used to evaluate liver functions, enzyme activities and phenotypes. The ratio of the urinary concentrations of different metabolites represents a sensitive index of hepatic xanthine oxidase, microssomal 3-methyl and 7-methyl demethylation as well as 8-hydroxylation. Caffeine is widely employed as a model drug to assess the hepatic CYP1A2 which catalyzes the metabolism of a number of clinically used drugs. The determination of human acetylator phenotype through the analysis of AFMU is the most widely used application of the use of caffeine metabolites for the evaluation of enzyme phenotypes.

Among the major metabolites that have been proposed as probes for the *in vivo* assessment of CYP1A2 activity, the metabolic ratio (AFMU + 1U + 1X) :

17U seems to be the most reliable (Rostami-Hodjegan *et al* 1996). This metabolic ratio has been reported to be decreased in patients with various types of liver disease (Bechtel *et al* 2000; Lelouet *et al.*, 2001) and may be used for monitoring the progress of liver disease. Xanthine oxidase, CYP2A6 and *N*-acetyl-transferase-2 activities have been assessed by the molar ratios 1U : (1U + 1X) (Kalow and Tang, 1991), 17U : 17X (Grant *et al* 1983) and AFMU : (AFMU + 1U + 1X) (Rostami-Hodjegan *et al* 1996), respectively.

Many analytical methods have been used for the quantitation of these caffeine metabolites. Common procedures involve HPLC with UV detection after liquid/liquid extraction of an acidified urine sample (Bendriss *et al* 2000). These procedures have some shortcomings in the extraction as well as the detection steps. Extraction is laborious and incomplete due to the hetero-geneous physico-chemical properties of the metabolites. Detection by UV is often insufficiently selective, as urine contains numerous UV-absorbing compounds which are not derived from caffeine but may overlay peaks of caffeine metabolites. Therefore, in the last ten years, due to its selectivity and sensitivity, LC-MS is becoming more and more the method of choice when one desires to analyze caffeine and its metabolites in biological fluids.

Baud-Camus *et al* (2001) reported the development and validation of a rapid and sensitive method using LC-ESI-MS for the simultaneous quantitative determination of caffeine metabolites 1U, 1X and AAMU in human urine. The sample preparation step of their method involved a simple dilution of urine samples. Chromatographic separation was achieved on a C18 column using a gradient of acetonitrile and 2mM ammonium formate (pH 3.0) as mobile phase. Ionization was achieved in an electrospray source operated in the SIM negative mode. The proposed method showed acceptable accuracy and precision, and improved specificity when compared with HPLC-UV for the investigation of *N*-acetylator phenotype of HIV-infected patients.

A different approach on the analysis of caffeine metabolites was introduced by Caubet *et al* (2004). They profited from the power and selectivity of MS detection to develop a method using LC-MS for the analysis of ^{13}C labeled caffeine and 14 of its metabolites in urine. The advantage of this method was the more accurate determination of CYP1A2 activity since it excluded the interference of background caffeine metabolites which are usually present in the urine of subjects, even after a 10 h fasting period. Disadvantages included a laborious sample preparation and a long analysis time (50 min). Sample extraction was performed by SPE, followed by membrane filtration and spiking with internal standards (β-hydroxyethyltheophylline and 1,9-dimethy-luric acid). Chromatographic separation was achieved on Eclipse XDB-C18 column using a gradient of a mixture of water–acetic acid–THF (996.5 : 1 : 2.5, *v/v/v*) (pH 3.3) and acetonitrile. ESI was used both in either positive or negative modes and quantification was achieved using SIM. The proposed method showed a 50-fold increase in selectivity and sensitivity in comparison to HPLC-UV.

Schneider *et al* (2003) developed an extractionless method for the determination of caffeine and 11 of its major urinary metabolites using LC-ESI-MS-MS. Sample preparation entailed a dilution step, followed by the addition of two internal standards (3-propylxanthine and 7β-hydroxypropyltheophylline) and centrifugation. Chromatographic separation was achieved on an Ultrasphere C18 column using a gradient of 0.05% acetic acid (*v/v*) with 3% methanol (*v/v*) and 1.5% of 2-propanol (*v/v*) (solvent A), methanol (solvent B), and 0.05% acetic acid (*v/v*) (solvent C). The MS detector was a triple quadrupole fitted with an ESI source in either positive or negative modes, depending on the analyte. MRM mode was utilized and parent and daughter ions were selected using intensity and selectivity as the criteria.

Weimann *et al* (2005) proposed a method for the analysis of caffeine and five of its major metabolites (1X, 1U, 17U, AAMU and AFMU) using LC-MS-MS. The advantage of their method in comparison to that developed by Schneider *et al* (2003) was the shorter analysis time (11.5 min *versus* 37 min). Even though Weimann *et al* (2005) acknowledge that fewer metabolites could be investigated using their methodology (5 *versus* 11), they stated that this limitation would not be a major disadvantage, since the analysis of the extra metabolites is not necessary for the calculation of enzymatic activities and phenotypes. Sample preparation only required dilution with buffer and centrifugation before injection into the LC-MS-MS. Separation of analytes was performed on an YMC-Pack C30 column using gradient elution with 0.5% acetic acid and acetonitrile. A triple quadrupole MS fitted with a pneumatically assisted ESI source operating in either positive or negative modes was used for detection. Quantification and confirmation of analytes was performed using MRM.

7.2.3.4   Other Fields

Caffeine analysis by LC-MS techniques is also relevant in other fields of investigation, such as metabolomics, doping control and pharmacokinetics. Metabolomics (also denominated metabolic profiling or metabonomics) is a term related with the analysis of low molecular weight compounds present in complex samples such as human biofluids (Nicholson *et al* 1999). Metabolomic investigations could provide deeper insights into mechanisms related with diseases, point out novel markers which could be used for diagnostic purposes and also offer greater understanding of the influence of lifestyle and dietary factors in relation to specific diseases and conditions (Goodacre *et al* 2004). Since caffeine is widely consumed by humans and it presents a myriad of physiological effects, its investigation by metabolomic methodologies is relevant. Due to its selectivity, sensitivity, speed and versatility, LC-MS and, more recently, UPLC-MS has become the method of choice for metabolomic studies. One of the key characteristics of such chromatographic methods is their capability of analyzing a wide variety of chemically diverse low molecular

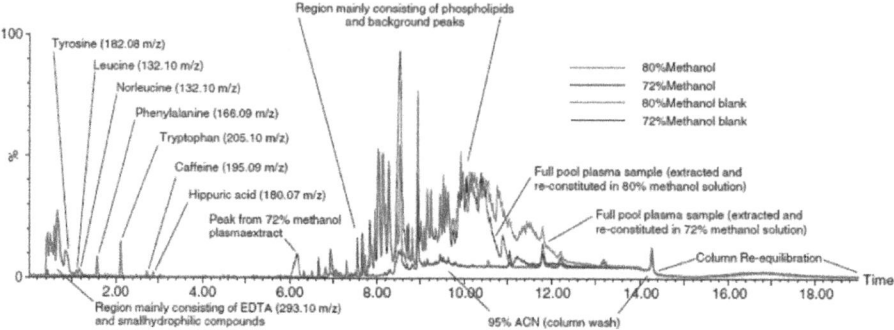

Figure 7.8 UPLC-MS spectra of pooled extracted plasma reported by Bruce *et al.* (2008).

weight compounds. Therefore, caffeine and its metabolites are only a few of many other compounds investigated when using such methodologies.

Bruce *et al* (2008) described a protocol designed to perform high-throughput metabolic profiling analysis on human blood plasma by UPLC-MS. They investigated different protocols of extraction, as well as UPLC columns with three different stationary phases (C8, C18, and phenyl). Gradient elution with a mixture of 0.1% formic acid and acetonitrile was used for chromatographic separation. A time-of-flight (ToF) MS equipped with an ESI source operating in positive ion mode was employed for detection and all mass spectral data from m/z 50 to 850 were acquired. Caffeine was observed as a positive ion with m/z 195.09 and retention time of 2.74 min (Figure 7.8).

Caffeine analysis in the background of doping has also been the subject of novel LC-MS methodologies. Even though it was removed from the World Anti-Doping Agency (WADA) prohibited list in 2004, caffeine is currently part of the WADA monitoring program in order to detect patterns of misuse of caffeine in sport. Badoud *et al* (2009) developed a multi-analyte high-throughput screening method based on UPLC coupled to hybrid quadrupole time-of-flight (Q-ToF) MS to identify 103 forbidden substances from various classes (such as stimulants, diuretics, narcotics, anti-estrogens) and WADA-monitored compounds, such as caffeine in urine. Due to the high sensitivity and selectivity of the method, urine samples could be simply diluted 2-fold prior to injection. Caffeine was observed as a positive ion with m/z 195.0882, retention time of 2.03 min and showed an adequate limit of detection, of 25 ng mL^{-1}.

Human pharmacokinetic studies that involve the simultaneous quantification of multiple agents require specific analytical methods such as LC-MS. When complicating factors such as the widely different polarities of the drugs and their wide range of concentrations found in plasma are taken into account, LC-MS-MS methods are the most adequate. Moreover, LC-MS-MS allows for simpler sample preparation protocols and more rapid run times. Cold medications usually combine an analgesic like paracetamol and an antihistamine like chlorpheniramine. Other pharmacological agents are also commonly found in these medications, such as a decongestant like pseudoephedrine, an antitussive like cloperastine and an antiviral like amantadine. Caffeine is also commonly included as a mild stimulant and vasodilator to counteract the sedative effects of the antihistamine and any systemic vasoconstriction due to the decongestant.

Feng *et al* (2009) developed and validated a rapid, simple and sensitive LC-MS-MS high performance liquid chromatography with positive ion electro-spray ionization tandem mass spectrometry to simultaneously determine paracetamol, amantadine hydrochloride, chlorpheniramine maleate and caffeine in human plasma. Sample preparation entailed methanol-induced protein precipitation and adding of tramadol hydrochloride as internal standard. Analytes were separated using a mobile phase comprised of methanol and water (80 : 20, v/v), containing 0.5% formic acid and on a C18 column. A triple quadrupole MS fitted with an ESI source was employed in the MRM mode. Quantification of caffeine was done by investigating the m/z 195.1→138.3 transition.

In the following year, Li *et al* (2010) developed and validated a rapid and sensitive method based on LC-MS-MS for the simultaneous quantification of paracetamol, pseudoephedrine, chlorpheniramine, cloperastine and caffeine in human plasma. Sample preparation involved liquid/liquid extraction and addition of diphenhydramine as an internal standard. Chromatographic separation was achieved in a run time of only 2.6 min, on a C18 column, using a mixture of formic acid : 10 mM ammonium acetate : methanol (1 : 40 : 60, v/v/v) for elution. Detection was carried out by a triple quadrupole MS fitted with an ESI source operating in the positive mode using MRM mode for quantification. The same m/z transition (195.1→138.3) used by Feng *et al* (2009) was employed in this method.

7.3   Concluding Remarks

The analysis of caffeine is usually fairly simple, although the complexity may increase depending on the investigated matrix. In major food sources, such as coffee and tea, caffeine analysis may be routinely and reliably performed by colorimetric methods or HPLC with UV detection. In biological fluids, major caffeine metabolites, such as paraxanthine and 1-methyl-uric acid, may also be analyzed using HPLC-UV. LC-MS methods are preferred especially for determining caffeine and metabolites at low levels and–or in very complex matrices.

Summary Points

- This chapter focuses on the different existing methods for analysis of caffeine and other methylxanthines in food, water and biological matrixes.

- The most common and abundant sources of caffeine are coffee (*Coffea* sp.), tea (*Camellia sinensis*) and cocoa (*Theobroma cacao*).
- The methodologies chosen for extraction and analysis of methylxanthines including caffeine depend on the complexity of the matrix and the amount of caffeine contained in it.
- Colorimetric or Isocratic HPLC-UV methods can be used for determination of caffeine and other methylxanthines in coffee and tea.
- LC-MS methods are more appropriate for matrixes containing very low amounts of caffeine due to their higher sensitivity.
- Reverse-phase columns are commonly used for chromatographic analysis of caffeine and related compounds.

Key Facts of Caffeine Analysis

- Caffeine is the most consumed psychoactive substance in the world.
- The most common and abundant sources of caffeine are coffee (*Coffea* sp.), tea (*Camellia sinensis*) and cocoa (*Theobroma cacao*).
- In addition to internationally known sources of caffeine, there are other local plant sources of caffeine such as mate (*Ilex paraguariensis*) and guaraná (*Paullinia cupana*) in South America, Yaupon Holly (*Ilex vomitoria*) in the United States and Mexico, and Miang (pickled tea) in Thailand.
- Caffeine was first isolated from coffee in 1820 by the German chemist Friedlieb Ferdinand Runge and in 1821 by a team of French chemists, Robiquet, Pelletier and Caventou.
- Caffeine analysis in major food sources, such as coffee and tea, may be routinely performed by colorimetric methods or HPLC with UV detection.
- When analytes are present from low to trace levels (ppb to low ppt) in samples like water, LC-MS methods are preferred for caffeine analysis.

Definition of Words and Terms

Atmospheric pressure chemical ionization: Chemical ionization technique used in mass spectrometry which takes place at atmospheric pressure.

Cytochrome P450: Large and diverse enzyme system responsible for the metabolism of xenobiotic substances, such as caffeine.

Desorption electrospray ionization: Atmospheric pressure ionization technique which ionizes gases, liquids and solids at open air under ambient conditions.

Diode array detector: Analytical equipment commonly used in liquid chromatography to detect and identify UV-Vis absorbing compounds according to their spectra.

Electrospray ionization: Most common ionization technique in mass spectrometry, which produces ions at atmospheric pressure through an electric discharge.

High performance liquid chromatography: Liquid chromatographic technique operated at high pressures used to separate, identify and quantify components present in complex mixtures.

Mass spectrometry: Analytical technique which measures the charge to mass ratio of charged compounds. Commonly used as a detector coupled to liquid and gas chromatography techniques.

Multiple reaction monitoring: Quantification technique used in tandem mass spectrometry in which mass analyzers selectively monitor single ions.

Single ion monitoring: Quantification technique used in mass spectrometry in which the mass analyzer selectively monitors a single ion.

Solid phase extraction: Sample clean-up or concentration process by which compounds dissolved in a liquid mixture are separated from interferants according to their physical and chemical properties.

List of Abbreviations

17U	1,7-dimethyl-uric acid
17X	1,7-dimethyl-xanthine
1U	1-methyl-uric acid
1X	1-methyl-xanthine
AAMU	5-acetylamino-6-amino-3-methyluracil
AFMU	5-acetylamino-6-formylamino-3-methyluracil
APCI	Atmospheric Pressure Chemical Ionization
API	Atmospheric Pressure Ionization
CID	Collision Induced Dissociation
CYP	Cytochrome P450
DAD	Diode Array Detector
DAPCI	Desorption Chemical Ionization
DART	Direct Analysis in Real Time
DESI	Desorption Electrospray Ionization
EI	Electron Ionization
ESI	Electrospray Ionization
FAB	Fast Atom Bombardment
FDA	US Food and Drug Administration
HPLC	High-Performance Liquid Chromatography
LC-MS	Liquid Chromatography-Mass Spectrometry
LTQ-MS	Linear Ion Trap-Mass Spectrometer
MRM	Multiple Reaction Monitoring
MS	Mass Spectrometry
MS-MS	Mass-mass spectrometry
MS^n	*Tandem* Mass Spectrometry
ODS	Octadecylsilane
PB-GD	Particle Beam-Glow Discharge
PB-GD-MS	Particle Beam-Glow Discharge-Mass Spectrometry
Q-ToF	Quadrupole Time-of-Flight

SIM   Single Ion Monitoring
SPE   Solid-Phase Extraction
ToF   Time-of-Flight
UPLC   Ultra-Performance Liquid Chromatography
UV   Ultraviolet
WADA   World Anti-Doping Agency

References

Aranda, M. and Morlock, G., 2006. Simultaneous determination of riboflavin, pyridoxine, nicotinamide, caffeine and taurine in energy drinks by planar chromatography-multiple detection with confirmation by electrospray ionization mass spectrometry. Journal of Chromatography A. 1131: 253–260.

Araque, P., Casanova, H., Ortiz, C., Henao, B. and Peláez, C., 2007. Insecticidal activity of caffeine aqueous solutions and caffeine oleate emulsions against *Drosophila melanogaster* and *Hypothenemus hampei*. Journal of Agricultural and Food Chemistry. 55: 6918–6922.

Arinobu, T., Hattori, H., Kumazaka, T., Lee, X., Mizutani, Y., Katase, T., Kojima, S., Omori, T., Kaneko, R., Ishii, A. and Seno, H., 2009. High-throughput determination of theophylline and caffeine in human serum by conventional liquid chromatography-mass spectrometry. Forensic Toxicology. 27: 1–6.

Arnaud, M. J. 1999. Caffeine: Chemistry and Physiological Effects. In Sadler, M. J., Strain, J. J. and Caballero, B., (ed.) Encyclopedia of Human Nutrition, 1st Edition. Elsevier, London, UK. Vol 1. 206–214.

Arora, D. S. and Ohlan, D., 1997. In vitro studies on antifungal activity of tea (*Camellia sinensis*) and coffee (*Coffea arabica*) against woodrotting fungi. Journal of Basic Microbiology. 37: 159–165.

Astill, C., Birch, M. R., Dacombe, C., Humphrey, P. G. and Martin, P. T., 2001. Factors affecting the caffeine and polyphenol contents of black and green tea infusions. Journal of Agricultural and Food Chemistry. 49: 5340–5347.

Badoud, F., Grata, E., Perrenoud, L., Avois, L., Saugy, M., Rudaz, S. and Veuthey, J.-L., 2009. Fast analysis of doping agents in urine by ultra-high-pressure liquid chromatography–quadrupole time-of-flight mass spectrometry I. Screening analysis. Journal of Chromatography A. 1216: 4423–4433.

Barone, J. J. and Roberts, H. R., 1996. Caffeine consumption. Food and Chemical Toxicology. 34: 119–129.

Baud-Camus, F., Marquet, P., Soursac, M., Davrinche, C. and Farinotti, R., 2001. Determination of *N*-acetylation phenotype using caffeine as a metabolic probe and high-performance liquid chromatography with either ultraviolet detection or electrospray mass spectrometry. Journal of Chromatography B. 760: 55–63.

Bechtel, Y. C., Lelouët, H., Brientini, M. P., David-Laroche, M., Miguet, J. P., Paintaud, G. and Bechtel, P. R., 2000. Caffeine metabolism differences in acute hepatitis of viral and drug origin. Therapie. 55: 619–627.

Begas, E., Kouvaras, E., Tsakalof, A., Papakosta, S. and Asprondini, E. K., 2007. *In vivo* evaluation of CYP1A2, CYP2A6, NAT-2 and xanthine oxidase activities in a Greek population sample by the RP-HPLC monitoring of caffeine metabolic ratios. Biomedical Chromatography. 21: 190–200.

Bell, L. N., Wetzel, C. R. and Grand, A. N., 1996. Caffeine content in coffee as influenced by grinding and brewing techniques. Food Research International. 29: 785–789.

Bendriss, E., Markoglou, N. and Wainer, I. W., 2000. Liquid chromatographic method for the simultaneous determination of caffeine and fourteen caffeine metabolites in urine. Journal of Chromatography B. 746: 331–338.

Benowitz, N. L., 1990. Clinical pharmacology of caffeine. Annual Review of Medicine. 41: 277–288.

Bruce, S. J., Jonsson, P., Antti, H., Cloarec, O., Trygg, J., Marklund, S. L. and Moritz, T. 2008. Evaluation of a protocol for metabolic profiling studies on human blood plasma by combined ultra-performance liquid chromatography/mass spectrometry: From extraction to data analysis. Analytical Biochemistry. 372: 237–249.

Casal, S., Oliveira, M. B. and Ferreira, M. A., (1998). Development of an HPLC/diode-array detector method for simultaneous determination of trigonelline, nicotinic acid, and caffeine in coffee. Journal of Liquid Chromatography and Related Technologies. 21: 3187–3195.

Casal, S., Oliveira, M. B. P. P., Alves, M. R. and Ferreira, M. A., 2000. Discriminate analysis of roasted coffee varieties for trigonelline, nicotinic acid, and caffeine content. Journal of Agricultural and Food Chemistry. 48: 3420–3424.

Castro, J., Pregibon, T., Chumanov, K. and Marcus, R. K., 2010. Determination of catechins and caffeine in proposed green tea standard reference materials by liquid chromatography-particle beam/electron ionization mass spectrometry (LC-PB/EIMS). Talanta. 82: 1687–1695.

Caubet, M., Comte, B. and Brazier, J., 2004. Determination of urinary ^{13}C-caffeine metabolites by liquid chromatography–mass spectrometry: the use of metabolic ratios to assess CYP1A2 activity. Journal of Pharmaceutical and Biomedical Analysis. 34: 379–389.

Chen, H., Liang, H., Ding, J., Lai, J., Huang, Y. and Qiao, X., 2007. Rapid differentiation of tea products by surface desorption atmospheric pressure chemical ionization mass spectrometry. Journal of Agricultural and Food Chemistry. 55: 10093–10100.

Clifford, M. N. and Ramirez-Martinez, J. R., 1990. Chlorogenic acids and purine alkaloids contents of maté (*Ilex paraguariensis*) leaf and beverage. Food Chemistry. 35: 13–21.

Cody, R. B., Laramée, J. A. and Durst, H. D., 2005. Versatile new ion source for the analysis of materials in open air under ambient conditions. Analytical Chemistry. 77: 2297–2302.

Conley, J. M., Symes, S. J., Kindelberger, S. A. and Richards, S. M., 2008. Rapid liquid chromatography–tandem mass spectrometry method for the determination of a broad mixture of pharmaceuticals in surface water. Journal of Chromatography A. 1185: 206–215.

Davis, W. C., Venzie, J. L., Willis, B., Coffee Jr, R. L., Arya, D. P. and Marcus, R. K., 2003. Particle beam glow discharge mass spectrometry: spectral characteristics of nucleobases. Rapid Communications in Mass Spectrometry. 17: 1749–1758.

Del Rio, D., Stewart, A. J., Mullen, W., Burns, J., Lean, M. E. J., Brighenti, F., and Crozier, A., 2004. HPLC-MS^n analysis of phenolic compounds and purine alkaloids in green and black tea. Journal of Agricultural and Food Chemistry. 52: 2807–2815.

Feng, S., Tian, Y., Zhang, Z., Zhang, J., Huang, M. and Chey, Y., 2009. Rapid simultaneous determination of paracetamol, amantadine hydrochloride, caffeine and chlorpheniramine maleate in human plasma by liquid chromatography/tandem mass spectrometry. Arzneimittelforschung. 59: 86–95.

Ford, M. J., Deibel, M. A., Tomkins, B. A. and Van Berkel, G. J., 2005. Quantitative thin-layer chromatography/mass spectrometry analysis of caffeine using a surface sampling probe electrospray ionization tandem mass spectrometry system. Analytical Chemistry. 77: 4385–4389.

Frary, C. D., Johnson, R. K. and Wang, M. Q., 2005. Food sources and intakes of caffeine in the diets of persons in the United States. Journal of the American Dietetic Association. 105: 110–113.

Gardinali, P. R. and Zhao, X., 2002. Trace determination of caffeine in surface water samples by liquid chromatography–atmospheric pressure chemical ionization–mass spectrometry (LC–APCI–MS). Environment International. 28: 521–528.

Goodacre, R., Vaidyanathan, S., Dunn, W. B., Harrigan, G. G. and Kell, D. B., 2004. Metabolomics by numbers: acquiring and understanding global metabolite data. Trends in Biotechnology. 22: 245–252.

Grant, D. M., Tang, B. K. and Kalow, W., 1983. Variability in caffeine metabolism. Clinical Pharmacology and Therapeutics. 33: 591–602.

Hieda, Y., Kashimura, S., Hara, K. and Kageura, M., 1995. Highly sensitive and rapid determination of theophylline, theobromine and caffeine in human plasma and urine by gradient capillary high-performance liquid chromatography-frit-fast atom bombardment mass spectrometry. Journal of Chromatography B. 667: 241–246.

Hollingsworth, R. G., Armstrong, J. W. and Campbell, E., 2002. Caffeine as repellent for slugs and snails. Nature. 417: 915–916.

Huerta-Fontela, M., Galceran, M. T. and Ventura, F., 2007. Ultraperformance liquid chromatography-tandem mass spectrometry analysis of stimulatory drugs of abuse in wastewater and surface waters. Analytical Chemistry. 79: 3821–3829.

Jones, J., Magri, R., Rios, R., Jones, M., Plate, C. and Lewis, D., 2011. The detection of caffeine and cotinine in umbilical cord tissue using liquid chromatography–tandem mass spectrometry. Analytical Methods. 3: 1310–1315.

Kalow, W. and Tang, B. T., 1991. Caffeine as a metabolic probe: Exploration of the enzyme-inducing effect of cigarette smoking. Clinical Pharmacology and Therapeutics. 49: 44–48.

Lelouët, H., Bechtel, Y. C., Paintaud, G., Brientini, M. P., Miguet, J. P., Bechtel, P. R., 2001. Caffeine metabolism in a group of 67 patients with primary biliary cirrhosis. International Journal of Clinical Pharmacology and Therapeutics. 39: 25–32.

Li, H., Zhang, C., Wang, J., Fawcett, J. P. and Gu, J., 2010. Simultaneous quantitation of paracetamol, caffeine, pseudoephedrine, chlorpheniramine and cloperastine in human plasma by liquid chromatography–tandem mass spectrometry. Journal of Pharmaceutical and Biomedical Analysis. 51: 716–722.

Lin, L., Chen, P. and Harnly, J. M., 2008. New Phenolic Components and Chromatographic Profiles of Green and Fermented Teas. Journal of Agricultural and Food Chemistry. 56: 8130–8140.

Matuszewski, B. K., Constanzer, M. L. and Chavez-Eng, C. M., 2003. Strategies for the assessment of matrix effect in quantitative bioanalytical methods based on HPLC-MS/MS. Analytical Chemistry, 75: 3019–3030.

Nathanson, J. A., 1984. Caffeine and related methylxanthines: Possible naturally occurring pesticides. Science. 226: 184–187.

Nicholson, J. K., Lindon, J. and Holmes, E., 1999. Metabonomics: Understanding the metabolic responses of living systems to pathophysiological stimuli *via* multivariate statistical analysis of biological NMR spectroscopic data. Xenobiotica. 29: 1181–1189.

Palumbo, M. J., Putz, F. E. and Talcott, S. T., 2007. Nitrogen fertilizer and gender effects on the secondary metabolism of yaupon, a caffeine-containing North American holly. Oecologia. 151: 1–9.

Pellegrini, M., Marchei, E., Rossi, S., Vagnarelli, F., Durgbanshi, A., García-Algar, Ó., Vall, O. and Pichini, S., 2007. Liquid chromatography/ electrospray ionization tandem mass spectrometry assay for determination of nicotine and metabolites, caffeine and arecoline in breast milk. Rapid Communications in Mass Spectrometry. 21: 2693–2703.

Pelozo, M. I. G., Cardoso, M. L. C. and de Mello, J. C. P., 2008. Spectrophotometric determination of tannins and caffeine in preparations from *Paullinia cupana* var. *sorbilis*. Brazilian Archives of Biology and Technology. 51: 447–451.

Perrone, D., Donangelo, C. M. and Farah, A., 2008. Fast simultaneous analysis of caffeine, trigonelline, nicotinic acid and sucrose in coffee by liquid chromatography–mass spectrometry. Food Chemistry. 110: 1030–1035.

Phromrukachat, S., Tiengburanatum, N. and Meechui, J., 2010. Assessment of active ingredients in pickled tea. Asian Journal of Food and Agro-Industry. 3: 312–318.

Power, F. B. and Chesnut, V. K., 1919. *Ilex vomitoria* as a natural source of caffeine. Journal of the Americal Chemical Society. 41: 1307–1312.

Ptolemy, A. S., Tzioumis, E., Thomke, A., Rifai, S. and Kellogg, M., 2010. Quantification of theobromine and caffeine in saliva, plasma and urine via liquid chromatography–tandem mass spectrometry: A single analytical protocol applicable to cocoa intervention studies. Journal of Chromatography B. 878; 409–416.

Rostani-Hodjegan, A., Nurminen, S., Jackson, P. R. and Tucker, G. T., 1996. Caffeine urinary metabolite ratios as markers of enzyme activity: a theoretical assessment. Pharmacogenetics. 6:121–149.

Sakairi, M. and Kambara, H., 1988. Determination of drugs in human serum by liquid chromatography/atmospheric pressure ionization mass spectrometry. Analytical Sciences. 4: 199–201.

Schneider, H., Ma, L. and Glatt, H., 2003. Extractionless method for the determination of urinary caffeine metabolites using high-performance liquid chromatography coupled with tandem mass spectrometry. Journal of Chromatography B. 789: 227–237.

Serra, A., Macià, A., Romero, M., Piñol, C. and Motilva, M., 2011. Rapid methods to determine procyanidins, anthocyanins, theobromine and caffeine in rat tissues by liquid chromatography-tandem mass spectrometry. Journal of Chromatography B. 879: 1519–1528.

Setchell, K. D. R., Welsh, M. B., Klooster, M. J., Balistreri, W. F. and Lim, C. K., 1987. Rapid high-performance liquid chromatography assay for salivary and serum caffeine following an oral load: An indicator of liver function. Journal of Chromatography A. 385: 267–274.

Takáts, Z., Wiseman, J. M., Gologan, B. and Cooks, R. G., 2004. Mass spectrometry sampling under ambient conditions with desorption electro-spray ionization. Science. 306: 471–473.

Tuomi, T., Johnsson, T. and Reijula, K., 1999. Analysis of nicotine, 3-hydroxy-cotinine, cotinine, and caffeine in urine of passive smokers by HPLC-tandem mass spectrometry. Clinical Chemistry. 45: 2164–2172.

Venzie, J. L., Castro, J., Krishna, M. V. B., Nelson, D. M. and Marcus, R. K., 2007. Electron-impact and glow-discharge ionization LCMS analysis of green tea tincture. Analytical and Bioanalytical Chemistry. 387: 321–333.

Wang, C., Shi, H., Adams, C. D., Gamagedara, S., Stayton, I., Timmons, T. and Ma, Y., 2011. Investigation of pharmaceuticals in Missouri natural and drinking water using high performance liquid chromatography-tandem mass spectrometry. Water Research. 45: 1818–1828.

Wang, D., Lu, J., Miao, A., Xie, Z. and Yang, D., 2008. HPLC-DAD-ESI-MS/MS analysis of polyphenols and purine alkaloids in leaves of 22 tea cultivars in China. Journal of Food Composition and Analysis. 21: 361–369.

Watson, J. 2003. Caffeine. In Caballero, B., Trugo, L. C. and Finglas, P. M., (ed.) Encyclopedia of Food Sciences and Nutrition, 2nd Edition. Elsevier, London, UK. Vol 2. 745–750.

Weimann, A., Sabroe, M. and Poulsen, H. E., 2005. Measurement of caffeine and five of the major metabolites in urine by high-performance liquid chromatography/tandem mass spectrometry. Journal of Mass Spectrometry. 40: 307–316.

Williams, J. P., Patel, V. J., Holland, R. and Scrivens, J. H., 2006. The use of recently described ionisation techniques for the rapid analysis of some common drugs and samples of biological origin. Rapid Communications in Mass Spectrometry. 20: 1447–1456.

Zhang, Y., Mehrotra, N., Budha, N. R., Christensen, M. L. and Meibohm, B., 2008. A tandem mass spectrometry assay for the simultaneous determination of acetaminophen, caffeine, phenytoin, ranitidine, and theophylline in small volume pediatric plasma specimens. Clinica Chimica Acta. 398: 105–112.

Zhao, X. and Metcalfe, C. D., 2008. Characterizing and compensating for matrix effects using atmospheric pressure chemical ionization liquid chromatography-tandem mass spectrometry: Analysis of neutral pharmaceuticals in municipal wastewater. Analytical Chemistry. 80: 2010–2017.

Zhu, X., Chen, B., Ma, M., Luo, X., Zhang, F., Yao, S., Wan, Z., Yang, D. and Hang. H., 2004. Simultaneous analysis of theanine, chlorogenic acid, purine alkaloids and catechins in tea samples with the help of multi-dimension information of on-line high performance liquid chromatography/electrospray–mass spectrometry. Journal of Pharmaceutical and Biomedical Analysis. 34: 695–704.

CHAPTER 8

Simultaneous Determination of Caffeine and Phenolic Compounds in Tea and Coffee

CAROLYNE B. FARIA[1], JULIANA M. PRADO[2], MAURICIO A. ROSTAGNO*[2], FLAVIO L. SCHMIDT[1] AND M. ANGELA A MEIRELES[2]

[1] Department of Food Technology, University of Campinas (UNICAMP), Rua Monteiro Lobato, 80, 13083-862 Campinas, São Paulo, Brazil; [2] LASEFI/ DEA, School of Food Engineering, University of Campinas (UNICAMP), Rua Monteiro Lobato, 80, 13083-862 Campinas, São Paulo, Brazil *E-mail: rostagno@fea.unicamp.br

8.1 Introduction

Tea and coffee are among the most consumed beverages worldwide, not only for their organoleptic properties but also due to their potential effects in the human organism. In recent decades, intensive research has been conducted on the phytochemical profiles of these beverages. Special interest has been paid to alkaloids and polyphenols, as several *in vitro* and *in vivo* studies using these compound classes are consistent with the potential health effects of tea and coffee.

Polyphenols are believed to be the main compounds responsible for the putative effects of tea and coffee in human health. The dominating phenolic compounds (Figure 8.1) present in coffee are hydroxycinnamic acids, which are also present in teas, although at lower concentrations. The characteristic phenolic compounds found in coffee are caffeic and ferulic acid derivatives,

Food and Nutritional Components in Focus No. 2
Caffeine: Chemistry, Analysis, Function and Effects
Edited by Victor R Preedy
© The Royal Society of Chemistry 2012
Published by the Royal Society of Chemistry, www.rsc.org

such as chlorogenic and feruloylquinic acids. The main polyphenols found in tea are catechins or flavones. Among tea polyphenols, catechins can be found in relatively high amounts. However, the most known and studied phytochemical found in tea and coffee is caffeine. Although other alkaloids, such as theophylline and theobromine, are also present in tea and coffee, caffeine is found in much higher concentrations. The concentrations of phenolic acids, catechins, flavones, flavonols and caffeine in several commercial tea and coffee samples are presented in Table 8.1.

8.2 Determination of Alkaloids and Polyphenols in Tea and Coffee

For the determination of phytochemicals from solid samples, several consecutive steps are usually necessary; if one of them is not properly followed, the overall performance of the analysis will be poor, errors will be introduced, and inconsistency in the results can thus be expected. Sample preparation can be a complex task involving multiple steps, such as drying, sieving, extraction of target compounds, pre-concentration and sample clean-up. The resulting sample should have a high concentration of target analytes free of interfering compounds from the matrix. The extract obtained from sample preparation is then analyzed by one of the several techniques available.

Simultaneous determination of different compound classes can be a difficult task because the concentration of most sample components may depend on several variables. The differences in composition, for instance, may result in interfering compounds detected in the final analysis step. In addition, it can result in lower extraction efficiency for a given class of compounds due to polarity differences between the phytochemicals present in the sample. Furthermore, differences in the matrix can influence the efficiency of an extraction method used for different samples. Therefore, it is of critical importance that all steps involved in the overall determination of polyphenols and alkaloids present in teas and coffee samples are optimized.

8.3 Sample Preparation

Sample preparation is a key component of the analytical process because it is responsible for the isolation of compounds of interest that will be further identified. Among the procedures used in sample preparation, extraction is of key importance in the analytical determination of polyphenols and alkaloids. There is no standard extraction method; each one possesses advantages and disadvantages. For the extraction of caffeine and phenolic compounds from coffee and tea, there are several solid/liquid extraction techniques available, including classical maceration and soxhlet extraction, and "modern" techniques such as ultrasound-assisted extraction (UAE) and supercritical fluid extraction (SFE).

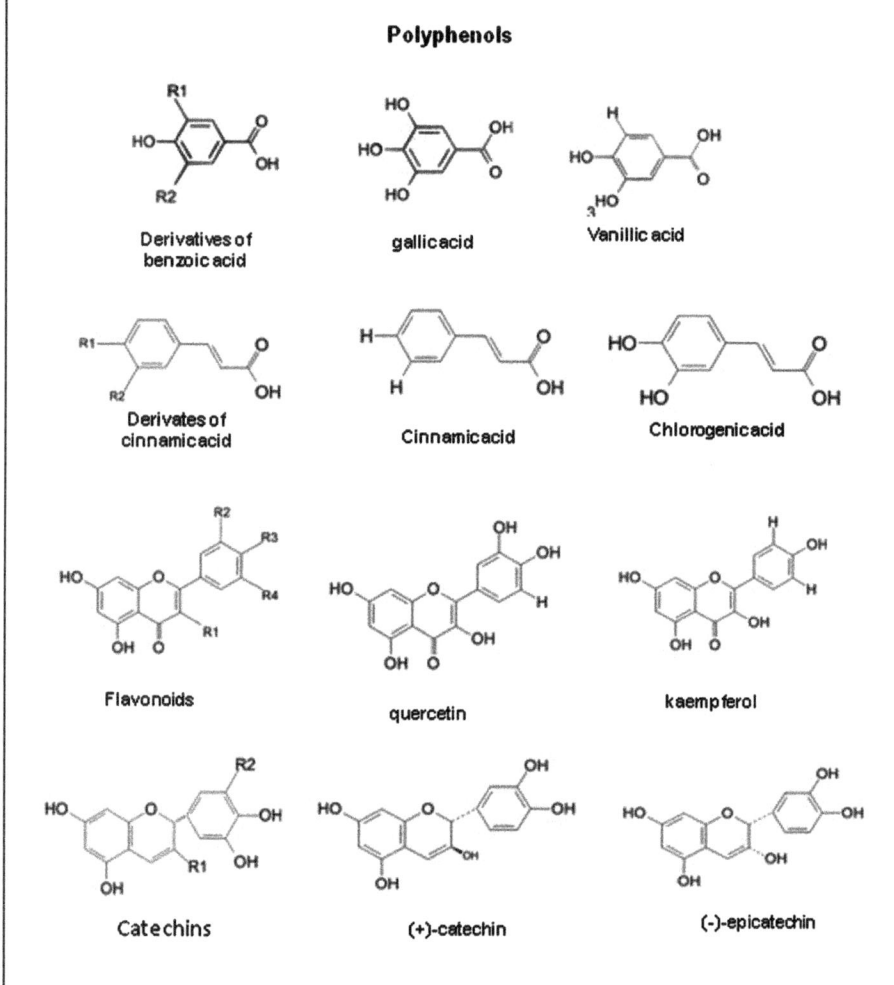

Figure 8.1 Chemical structure of major polyphenol classes and alkaloids. The chemical structure of compounds can indicate their relative polarity and their solubility in the extraction solvent.

Table 8.1   Concentration (mg g^{-1}) of phenolic acids, catechins, flavones, flavonols and caffeine determined in different samples.

Compound	Green Tea	White Tea	Black Tea	Instant Coffee
Caffeine	26.01	27.99	17.82	13.92
Gallic acid	0.86	1.12	1.63	—
Protocatechuic acid	—	—	0.08	–
Chlorogenic acid	1.39	0.49	0.24	12.09
Caffeic acid	—	—	—	0.91
Total phenolic acids	2.25	1.61	1.95	13.00
(–) Gallocatechin	5.86	2.22	0.15	—
(–) Catechin	6.28	5.67	0.25	—
(–) Epigallocatechin	13.28	8.91	0.32	—
(–) Epicatechin	2.44	1.92	0.84	—
(–) Epigallocatechin gallate	6.12	5.83	—	—
(–) Epicatechin gallate	2.96	4.39	0.11	—
Total catechins	36.94	28.94	1.67	—
Myricetin-3-O-rhamnoside	0.50	0.37	0.34	—
Quercetin-3-O-rutinoside	3.52	2.49	1.71	—
Quercetin-3-O-glucopyranoside	1.10	0.48	0.38	—
Kaempferol-3-O-rutinoside	0.89	0.30	0.33	—
Kaempferol-3-O-glucoside	0.76	0.43	0.38	—
Total flavonols	6.77	4.07	3.14	—

This table gives an example of the relative distribution of main polyphenols and caffeine in tea and coffee. Adapted from Rostagno *et al* 2011 with permission from Elsevier.

In addition to the technique, extraction efficiency is also a function of process conditions. Several factors affect the concentration of the desired components in the extract, such as solvent type, temperature, sample to solvent ratio, contact time, and particle size. When determining the content of compounds in a sample, an extraction method with optimized conditions should be employed. A survey of extraction methods and process conditions available for caffeine and polyphenol recovery from coffee and tea is presented in Table 8.2.

Conventional methods that employ only heating and–or stirring can be used to extract caffeine and phenolic compounds; however, the disadvantages are the loss of polyphenols due to thermal degradation and the long extraction time. To overcome these drawbacks, various novel extraction techniques have been developed, such as UAE, microwave-assisted extraction (MAE), SFE and others. These techniques, in most cases, are an efficient alternative to reduce the process time, its environmental impact and its cost.

It has been reported that UAE is of higher efficiency than simple infusion at lower temperature for the extraction of tea; it also presents better results compared to stirring (Wang *et al* 2008). On the other hand, although UAE with methanol is theoretically a more preferable method for polyphenol recovery compared to hot water maceration with stirring, UAE appears to be less efficient in the case of non-fermented teas (Nováková *et al* 2010).

Table 8.2 Sample preparation for obtaining caffeine and phenolic compounds from coffee and tea.

Sample	Operational conditions	Compounds analyzed	Reference
Tea (green)	1) m = supercritical fluid extraction; s = CO_2, CO_2 : water (0.911 : 0.089, w/w), CO_2 : ethanol (0.907 : 0.093, w/w), CO_2 : water : ethanol (0.892 : 0.089 : 0.019-0.906 : 0.005 : 0.089, w/w/w); So/Sa = 10 (w/w); T = 60 °C; P = 31 MPa 2) m = soxhlet; s = water, ethanol : water (95 : 5, v/v); T = 100 °C; t = 12 h	(CAF), (EC), (ECG), (EGC), (EGCG), (GA)	Chang et al 2000
Coffees (instant; light, medium and dark roasts) Teas	1) s = methanol : water (70 : 30, v/v); So/Sa = 2 × 25 (v/w); T = 70 °C; t = 2 × 10 min 2) s = water; So/Sa = 75-90 (v/w); T = boiling; t = 25-240 s; for teas only 3) m = percolation; s = water; So/Sa = 25-50 (v/w); for coffee only	(CAF), catechins, chlorogenic acids, flavone glycosides, flavonol glycosides, (GA), (KAEM), (MYR), (QCE), (RUT), (TB), theaflavins, theogallin	Lakenbrink et al 2000
Teas (green, black, Oolong)	m = shaking each 2 min for 30 s; s = water, methanol : water (20 : 80-80 : 20, v/v), ethanol : water (70 : 30, v/v); So/Sa = 100 (v/w); T = 60-100 °C; t = 5-10 min	(CAF), (C), (−)-(EC), (−)-(ECG), (−)-(EGC), (EGCG), (GA)	Khokhar and Magnusdottir 2002
Teas (green, black, Oolong, Pu-erh)	s = 3 × methanol : water (80 : 20, v/v) + 2 × methanol : water : HCl (80 : 20 : 0.15, v/v/v); So/Sa = 5 × 5-11 (v/w); t = 5 × 3 h	(p-ASA), (CAF), (−) (CG), (3,5-HBA), (−)-(EC), (−)-(ECG), (−)-(EGC), (−)-(EGCG), (GA), (MYR), (3,4,5-MCNA)	Zuo et al 2002
Teas (green, black, Oolong, Pu-erh)	m = steeping; s = distilled water : ethanol (100 : 0-5 : 95, v/v); T = boiling; t = 30 min	(CAF), (+)-(C), (−)-(EC), (−)-(EC3G), (−)-(EGC), (−)-(EGC3G), (−)-(GC3G)	Lin et al 2003

Table 8.2 (*Continued*)

Sample	Operational conditions	Compounds analyzed	Reference
Tea (green)	1) m = microwave-assisted extraction; s = ethanol : water (0 : 100–100 : 0, v/v), acetone : water (50 : 50–100 : 0, v/v), methanol : water (50 : 50–100 : 0, v/v); So/Sa = 10–25 (v/w); t = 0.5–8 min; pre-leaching t = 0–90 min 2) m = ultrasound-assisted extraction; s = ethanol : water (50 : 50, v/v); So/Sa = 20 (w/w); T = 20–40 °C; t = 90 min 3) m = extraction at room temperature; s = ethanol : water (50 : 50, v/v); So/Sa = 20 (v/w); T = 20–40 °C; t = 20 h 4) m = soxhlet; s = ethanol : water (50 : 50, v/v); So/Sa = 20 (v/w); T ~85 °C; t = 45 min	(CAF), polyphenols	Pan et al 2003
Coffee (bean) Teas (green, Oolong, black)	m = ultrasound-assisted extraction; s = methanol : water : acetic acid (90 : 10 : 0.5, v/v/v); So/Sa = 3 × 40 (v/w); t = 3 × 1 min	(CA), (CAF), (+)-(C), (−)-(CG), (CHLA), cinnamic acids, (−)-(EC), (−)-(ECG), (−)-(EGC), (−)-(EGCG), (GA), (−)-(GC), (−)-(GCG), isovitexin, (KAEM3-O-G), (KAEM3-O-R), (MYR3-O-R), (QCE3-O-RM), (TF), (TF3G), (TF3'G), (TF33'DG)	Sakakibara et al 2003
Tea	1) m = blending; s = distilled water; So/Sa = 12 (v/w); T = 100 °C; t = 4 min 2) m = blending; s = methanol; So/Sa = 12 (v/w); T = 20 °C; t = 4 min	(CAF), (CA), (C), (CG), (CHLA), (p-CMA), coumarin, (EC), (ECG), (ECDG), (EGC), (EGCG), (EGCDG), (GA), (GC), (GCG), (TB), (KAEM), 3-(p-hydroxyphenyl)-propionic acid	Yao et al 2004

Table 8.2 (*Continued*)

Sample	Operational conditions	Compounds analyzed	Reference
Tea (green)	1) m = stirring; s = water, acetone : water (25 : 75–100 : 0, v/v), methanol : water (25 : 75–100 : 0, v/v), ethanol : water (25 : 75–100 : 0, v/v), acetonitrile : water (50 : 50–100 : 0, v/v); So/Sa = 20 (v/w); T = 70–100 °C for water and boiling point for other solvents; t = 2 h; h = 3.96% 2) m = stirring; s = water; So/Sa = 40–100 (v/w); T = 80–95 °C; t = 10–240 min; h = 3.96% 3) m = stirring; s = water; So/Sa = 4 × 40 (v/w); T = 95 °C; t = 4 × 10 min; h = 3.96% 4) m = supercritical fluid extraction; s = CO_2; h = 15–50% (w/w); T = 50–80 °C; P = 225–350 bar; h = 3.96%	(KAEM), (CAF), (EGCG), (EC), (ECG), (EGC), (MYR), (QCE)	Perva-Uzunalic et al 2006
Tea (green)	1) m = infusion; s = distilled water; So/Sa = 100 (v/w); T = 85 °C; t = 15 min 2) m = ultrasound-assisted extraction; s = distilled water; So/Sa = 100 (v/w); T = 60 °C; t = 40 min; PW = 250 W	(CAF), polyphenols	Xia et al 2006
Tea (Oolong)	m = ultrasound-assisted extraction; s = acetonitrile : water (1 : 1, v/v); So/Sa = 100 (v/w); T = RT; t = 20 min; PW = 250 W, f = 40 kHz	(CAF), (+)-(C), (–)-(CG), (–)-(EC), (–)-(ECG), (–)-(EGC), (–)-(EGCG), (–)-(GC), (–)-(GCG), (TB)	Wang et al 2008
Teas (green, green and herbal blend)	m = stirring; s = water; So/Sa = 1–3 × 100 (v/w); T = 60–100 °C; t = 3–30 min	(CAF), (+)-(C), (CHLA), (p-CMA), (–)-(EC), (–)-(ECG), (–)-EGCG), (FA), (GA), (+)-(GC), (–)-(GCG), (PCCA), (TB), (TP), (VA)	Komes et al 2010
Teas (green, fermented, blend)	1) m = maceration with mild stirring; s = water; So/Sa = 100 (v/w); T = 90 °C; t = 5 min 2) m = ultrasound assisted extraction; s = methanol; So/Sa = 50 (v/w); t = 20 min	(CA), (CAF), (C), (CG), (CHLA), (EC), (ECG), (EGC), (EGCG), (FA), (GA), (GC), (GCG), (PCCA), (QCI), (RUT), (VA)	Nováková et al 2010

Table 8.2 (*Continued*)

Sample	Operational conditions	Compounds analyzed	Reference
Tea (green)	1) m = pressurized liquid extraction; s = ethanol : water (1 : 1, v/v); So/Sa = 20 (v/w); T = RT; P = 100–600 MPa; t = 15 min; d <40 mesh 2) m = stirring; s = ethanol; So/Sa = 20 (v/w); T ~ 85 °C; t = 2 h; d <40 mesh	(CAF), (EC), (ECG), (EGC), (EGCG), (GA)	Xi et al 2010
Coffee (instant) Teas (black, green, white)	m = ultrasound-assisted extraction; s = methanol : water (50 : 50, v/v) + methanol : water (75 : 25, v/v) + methanol; So/Sa = 30 + 30 + 30 (v/w); T = 60 °C; t = 3 × 30 min; f = 25 kHz	(CA), (CAF), (–)-(C), (CHLA), (p-CMA), (–)-(EC), (–)-(ECG), (–)-(EGC), (–)-(EGCG), (GA), (–)-(GC), (–)-(GCG), (KAEM3-O-R), G), (KAEM3-O-R), (MYR3RM), (PCCA), (QCE3GP), (QCE3R)	Rostagno et al 2011

This table consists of a compilation of data available in the literature and provides an outline of conditions used for the extraction of alkaloids and polyphenols from tea and coffee. Method (m), solvent (s), solvent to sample ratio (So/Sa), temperature (T), pressure (P), time (t), particle size (d), sample moisture content (h), rotation (r), power (PW), frequency (f). Compound identification: Caffeine (CAF); catechin (C); catechin gallate (CG); (EC) epicatechin; epicatechin gallate (ECG); epicatechin-3-gallate (EC3G); epicatechin digallate (ECDG); gallocatechin (GC); gallocatechin gallate (GCG); gallocatechin-3-gallate (GC3G); epigallocatechin (EGC); epigallocatechin gallate (EGCG); epigallocatechin-3-gallate (EGC3G); epigallocatechin digallate (EGCDG); epigallocatechin 3-O-(3-O-methyl)-gallate (EGCG'' 3Me); theobromine (TB); chlorogenic acid (CHLA); theophylline (TP); gallic acid (GA); caffeic acid (CA); ferulic acid (FA); vanillic acid (VA); rutin (RUT); myricetin (MYR); quercitrin (QCI); quercetin (QCE); kaempferol (KAEM); quercetin-3-O-glucopyranoside (QCE3GP); quercetin-3-O-rhamnoside (QCE3-O-RM); quercetin-3-O-rutinoside (QCE3R); kaempferol-3-O-rutinoside (KAEM3-O-R); kaempferol-3-O-glucoside (KAEM3-O-G); theaflavin (TF); theaflavin-3-gallate (TF3G); theaflavin-3'-gallate(TF3'G); theaflavin-3,3'-digallate (TF33G); myricetin-3-O-rhamnoside (MYR3RM); myricetin-3-O-rutinoside (MYR3-O-R); 3,5-dihydroxybenzoic acid (3,5-HBA); 3,4,5-trimethoxycinnamic acid (3,4,5-MCNA); p-anisic acid (p-ASA); theacrine (Teac); 5-caffeoylquinic acid (5-CQA); p-coumaric acid (p-CMA); protocatechuic acid (PCCA).

MAE has been reported to be an efficient technique for the extraction of phytochemicals. Illustratively, the extraction of tea polyphenols and caffeine with MAE for 4 min was more efficient than extraction at room temperature for 20 h, UAE for 90 min and soxhlet for 45 min (Pan *et al* 2003).

Another emerging technology available for the extraction of phytochemicals from natural products is SFE. Although SFE has been proposed as an alternative green method for recovering bioactive compounds from natural sources, when using pure CO_2, the process is not able to reach the recovery of caffeine and polyphenols achieved with soxhlet using methanol, ethanol or water. The addition of 10% (w/w) of 70% aqueous ethanol as co-solvent, on the other hand, improves 4.4-fold the SFE efficiency (Chang *et al* 2000).

Pressurized liquid extraction (PLE) has also been studied recently. PLE presented results similar to stirring, UAE and soxhlet for extraction of green tea, with excellent advantages, such as shorter extraction time, higher yield, lower energy consumption and eco-friendliness (Xi *et al* 2010).

Extraction with methanol usually is the most efficient (Lakenbrink *et al* 2000; Yao *et al* 2004). Nevertheless, it has been reported that aqueous methanol (70%) as solvent leads to a higher extraction efficiency of flavonoids and caffeine from tea and coffee than pure water (Lakenbrink *et al* 2000). On the other hand, relatively low recovery has been reported when UAE with aqueous methanol (90%) was employed (Sakakibara *et al* 2003). Although methanol leads to higher yields, hot water extraction better represents the product ingested by the consumers, which should be more interesting for studies dealing with the influence of coffee and tea consumption on human health. Better results were obtained for the extraction of catechins and caffeine using pure water than its mixture with ethanol or methanol (Khokhar and Magnusdottir 2002).

Acetone leads to a higher extraction of polyphenols than methanol, water and ethanol. As for caffeine, methanol leads to a higher extraction than pure water, ethanol and acetone. On the other hand, when water is added to ethanol, the ethanol : water (1 : 1 v/v) solution leads to higher extractions of polyphenols and caffeine than methanol, ethanol and water (Pan *et al* 2003). Ethanol and water (2 : 3 v/v) also proved to be more efficient than these solvents separately (Wang *et al* 2010). Thus, there is no consensus in the literature about the best extraction solvent for recovering caffeine and phenolic compounds from coffee and tea, but there is an indication that aqueous ethanol, a green solvent, could be used to substitute toxic solvents.

There are also reports on the influence of particle size, solvent to sample ratio, time and temperature on extraction efficiency. Temperature and time should be balanced, because prolonged extraction times at high temperatures can lead to degradation. Thus, it is advisable to use either a combination of high temperature (95 °C) and short extraction times (5–10 min), or lower temperatures (60 or 80 °C) and longer extraction times (20 min) to avoid major degradation of catechins during extraction (Perva-Uzunalic *et al* 2006). In one study, UAE for 10–60 min was tested for tea extraction, and 20 min was chosen as the best processing time (Wang *et al* 2008). In another study, the extraction of polyphenols and caffeine increased as MAE time increased, reaching a maximum at 4 min, with a decrease in caffeine extraction for longer extraction times (Pan *et al* 2003).

Altogether, it is a difficult task to ensure quantitative recoveries are being achieved. The differences in sample matrices and the profile and concentration of phytochemicals present can influence the ability of certain processes and their parameters to efficiently extract all compounds of interest, which means that the extraction method and the optimal conditions for a particular sample cannot always be extrapolated to other cases.

One conservative approach when quantifying compounds in coffee and tea is multiple consecutive extractions that can include different extraction methods and–or solvents. Up to four steps of stirring were used to extract catechins and caffeine from green tea, and their cumulative recovery increased with each subsequent extraction step (Perva-Uzunalic *et al* 2006).

8.4   Analysis of Alkaloids and Polyphenols in Tea and Coffee Samples

The last step in the analytical determination of the sample components is usually carried out by instrumental analysis. High performance liquid chromatography (HPLC) is the most common technique because it allows simultaneous determination of both compound classes, requires simple pre-analysis sample preparation and is highly sensitive and reproducible.

However, the correct separation of compounds present in complex samples depends on several factors. Briefly, separation is determined by the number of target compounds, column dimensions and characteristics, temperature, mobile phase flow rate and composition, among other factors. In general, the separation of a large number of alkaloids and polyphenols can be achieved in relatively short times when employing optimal chromatographic conditions.

Most mobile phases are composed of water and methanol or acetonitrile, with small amounts of acid (Table 8.3). Both methanol and acetonitrile have advantages and disadvantages from analytical, economic and environmental points of view. In general, it is feasible to assume that separation of polyphenols and alkaloids is possible using both types of solvents, but the ability of each mobile phase to effectively separate all sample components will depend on several factors, such as the number and type of target analytes and the analysis conditions.

Separation is usually performed in gradient mode, although several isocratic separations have been reported. Gradient elution is generally used when a larger number of compounds are analyzed, as higher resolution and separation efficiency are required. Table 8.3 presents a survey of HPLC methods for the separation of caffeine and other compounds from coffee and tea. There are several reports of complex separations of the major polyphenols and alkaloids

Table 8.3 Survey of methods for the simultaneous analysis of alkaloids and polyphenols from tea and coffee.

Sample	Target analytes	Chromatographic conditions	Elution	Detection	Ref
Tea (Oolong, Pu-erh, green, black)	(GA), (3,5-HBA), (3,4,5-MCNA), (MYR), (CAF), (p-ASA), (CG), (EGC),(EGCG) and (ECG)	Injection volume: 10 uL; Solvent: water–acetic acid, (97 : 3 v/v) (solvent A); methanol (solvent B); Column: Alltech Adsorbosil C18 (250 mm x 4.5 mm, 5 μm); Temperature: n. e.; Flow rate: 1 mL min^{-1}; Run time: 3,4,5-trimethoxycinnamic acid, approx. 25 min; Equilibration time: 2 min	Gradient: 0–1 min, 0% B; 1–27min, 0–63% B.	UV: 280 nm	Zuo et al 2002
Tea (black, white, green)	(CAF), (GA), (PCCA), (CHLA), (CA), (p-CMA), (GCG), (EC), (C), (EGC), (GC), (EGCG), (ECG), (MYR3RM), (QCE3R), (QCE3GA), (KAEM3R), (KAEM3G), (AP-7-O-G), (Lut-7-O-G)	Injection volume: 10 μL	Gradient	UV: 270nm, 280 and 320nm	Rostagno et al 2011

Table 8.3   *(Continued)*

Sample	Target analytes	Chromatographic conditions	Elution	Detection	Ref
Coffee (instant)		**Solvent:** water (1% phosphoric acid) (solvent A) and acetonitrile (1% phosphoric acid) (solvent B) **Column:** Kinetex C18, (2.6μm, 100 × 4.6mm) Temperature: 55 °C **Flow rate:** 2.2 mL min^{-1} **Run time:** kaempferol-3- O-glucoside, 4.66 min Equilibration time: 3 min	0 min, 5% B; 0.5 min, 10% B; 2.0 min, 12.5% B; 3.0 min, 15% B; 4.0 min, 80% B; 5min, 100% B; 6.0 min, 100% B; 7.0 min, 5% B	Fluorescence UV: 270nm, 280 and 320nm Fluorescence	
Tea	Esculin, Protocatechuic acid, (GA), (PCCA), (CHLA), (VA), (CA), (SA),(FA), (CNA), (C), (CG), (EC), (ECG), (EGC), (EGCG), (GC), (GCG), (RUT), (MYR), (QCI), (QCE), (NAR), (LUT), (HES), (KAEM), (AES), (SCOP), (4OH-COUM), (6M-COUM), (O-CMA) and(CAF)	Injecton Volume: 2 μL **Solvent:** 0.1% formic acid (solvent A) and methanol (solvent B) **Column:** BEH C18 (100 mm × 2.1 mm, 1.7 μm) Temperature: 25 °C **Flow rate:** initial 0.45mL min^{-1} **Run time:** Kaempferol, approx. 18 min Equilibration time: n. e.	Gradient: 0–4 min, (88.5% A : 11.5% B); 4–20 min, linear gradient elution from 88.5% A : 11.5% B to 50% A : 50% B.	UV: 280nm	Nováková *et al* 2010

Table 8.3 (Continued)

Sample	Target analytes	Chromatographic conditions	Elution	Detection	Ref
Tea (green, white)	(C), (EC), (EGC), (-)-(EGC3G), (GA), (CAF), (TB), (TP), (APG), (p-CMA), (CA), (CHLA), (KAEM), (QCE, (KAEM-3-O-p-CMG), (QCE3G), (KAEM3G), (MYR)	Injecton Volume: 5 µL **Solvent:** (H$_2$O 0.1% formic acid) (solvent A) and (acetonitrile 0.1% formic acid) (solvent B) **Column:** Acquity HSS C18 (1.8 µm, 2.1 x 100 nm) Temperature: 40 °C **Flow rate:** 0.5 mL min^{-1} **Run time:** Kaempferol, 13.33 min Equilibration time: n. e.	**Gradient:** 0–12 min, 5–26% B; 12–14 min, 26–65% B.	MS	Zhao et al 2011
Coffee and Beverage cola	(CAF), (TP) and (TB)	Injection volume: 20 µL **Solvent:** acetonitrile–water (10 : 90 v/v) **Column:** RP-18$_e$ monolithic; Chromolithi SpeedRod (50 mm x 4.6 mm i.d.) Temperature: 25 °C **Flow rate:** 3.0 mL min^{-1} **Run time:** caffeine, 0.547 min Equilibration time: n. e.	Isocratic	UV: 274 nm	Tzanavaras and Themelis 2007

This table consists of a compilation of data available in the literature and provides actual examples of separations of alkaloids and polyphenols by high performance liquid chromatography. Compound abbreviations as shown in Table 8.2.

using different conditions. Without doubt, caffeine is the main focus, but other alkaloids and different phenolic classes can be simultaneously analyzed.

In general, flow-rates used in the methods are in the $1.0–1.2$ mL min^{-1} range, but flow rates as high as 2.2 mL min^{-1} and as low as 0.2 mL min^{-1} have been reported. Flow-rate is limited by the system backpressure, which is influenced by the mobile and stationary phases. For example, due to its higher viscosity, methanol produces a higher column backpressure than acetonitrile, which limits flow-rates (Figure 8.2). As can be seen, the correct choice of mobile and stationary phases and temperature can expand the range of workable flow-rate, which can be explored to reduce analysis time.

Conventional RP-C18 columns of 5 µm particles are the most used stationary phases for the separation of alkaloids and phenolics. Usually, relatively long columns are necessary to separate a greater number of compounds. Higher performance can be achieved with columns of smaller particles ($1.7–3.5$ µm), but systems that can withstand higher pressures are required (ultra-high performance liquid chromatography – UHPLC). It was recently reported a UHPLC method for the determination of different phenolics and caffeine in tea and tea extracts (Nováková *et al* 2010). The separation of 30 compounds was achieved in less than 19 min. In a similar way, the separation of a large number of phenolics and alkaloids in teas was reported using a similar column (Zhao *et al* 2011). Chromatograms of both methods are shown in Figures 8.3 and 8.4. However, in both cases very low flow-rates were used due to the high backpressure generated by the small particle column.

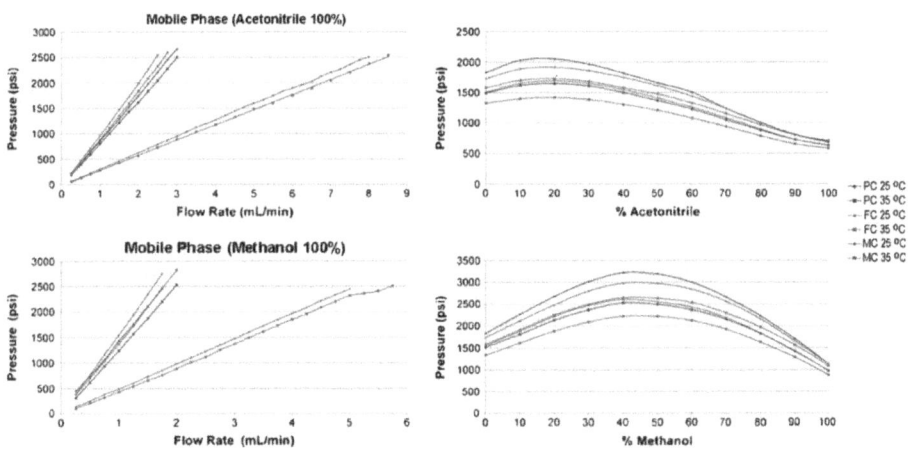

Figure 8.2 System pressure using different columns (PC: conventional particle column, FC: fused-core column, MC: monolithic column) at different temperatures, flow rates, and mobile phase compositions. Illustrates the effect of the column used and how, depending on its characteristics, higher flow-rates can be used to reduce analysis time. Reproduced from Manchon *et al* 2011 with permission from Springer.

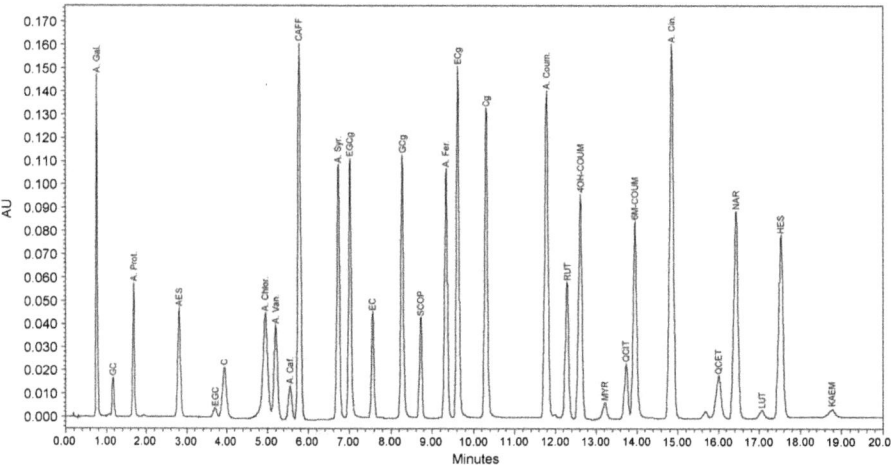

Figure 8.3   Chromatogram (UV 280 nm) of a standard mixture of 29 phenolic compounds and caffeine. Chromatographic conditions are described in Table 8.3. Illustrates a good example of the separation of alkaloids and polyphenols by UPLC. Reproduced from Nováková *et al* 2010 with permission from Elsevier.

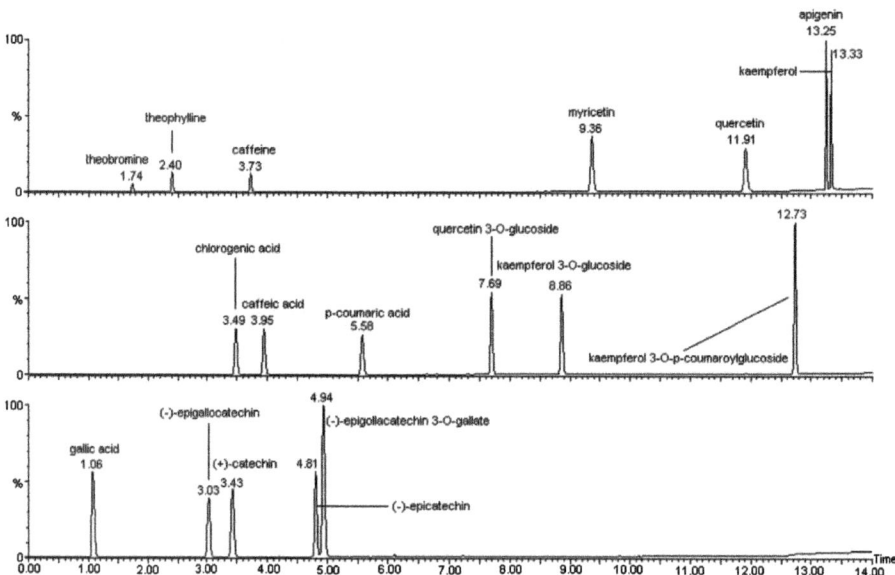

Figure 8.4   Chromatograms of mixed standards of phenolics and alkaloids (total scan 200–400 nm). Chromatographic conditions are described in Table 8.3. Illustrates a good example of the separation of alkaloids and polyphenols by UHPLC. Reproduced from Zhao *et al* 2011 with permission from Elsevier.

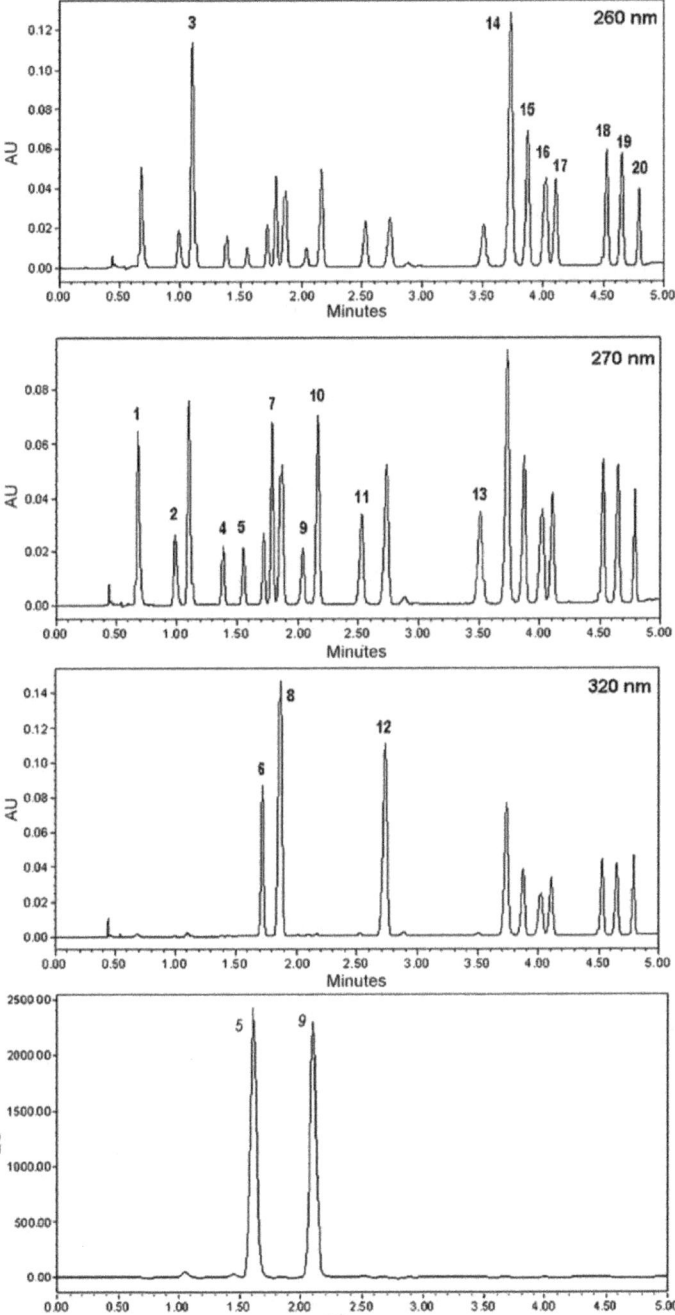

Figure 8.5 Chromatogram of a standard mixture of 19 polyphenols and alkaloids using HPLC-DAD-fluorescence. Chromatographic conditions are described in Table 8.3. Illustrates a good example of the separation of alkaloids and polyphenols using a fused-core column. Reproduced from Rostagno *et al* 2011 with permission from Elsevier.

New stationary phases are being developed to reduce the impact of particle size on the column pressure drop. Currently, columns made of particles with a porous shell fused to a solid core are proving to be an efficient way to improve chromatographic performance while generating lower pressure than conventional stationary phases. As an example of the enormous potential of this new column technology, an analysis method was recently reported for the determination of 19 phenolics and alkaloids that takes less than 5 minutes (Rostagno *et al* 2011). A representative chromatogram obtained with this method is shown in Figure 8.5.

The high efficiency of this type of stationary phase and its potential for the separation of a large number of tea and coffee components in short times is clear. The use of relatively high temperatures and acetonitrile as the mobile phase allows using a high flow-rate, which is in great part responsible for achieving an extremely fast separation.

However, there are other alternatives to particle-packed columns available, such as monolithic columns. Because it is a relatively new technology, only a few analytical monolithic columns are commercially available. Among the few applications available, a short monolithic column was used for the fast quantitation of theophylline and theobromine in coffee and cola beverages (Figure 8.6) (Tzanavaras and Themelis 2007). Separation was achieved in less than one minute with an isocratic elution and a high flow-rate (3 mL min^{-1}). It is clear that monolithic columns are capable of fast separations, but the range of applications is still limited.

Altogether, column technology is still evolving, and it is reasonable to assume that these new stationary phases will keep improving chromatographic performance in terms of resolution, speed and organic solvent consumption compared to current chromatographic systems. The increase in chromatographic performance is also linked to the number of compounds that can be

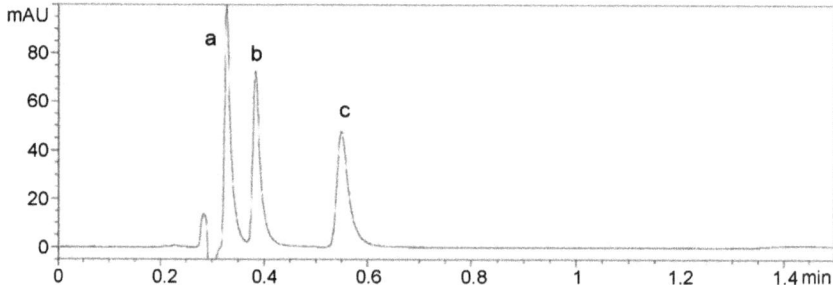

Figure 8.6   Chromatogram of alkaloids by HPLC using a monolithic column. Peak identification: caffeine (c), theophylline (b) and theobromine (a). Chromatographic conditions are presented in Table 8.3. Illustrates a good example of the separation of alkaloids using a monolithic column. Reproduced from Tzanavaras and Themelis 2001 with permission from Elsevier.

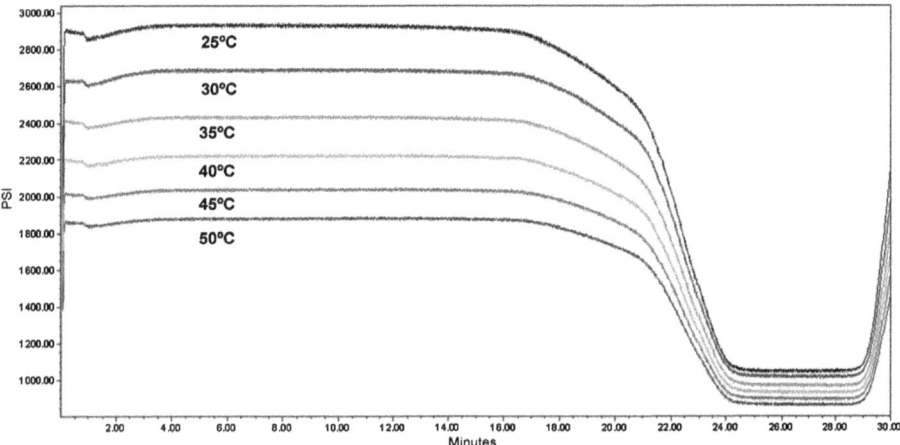

Figure 8.7   System back pressure using different column temperatures. This figure illustrates the effect of temperature on system pressure and the potential of increasing flow-rate by the use of higher temperatures. Reproduced from Manchon *et al* 2011 with permission from Elsevier.

simultaneously determined in a single run, which can be expected to increase as stationary phase technology advances.

Column temperature can have a great impact not only on the separation achieved by the column but also in analysis time. Higher temperatures reduce mobile phase viscosity, decreasing system backpressure, which can be used to increase the flow-rate and reduce analysis time. Figure 8.7 illustrates the influence of temperature on the system backpressure. Increasing the temperature from 25 to 50 °C may reduce pressure by more than 2-fold, implying that a proportional increase in the flow-rate is achievable. Although the use of higher temperatures can be advantageous in the case of polyphenols and alkaloids, columns are normally kept at ambient or slightly above ambient temperatures.

Detection of target analytes is often carried out by UV and UV-diode array detection (DAD) in the range of 270–320 nm. Mass spectrometry is also used but to a smaller extent. Multiple UV detection wavelengths are also used to improve sensitivity and to selectively detect different compound classes present. Fluorescence detection is also sometimes explored and can be especially useful for the determination of catechins. This type of detection is very sensitive and can provide low detection and quantitation limits. Moreover, selective detection also minimizes interference from some peaks and improves separation. There are several examples where selective detection was used for the separation of different compound classes and a general overview of the results indicate that this is the most adequate approach for the simultaneous determination of alkaloids and polyphenols. It is clear that using a combination of different detectors (fluorescence, MS, *etc.*), and UV wavelengths is the best approach for the separation, identification and quantification of phenolics and alkaloids in tea and coffee samples.

8.5   Conclusions

Accurate and reliable data on the composition of coffee and tea are crucial for the correct assessment of their potential health implications. Understanding the complexities of phytochemicals through extensive laboratory analysis is fundamental for the proper utilization of their potential health benefits. Among phytochemicals present in these foods, phenolics and alkaloids are the focus of intensive research. However, their determination in complex matrices, such as tea and coffee, is a multifarious task, requiring the consideration of many aspects that can influence their determined concentration. Sample preparation is an important but often neglected step in the determination of alkaloids and phenolics in tea and coffee. Special attention should be paid to this step because several factors can influence the extractability of target analytes, from the extraction technique to the operational conditions, so that quantitative recovery of compounds of interest is achieved. Unfortunately, there is not a universal extraction method, and each sample should be studied individually. In contrast, the analysis step of the whole process is advancing, and our ability to identify and quantify a larger number of compounds is increasing. Therefore, it is possible to perform a comprehensive analysis of the composition of these largely consumed products. Also, analysis time is clearly benefiting from recent technological advances, and it is now possible to have large amounts of compositional data in relatively short times.

Summary Points

- This chapter presents the necessary steps for the simultaneous determination of caffeine and phenolic compounds in tea and coffee samples.
- It is important to have accurate and reliable data about the composition of coffee and tea for the correct assessment of their potential health implications.
- The determination of alkaloids and phenolic compounds in complex matrices, such as tea and coffee, is a multifarious task requiring the consideration of many aspects that can influence their determined concentration.
- In the sample preparation step, the factors that influence the extractability of the target analytes should be considered: the extraction technique used and the optimization of the operational conditions.
- The analysis step is advancing and our ability to identify and quantify a larger number of compounds in a shorter time using high-performance liquid chromatography is increasing.

- Several aspects influencing the extraction and analysis of alkaloids and phenolics are discussed.
- The large amount of information that can be acquired should be used in conjunction with nutritional and biological observations to provide a clear view of the mechanisms of action and to point out directions for future research in this area.

Key Facts of the Chemical Composition of Tea and Coffee

- Tea and coffee are among the most consumed beverages worldwide; they are consumed for their organoleptic properties and potential effects in the human organism.
- The alkaloids and polyphenols in their chemical composition are likely responsible for their potential health effects.
- To understand these health effects, it is important to correctly identify and quantify the bioactive phytochemicals present in tea and coffee.
- The main phenolic compounds present in coffee are hydroxycinnamic acids (coumaric, caffeic and ferulic acids) and their derivatives (chlorogenic and feruloylquinic acids).
- The main polyphenols found in tea are flavonols (catechins) or flavones.
- The most known and studied bioactive phytochemical found in tea and coffee is caffeine, an alkaloid.
- Caffeine is largely known for its stimulating effects on the central nervous system.
- Other alkaloids, such as theophylline and theobromine, are also present in tea and coffee.

Key Facts of Phenolic Compounds

- Phenolic compounds are molecules containing one or more phenolic groups.
- They occur naturally in the plant kingdom.
- A group of natural phenols called flavonoids are of most popular interest because they have potential health benefits.
- A group of flavonoids called catechins is responsible for the majority of flavonoids in growing tea leaves.
- In terms of taste, catechins are bitter.
- When the tea leaves are fermented to produce black tea, most of their catechins are oxidized and turn into theaflavins and thearubigins.
- Studies have pointed that theaflavins have potential health beneficial effects.

- Thearubigins are responsible for the red color of black tea, while theaflavins contribute a yellowish orange color; typically, a finer black tea has greater theaflavins content and thus a more orange tone.

Key Facts of Determination of Caffeine and Phenolic Compounds in Tea and Coffee

- Polyphenols may be simultaneously determined with caffeine and other alkaloids.
- The basic steps necessary for the determination of the profile and concentration of the phytochemicals present in tea and coffee are sample preparation and analysis.
- Among the procedures used in sample preparation, extraction is of key importance.
- There is an indication that aqueous ethanol, a green solvent, could be used to substitute toxic solvents for the extraction of tea and coffee phytochemicals.
- For classical extraction techniques, it is advisable to use either a combination of high temperature (95 °C) and short extraction times (5–10 min), or lower temperature (60 or 80 °C) and longer extraction times (20 min) to avoid the major degradation of catechins during extraction.
- Of the novel extraction techniques, ultrasound-assisted extraction is the most frequently used, but pressurized liquid extraction seems to be faster, more efficient, more eco-friendly and with lower energy consumption.
- The most used method of instrumental analysis of tea and coffee polyphenols and alkaloids is HPLC.
- The analysis of a few compounds is usually achieved in short times with relatively simple methods; the analysis time increases proportionally with the increase of components to be determined.
- Separation efficiency is increasing and it is currently possible to simultaneously determine more than 20 phenolic compounds with caffeine.

Definition of Words and Terms

Phytochemicals: The term "phytochemicals" refers to a wide range of compounds produced by plants that are not considered essential nutrients for the normal development of plants.

Phenolic compounds: These compounds present one or more hydroxyl radicals linked to an aromatic ring; several different radicals, such as carboxyl, methoxyl and other non-aromatic cyclic groups, can be present. Polyphenols can be separated in several classes according to the type and number of phenolic rings and in subclasses according to specific substitutions in the basic structure, associations with carbohydrates and polymerized forms. Among these classes, non-flavonoid compounds include the subclasses of benzoic acids, hydroxycinnamic acids, hydrolyzable tannins and lignins, among others. The flavonoid class includes flavanols, flavonones, flavones, isoflavones, proanthocyanins and anthocyanins.

Extraction: Sample preparation procedure used to separate target analytes from the sample matrix to be later analyzed. Usually involves contact between the sample and some type of solvent.

Ultrasound-assisted extraction: This technique is based on the cavitation of the bubbles created by the passage of ultrasonic waves (>20 kHz) through the extraction medium. It enhances contact between the solvent and sample, improving extraction efficiency. It also may affect sample structure, which can release target analytes to the extraction medium.

Supercritical fluid extraction: This technique consists of separating sample components using a supercritical fluid as the extracting solvent. The most commonly used supercritical fluid is carbon dioxide (CO_2), which is capable of dissolving non-polar or moderately polar compounds.

Supercritical fluids: Any substance submitted to temperature and pressure above their thermodynamic critical point.

Critical point: In thermodynamics, a critical point specifies the conditions (temperature and pressure) at which a phase boundary ceases to exist.

Pressurized liquid extraction: Extraction based on a combination of elevated temperature (50–200 °C) and moderate pressures (100–140 atm) of liquid solvents. High temperature is used to improve extraction efficiency, while pressure is used to maintain the solvent in the liquid phase.

Microwave-assisted extraction: This technique is based on the molecular movement caused by electromagnetic waves with wavelengths ranging from 1 mm to 1 m, or frequencies between 300 MHz and 300 GHz.

High performance liquid chromatography: It is an analytical technique based on the selective retention of analytes by a stationary phase that is carried by the mobile phase. Once analytes are separated, they are individually detected by different types of detectors. There is usually a correlation between the signal generated and the concentration.

Isocratic: A separation in which the mobile phase composition remains constant throughout the procedure.

Gradient: A separation in which the mobile phase composition does not remains constant throughout the procedure.

List of Abbreviations

DAD	Diode array detection
HPLC	High-performance liquid chromatography
MAE	Microwave-assisted extraction
PLE	Pressurized liquid extraction
SFE	Supercritical fluid extraction
UHPLC	Ultra high performance liquid chromatography
UAE	Ultrasound-assisted extraction

References

Chang, C. J., Chiu, K.-L. and Chen. Y.-L., 2000. Separation of catechins from green tea using carbon dioxide extraction. Food Chemistry. 68:109–113.

Khokhar S. and Magnusdottir S. G. M., 2002. Total phenol, catechin, and caffeine contents of teas commonly consumed in the United Kingdom. Journal of Agricultural and Food Chemistry. 50:565–570.

Komes D., Horžić D., Belščak A., Ganič K.K. and Vulič I., 2010. Green tea preparation and its influence on the content of bioactive compounds. Food Research International. 43:167–176.

Lakenbrink C., Lapczynski S., Maiwald B., Engelhardt U. H., 2000. Flavonoids and other polyphenols in consumer brews of tea and other caffeinated beverages. Journal of Agricultural and Food Chemistry. 48:2848–2852.

Lin Y.-S., Tsai Y.-J., Tsay J.-S. and Lin J.-K., 2003. Factors affecting the levels of tea polyphenols and caffeine in tea leaves. Journal of Agricultural and Food Chemistry. 51:1864–1873.

Manchón, N., D'Arrigo, M., García-Lafuente, A., Guillamón, E., Villares, A., Ramos, A., Martínez, J. A. and Rostagno, M. A., 2010. Fast analysis of isoflavones by high-performance liquid chromatography using a column packed with fused-core particles. Talanta. 82:1986–1994.

Manchón, N., D'Arrigo, M., García-Lafuente, A., Guillamón, E. and Villares, A., 2011. Comparison of different types of stationary phases for the analysis of soy isoflavones by HPLC. Analytical and Bioanalytical Chemistry. 400:1251–1261

Nováková L., Spáčil Z., Seifrtová M., Opletal L. and Solich P., 2010. Rapid qualitative and quantitative ultra high performance liquid chromatography method for simultaneous analysis of twenty nine common phenolic compounds of various structures. Talanta. 80:1970–1979.

Pan X., Niu G. and Liu H., 2003. Microwave-assisted extraction of tea polyphenols and tea caffeine from green tea leaves. Chemical Engineering and Processing. 42:129–133.

Perva-Uzunalić A., Škerget M., Knez Ž., Weinreich B., Otto F. and Grüner S., 2006. Extraction of active ingredients from green tea (*Camellia sinensis*): Extraction efficiency of major catechins and caffeine. Food Chemistry. 96:597–605.

Rostagno M.A., Manchón N., D'Arrigo M., Guillamón E., Villares A., García-Lafuente A., Ramos A. and Martínez J. A., 2011. Fast and simultaneous determination of phenolic compounds and caffeine in teas, mate, instant coffee, soft drink and energetic drink by high-performance liquid chromatography using a fused-core column. Analytica Chimica Acta. 685:204–211.

Sakakibara H., Honda Y., Nakagawa S., Ashida H. and Kanazawa K., 2003. Simultaneous determination of all polyphenols in vegetables, fruits, and teas. Journal of Agricultural and Food Chemistry. 51:571–581.

Tzanavaras, P. D. and Themelis, D. G., 2007. Development and validation of a high-throughput high-performance liquid chromatographic assay for the determination of caffeine in food samples using a monolithic column. Analytica Chimica Acta. 581:89–94.

Wang D., Lu J., Miao A., Xie Z. and Yang, D., 2008. HPLC-DAD-ESI-MS/MS analysis of polyphenols and purine alkaloids in leaves of 22 tea cultivars in China. Journal of Food Composition and Analysis. 21:361–369.

Xi, J., Zhao, S., Lu, B., Zhang, R., Li, Y., Shen, D. and Zhou, G., 2010. Separation of major catechins from green tea by ultrahigh pressure extraction. International Journal of Pharmaceutics. 386:229–231.

Xia, T., Shi, S. and Wan, X., 2006. Impact of ultrasonic-assisted extraction on the chemical and sensory quality of tea infusion. Journal of Food Engineering. 74:557–560.

Yao, L., Jiang, Y., Datta, N., Singanusong, R., Liu, X., Duan, J., Raymont, K., Lisle, A. and Xu, Y., 2004. HPLC analyses of flavanols and phenolic acids in the fresh young shoots of tea (*Camellia sinensis*) grown in Australia. Food Chemistry. 84:253–263.

Zhao, Y., Chen, P., Lin, L., Harnly, J.M., Yu, L. and Li Z., 2011. Tentative identification, quantitation, and principal component analysis of green pu-erh, green, and white teas using UPLC/DAD/MS. Food Chemistry. 126:1269–1277.

Zuo, Y., Chen, H. and Deng, Y., 2002. Simultaneous determination of catechins, caffeine and gallic acids in green, oolong, black and pu-erh teas using HPLC with a photodiode array detector. Talanta 57:307–316.

CHAPTER 9

Analysis of Caffeine in Dietary Products by Multiple Injection Capillary Electrophoresis

AHMAD AMINI[1,2]

[1] Medical Products Agency (MPA), Laboratory, Box 26, Dag Hammarskjölds väg 42, 751 03 Uppsala, Sweden; [2] Uppsala University, Division of Analytical Pharmaceutical, Chemistry, Biomedical Centre, Box 574, 751 03 Uppsala, Sweden
E-mail: ahmad.amini@mpa.se

9.1   Introduction

Caffeine (1,3,7-trimethylxanthine) is found in certain leaves, beans and many fruits worldwide. Tea, coffee, chocolate, Coca-Cola® and energy drinks are common sources of caffeine in our diet (Ashton 1987; Astrup *et al* 1990; Curatolo and Robertson 1983). However, both extracted and synthetic caffeine may be added to other commercially available products. Caffeine is considered to be the most commonly consumed stimulant worldwide (Juliano and Griffiths 2004). Therefore, development of accurate, rapid and cost-efficient analytical methods for the assay of caffeine is important. The determination of caffeine is of interest from several points of view. It can be performed to measure the caffeine content of commercial products, monitor the caffeine intake as well as to evaluate the product quality, stability and shelf life (Horie and Kohata 1998; Walker *et al* 1997).

Food and Nutritional Components in Focus No. 2
Caffeine: Chemistry, Analysis, Function and Effects
Edited by Victor R Preedy
© The Royal Society of Chemistry 2012
Published by the Royal Society of Chemistry, www.rsc.org

Capillary electrophoresis (CE) has demonstrated to be an efficient and a fast technique for the analysis of a wide range of analytes. Different CE modes, such as capillary zone electrophoresis (CZE), micellar electrokinetic chromatography (MEKC), capillary gel electrophoresis (cGE), capillary iso-electrical focusing (cIEF) and multiple-injection capillary electrophoresis have been applied for the separation of enantiomers (Amini 2001), analysis of biological samples (Heller 1997), food, botanical products and beverages (Lindeberg 1996).

Caffeine is a xanthine alkaloid (Figure 9.1(A)) with very weak protolytic properties, *i.e.*, it will not be ionized at almost any pH (Jimidar *et al* 1993). There are many publications, reporting rapid assay of caffeine at high pH values where the caffeine due to the lack of electrophoretic mobility is quickly swept towards the detection site by the electroosmotic flow (EOF). EOF is generated by the negatively charged silanol groups on the inner surface of the capillary wall. The increased ionization of silanol groups at higher pH values

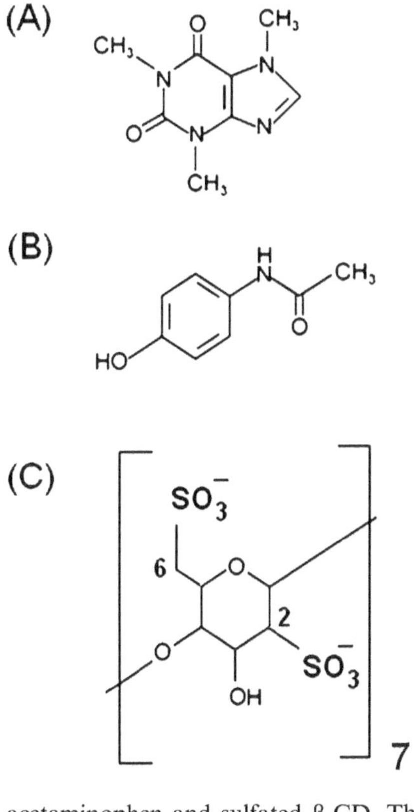

Figure 9.1 Caffeine, acetaminophen and sulfated β-CD. The chemical structure of caffeine (A), acetaminophen (B) and sulfated β-cyclodextrin (C). The number of sulfonyl groups per β-CD molecules varies between 1 and 14. The sulfonyl groups are linked to the primary hydroxyl group (C6) and the secondary hydroxyl group (C2) of the sugar units of the cyclodextrin.

results in stronger EOF. The bulk electroosmotic flow is a non-selective mobility which sweeps all uncharged compounds, irrespective of their molecular size, towards the detection window with the same velocity. A quantitative method requires selectivity and specificity, *i.e.*, the method must be able to measure accurately the quantity of caffeine in the presence of other neutrally charged compounds such as theobromine and theophylline (Lee 2000; Vogt and Conradi 1997).

In order to develop selective methods, micellar electrokinetic chromatography (MEKC) as a powerful separation mode of CE was employed for the assay of caffeine (Cianchino *et al* 2008; Vogt and Conradi 1997; Horie and Kohata 1998; Krogh *et al* 1994; Morneau *et al* 1999; Thompson *et al* 1995; Thanh *et al* 1997; Zhao and Lunte 1997; Jumppanen *et al* 1993; Pedersen-bjergaard *et al* 2000; Geldart and Brown 1993; Zhang *et al* 2005; Jimidar *et al* 1993). Sodium dodecyl sulfate (SDS) as an anionic surfactant at a concentration above its critical micelle concentration (CMC), was added to the BGE. The separations were performed at alkaline conditions to generate a strong cathodic EOF (Cianchino *et al* 2008; Horie and Kohata 1998; Lilley and Wheat 1993; Krogh 1994; Vogt and Conradi 1997; Conte *et al* 1996; Aucamp *et al* 2000; Walker *et al* 1997; Cianchino 2008; Thompson *et al* 1995; Thanh *et al* 1997; Pedersen-bjergaard *et al* 2000; Zhao and Lunte 1997; Larger P. J. 1998; Jumppanen *et al* 1993; Thormann *et al* 1992; Emre and Özaltın 2007). In MEKC, separation of neutral species such as caffeine, is based on their hydrophobicity. The migration velocity of these analytes in the capillary is determined by the vector sum of the electrophoretic mobility of the micelles (μ_{mc}) and that of the EOF. The counter-current mobility of the negatively charged SDS micelles with respect to the EOF has been considered as a reason for the high separation efficiency of MEKC (Muijselaar *et al* 1997).

This paper presents a new separation method based on multiple-injection CE for the assay of caffeine in dietary products. In multiple-injection CE, several samples are sequentially introduced in the capillary. The applied samples are electrophoresed for a short period of time (t_{PE}) immediately after the injection in order to be separated from the previously injected sample plugs. As a result the applied plugs are sandwiched between two segments of the BGE, which provides appropriate space for the expansion of the applied zones during the separation. After the final injection, the samples are electrophoresed for a longer period of time to complete the separation of the applied samples. The use of multiple-injection CE reduces the average analysis time for each sample to a fraction of what it is in the single-injection mode. It also brings additional benefits, such as a higher precision since several samples are analyzed under the same conditions (Lodén *et al* 2009).

Sulfated β-cyclodextrin (S-β-CD) was added to the BGE for separation of caffeine and other components in the analyte samples. Acetaminophen was added to the samples as an injection marker (IM) to compensate for injection volume fluctuations (Amini *et al* 2008). The quantifications were carried out by the use of calibration curves accomplished by available caffeine reference standard.

9.2 Methodological Considerations

Capillary electrophoresis experiments were performed on a ProteomeLab PA 800 system (Beckman Coulter, CA, USA), equipped with a photodiode array (DAD) system that monitored the wavelengths between 190 nm and 400 nm. Detection was performed at 270 nm. The background electrolyte (BGE) was prepared by adjusting the pH of 50 mM phosphoric acid to 7.0 with 1.0 M NaOH, followed by the addition of 15 mg mL^{-1} sulfated β-cyclodextrin, 10% (v/v) ACN and 20 mM NaCl. A 75 cm fused silica capillary (65 cm effective length) × 50 μm I.D. (O.D. 375 μm) from Polymicro Technologies, Phoenix, AZ, USA was used. The capillary was cut to the desired length using an SGT Shortix Capillary Column Cutter (Middelburg, The Netherlands).

Before the first injection, the capillary was consecutively preconditioned with water (5 min), 0.1 M HCl (5 min) and 0.1 M NaOH (10 min), followed by water and BGE for 10 min at 60 psi (413 kPa). Injections of standards and samples were performed at 0.1 psi (0.69 kPa) for 10 seconds. Between the injections in the multiple-injection CE mode I, the applied plugs were subjected to partial electrophoresis for 2.2 min (t_{PE}) at 25 kV. The ramp time was set at 0.17 minutes. Following the last sample injection, a constant voltage of 25 kV was applied across the capillary for 15 minutes. The observed electric current was 60 μA. Temperature of the capillary coolant was set to 20 °C. After each run the capillary was rinsed with NaOH (3 min), water (3 min) and with the BGE for 5 min. All solutions were filtered through 0.22 μm filters before use.

Reference standards of caffeine and acetaminophen were supplied by the European Directorate for the Quality of Medicines (EDQM, Strasburg, France). Sulfated β-CD was purchased from Sigma-Aldrich Chemie (Steinheim, Germany). All solutions were prepared using double distilled water supplied by a Maxima water purification system from USF Elga (Bucks, UK) and were filtered through 0.22 μm PVDF syringe filters (Millipore, Cork, Ireland) before use. Different tea products, chocolate, Coca-Cola®, and coffee were purchased from local grocery stores in Uppsala, Sweden.

9.3 Separation of Caffeine and Acetaminophen by Capillary Electrophoresis

The analysis of caffeine has usually been performed at high pH values, where caffeine as a neutral molecule rapidly migrates in the capillary by the bulk electroosmosis (Cianchino *et al* 2008; Vogt and Conradi 1997; Horie and Kohata 1998; Krogh *et al* 1994; Morneau *et al* 1999). In this study the separations were performed at pH 7.0. This pH was chosen because it provided closely migrating, but base-line resolved peaks of caffeine and acetaminophen

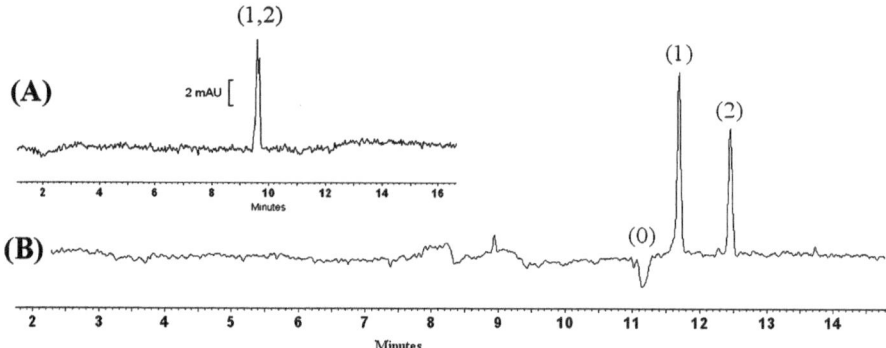

Figure 9.2   Capillary electrophoresis separations of caffeine and acetaminophen for different conditions. Representative electropherograms of caffeine and acetaminophen in the absence (A) and presence (B) of 15 mg mL^{-1} sulfated β-cyclodextrin (S-β-CD). The separations were performed in a fused silica capillary of dimensions: 50 μm × 65 × 75 cm; Applied voltage 25 kV; Hydrodynamic injection (0.1 psi) for 10 seconds; BGE 50 mM phosphoric acid (pH 7.0) containing 10% ACN (v/v %), 15 mg mL^{-1} S-β-CD and 20 mM NaCl; capillary temperature 20 °C; UV detection at 270 nm. Peaks: 0, water; 1, caffeine; 2, acetaminophen.

in the presence of sulfated β-cyclodextrin (see Figure 9.2), which are excellent conditions for the conversion of the method to multiple-injection CE mode I (Amini *et al* 2008). Acetaminophen (Figure 9.1(B)) was selected as an injection marker based on the fact that acetaminophen as a very weak acid (pK$_a$ ≈ 9) co-migrated with caffeine at the selected pH, Figure 9.2 (A). In addition, acetaminophen like caffeine exhibited a satisfactory UV absorption at 270 nm. Detection at this wavelength contributed to the specificity of the method since spectral interferences of background and other substances present in the extract samples were suppressed. In order to further prevent the matrix effects, as well as to improve the solubility of caffeine, the extract was diluted. The extent of dilution was dependent on the caffeine concentration in the sample. These measures also facilitated the application of the multiple-injection technique, where interferences are not desirable.

On the basis of previous experience sulfated β-CD (S-β-CD), as an effective complexing agent, was added to the BGE for the separation of the IM and caffeine (Amini *et al* 2006), Figure 9.1 (C). β-cyclodextrin is a cyclic oligosaccharide containing seven D-(+)-glucose units linked by α-1,4-glycosidic bonds, Figure 9.1 (C). It possesses a central hydrophobic cavity and a hydrophilic outside. The hydrophobic cavity of the CD is able to form inclusion complexes with hydrophobic compounds of appropriate size.

As Figure 9.2 (B) shows, base-line separation was achieved in less than fifteen minutes in the presence of 15 mg mL^{-1} S-β-CD. This concentration was considered to be adequate for the separation and conversion of the method to multiple-injection CE mode I (Amini *et al* 2008). This multiple-injection CE mode is considered to be suitable for compounds that exhibit small migration time difference (Δt_{mig}) and large $t_{mig2} : \Delta t_{mig}$ ratio, see Figure 9.2 (B) (Amini *et al* 2008). These requirements were met under the chosen separation conditions, *i.e.*, S-β-CD at 15 mg mL^{-1} dissolved in 50 mM phosphate buffer, pH 7.0. Higher S- β -CD concentrations resulted in increased resolution between the peaks but also in excessive analysis time, high current and reduced sample capacity.

In the presence of S-β-CD, the IM and caffeine were separated due to the difference in their effective mobilities ($\Delta\mu_{eff}$). In contrast to the non-specific mobility of the EOF, the effective mobility is a selective mobility. Eqn (1) shows that the analyte with stronger interactions with the negatively charged CD migrates slower. As evidenced by the electropherogram in Figure 9.2 (B), the CD had stronger influence on the migration time of acetaminophen than caffeine.

The negatively charged CD migrates electrophoretically towards the anode, *i.e.*, opposite to the direction of the EOF. However, the caffeine and the IM zones migrated towards the detection site near the cathodic end of the capillary by a combination of the electrophoretic mobility of S-β-CD and the bulk EOF. The contribution of each of these two mobilities to the overall mobility is described by eqn (1).

The separation method was further optimized by addition of 10% acetonitrile and 20 mM NaCl to the BGE, as well as by reducing the capillary temperature to 20 °C. It has been suggested that addition of acetonitrile at appropriate concentration, *e.g.* 10%, disrupts the complex between caffeine and polyphenols present in the extract samples (Larger *et al* 1998; Collier 1972) thereby solubilizing the polyphenols, which results in the sharpening of the caffeine peak. Acetonitrile changes the apparent electrophoretic mobility (μ_{app}) of the analytes by reducing the EOF as well as their effective electrophoretic mobility (μ_{eff}) by affecting the CD-analyte interactions.

In the presence of S-β-CD the apparent mobility (μ_{app}) of the analytes is determined by the vector sum of the effective electrophoretic mobility (μ_{eff}) and the electroosmotic mobility (μ_{eo}), eqn (1).

$$\mu_{app} = \mu_{eo} - \mu_{eff} \tag{1}$$

and

$$\mu_{eff} = \delta\ \mu_{ep(S-β-CD)} \tag{2}$$

where δ is a coefficient indicating the affinity of the analyte for the CD, see eqn (3). The effective mobility of the analyte is equal to the product of δ and $\mu_{ep(CD)}$, eqn (2). The negative sign in eqn (1) indicates anodic electrophoretic mobility of the S-β-CD. The value of δ is small since the apparent migration mobilities were only 5 to 10% smaller than the electroosmotic flow, indicating that the EOF had a huge contribution to the apparent mobility of the analytes, *i.e.*, the mobility was predominately determined by the cathodic EOF.

$$\delta = [\text{Analyte-S-β-CD}] / C \tag{3}$$

where C is the total concentration of the analyte. By combining equations (2) and (3) the concentration of the analyte- S-β-CD complex, can be calculated, eqn (4).

$$[\text{Analyte-S-}\beta\text{-CD}] = (C \ \mu_{\text{eff}}) / \mu_{\text{ep(S-}\beta\text{-CD)}} \qquad (4)$$

However, the electrophoretic mobility of S-β-CD was difficult to determine, since the sulfated β-CD showed considerable molecular heterogeneity in terms of sulfation degree, *i.e.*, the β-CD molecules were substituted with 1 to 14 sulfate groups, as determined by mass spectrometry and nuclear magnetic resonance spectroscopy (Amini *et al* 2004). The NMR results indicated that S-β-CD composed to a large extent of sugar moieties sulfated at both primary (C6) and secondary hydroxyl groups, Figure 9.1 (C).

9.4   Conversion of the Single Injection Method to Multiple Injection Mode I

Since the caffeine and the IM migrated close to each other and the migration time of the IM was about eight times longer than the difference in migration times (Δt_{mig}), the application of the first multiple-injection CE mode was possible (Amini *et al* 2008). The Δt_{mig} was determined to approximately 1.5 minutes by calculating the difference between the migration times of the last migrating peak ($t_{\text{mig(2)}}$) and the first migrating peak, *i.e.*, EOF ($t_{\text{mig(o)}}$), see Figures 9.2 and 9.3 (B). The shortest time period for the t_{PE} was determined to approximately 1.6 minutes, eqn (5) (Amini *et al* 2008).

$$t_{\text{PE}} \geq \Delta t_{\text{mig}} + 12\sigma \qquad (5)$$

where σ is the standard deviation for the broadest peak, which may be calculated by dividing the width at half of the peak height by 2.35 (Mills *et al* 1997). In order to increase the distance between the adjacent plugs, the t_{PE} was extended to 2.2 minutes. The number of injections was thus determined as five using eqn (6), Figure 9.3 (B).

$$n_{\text{inj}} = t_{\text{mig(2)}} / t_{\text{PE}} \qquad (6)$$

and

$$n_{\text{s}} = [t_{\text{mig(2)}} / (\Delta t_{\text{mig}} + 12\sigma)] \geq n_{\text{inj}} \qquad (7)$$

The sample capacity (n_{s}) refers to the maximum number of injections which can be performed in multiple-injection CE. The value of 7 was calculated for the sample capacity (n_{s}) by means of equation (7), where ($\Delta t_{\text{mig}} + 12\sigma$) = t_{PE} = 1.6 minutes (Amini *et al* 2008). The difference between the sample capacity (n_{s}) and n_{inj} (Δn) provides the possibility of compensating for variations in the migration times and Δt_{mig}, by optimizing the t_{PE}, see section 9.3.2.5.

$$\Delta n = (n_{\text{s}} - n_{\text{inj}}) \qquad (8)$$

(A)

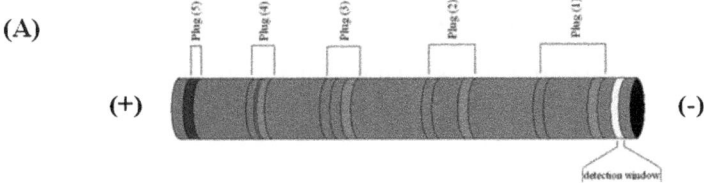

(+)   (-)

(B)

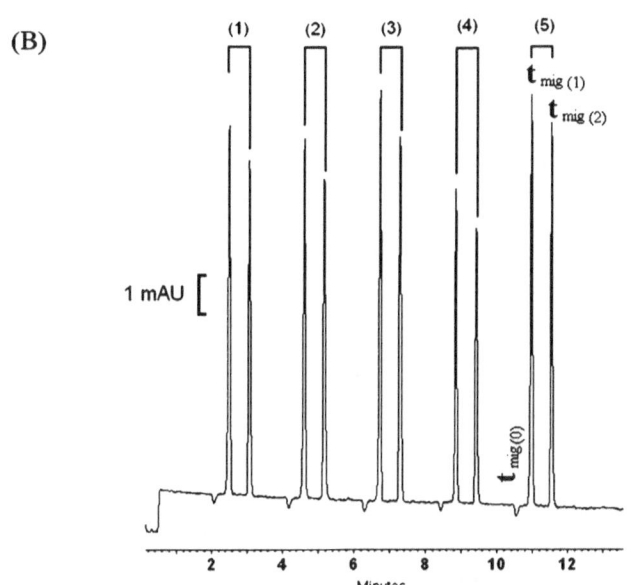

Figure 9.3 Multiple- injection capillary electrophoresis mode I. Schematic illustration of the multiple-injection technique mode I (A), applied to the assay
of caffeine in the dietary products (B). Five analyte samples, consisting of
caffeine and acetaminophen were analyzed simultaneously (B). The
hydrodynamically injected sample plugs were subjected to the partial
electrophoresis for 2.2 minutes in order to be isolated from the previously
injected plugs. Operation conditions 15 mg mL^{-1} sulfated β-cyclodextrin
(S-β-CD). The separations were performed in a 75 cm fused silica
capillary of dimensions: 50 μm × 65 cm (effective length); Applied
voltage 25 kV; Hydrodynamic injection (0.1 psi) for 10 seconds; BGE 50
mM phosphoric acid (pH 7.0) containing 10% ACN (v/v %), 15 mg m S-
β-CD and 20 mM NaCl; capillary temperature 20 °C; UV detection at
270 nm. Peaks: 0, water; 1, caffeine; 2, acetaminophen.

The developed multiple-injection CE method was considered to be a simple
and fast method for the assay of caffeine. The method was therefore validated
according to the ICH guide line with respect to linearity, precision,
intermediate precision, reproducibility, accuracy, robustness, limit of detection
(LOD) and limit of quantification (LOQ).

9.5   Method Validation

The purpose of validation is to ensure that the method measures accurately and consistently the amount of caffeine present in different samples.

9.5.1   Specificity and Selectivity

Specificity is defined as the ability of the method to detect caffeine that is free from any interference. The procedure is selective when it allows determination of caffeine independent of other components present in the analyte sample. Specificity is therefore the ultimate of selectivity, where no interference occurs. Specificity of the method was confirmed by comparing the electropherograms obtained from the multiple-injection CE analysis of a dietary sample and a standard sample, containing both caffeine and acetaminophen as injection marker. The purpose was to identify the caffeine peak and to ensure that the IM did not interfere with any other component in the sample. However, because of the lack of blank samples the purity and identity of the caffeine peaks were confirmed by comparing the obtained ultra violet (UV) spectra with the caffeine UV spectrum, Figure 9.4. The obtained UV spectral data revealed that the caffeine did not co-elute with any other compound in the sample and was thus free from any interference. The analysis of the dietary samples showed further that no interfering peaks were appeared at the elution position of the IM.

The high specificity of the multiple-injection CE method was attributed to the high concentration of caffeine in the dietary samples. The caffeine concentration was much higher than other co-extracted compounds, which provided the possibility of diluting the extract so that UV interferences from these compounds became negligible. The employed detection wave length was also a contributory factor.

9.5.2   Linearity

Stock solutions of caffeine and acetaminophen (1.8 mg mL^{-1} and 0.9 mg mL^{-1}, respectively) were prepared in water. The exactly weighed caffeine powder was dissolved in water containing 0.9 mg mL^{-1} acetaminophen to yield a concentration of 1.0 mg mL^{-1} caffeine. Twenty one standard solutions of caffeine at concentrations reported in Table 9.1 were prepared. The standard samples were analyzed to investigate the concentration–UV response relationships. A standard curve, comprised of five points, was generated for each run. Thus, a standard curve comprising all the standard concentrations was constructed by at least four multiple-injection CE runs. The range chosen for the calibration curve was based on the concentration of caffeine in the sample. Peak area ratios, *i.e.*, Area$_{caffeine}$: Area$_{actaminophen}$, were plotted *versus* the corresponding standard concentrations. It is important to note that the peak areas were not normalized with respect to their migration times. This is usually done to

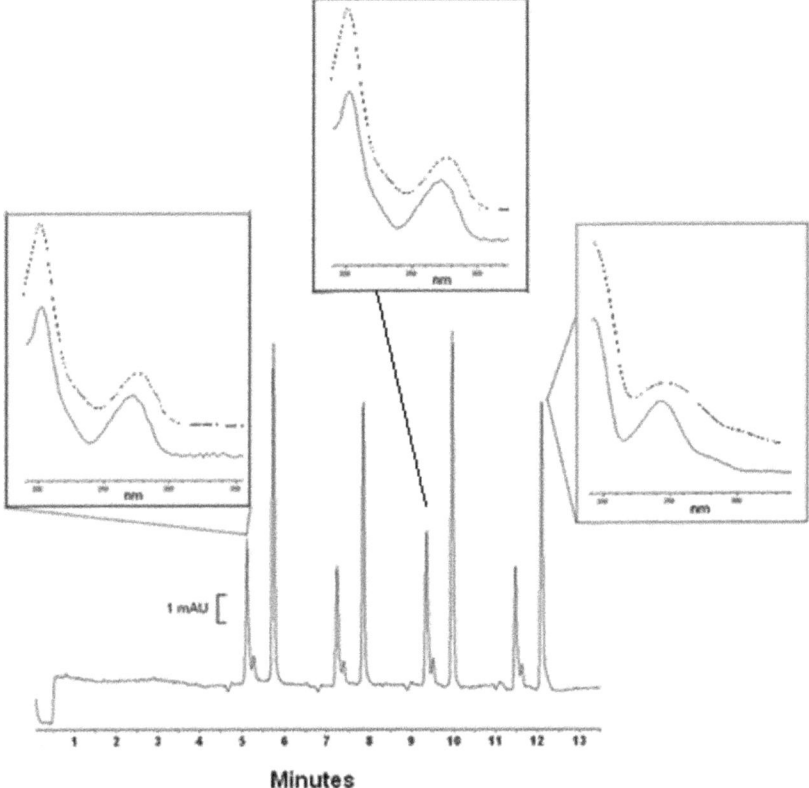

Figure 9.4   Multiple-injection capillary electrophoresis analysis of a coffee product. Multiple-injection CE mode I analysis of a coffee product. After the solid/liquid extraction to hot water, four replicate injections were performed, and the purity and identity of the peaks were confirmed by comparing the UV spectrum of each peak with those of the caffeine and the acetaminophen reference standards. The dashed lines are the caffeine and acetaminophen reference spectra. The small peak beside the large caffeine peak was not identified. Operation conditions as those given in Figure 9.2.

compensate for run-to-run fluctuations in the analyte migration velocity which is mainly caused by variations in the EOF (Altria 1993). The possibility to use un-normalized peak areas for the quantifications points out another advantage of multiple-injection CE mode compared to conventional single-injection mode: in multiple-injection CE mode the analyte in all the applied samples passes the detector with the same velocity, since the EOF remains almost constant during the separation.

Linear regression data tabulated in Table 9.2 show that there is a linear relationship, as indicated by the correlation coefficients ($R^2 > 0.990$), between the caffeine concentration and the detector response, *i.e.*, *Response = Slope × Concentration ± Intercept*. The agreement between the slopes of the standard curves constructed during a two-month period was excellent, Table 9.2. The

Table 9.1 Preparation of the caffeine standard solutions for construction of calibration graphs. Acetaminophen was added in the samples as the injection marker to compensate for injection volume fluctuations. Acetaminophen was dissolved in water at a concentration of 0.9 mg mL^{-1}. The stock solution of caffeine at 1 mg mL^{-1} was made in the acetaminophen solution. The caffeine stock solution was diluted to different concentrations in acetaminophen solution.

Sample number	[caffeine] $_{standard\ solution}$ ($\mu g\ mL^{-1}$)	$V_{Caffeine}$ (1 mg mL^{-1}) (μL)	$V_{Acetaminophen}$ (0.9 mg mL^{-1}) (μL)	V_{Total} (μL) ($V_C + V_A$)
1	900	900	100	1000
2	800	800	200	1000
3	700	700	300	1000
4	650	650	350	1000
5	600	600	400	1000
6	550	550	450	1000
7	500	500	500	1000
8	450	450	550	1000
9	400	400	600	1000
10	350	350	650	1000
11	300	300	700	1000
12	250	250	750	1000
13	200	200	800	1000
14	125	125	875	1000
15	100	100	900	1000
16	90	90	910	1000
17	80	80	920	1000
18	75	75	925	1000
19	70	70	930	1000
20	50	50	950	1000
21	25	25	975	1000

observed none-zero intercept, being either positive or negative, was due to random errors. These errors could be due to random peak intensity variations, caused by fluctuations in the UV light intensity, as well as errors in the small volumes used to prepare the standard solutions (see Table 9.1) and adsorption of the analytes onto the capillary wall.

The limits of detection and quantification, LOQ and LOD, respectively, were determined to be approximately 10 and 25 μg mL^{-1}. The LOQ and LOD were determined according to the signal to noise ratio approach, where signal-to-noise ratios of three and ten correspond to the detection and assay limits, respectively. The determined LOQ was shown to be low enough for the determination of caffeine in the examined dietary products.

9.5.3 Precision

Precision of the method was studied to define the degree of variation between a series of multiple measurements from the same, as well as different samples.

Table 9.2 Linear regression data of a series of caffeine standards analyzed on different days (n ≥6) and in different capillaries. Separation conditions as those given in Figure 9.3. Only five standard solutions were analyzed within each multiple-injection CE run.

Occasion	[a]Concentration range (μg mL[-1])	Slope (RSD)	[b]Intercept (RSD)	R[2] (RSD)
Day 1	500–900 (5 points)	0.0023 ± 4.3%	0.071 ± 71.2%	0.996 ± 0.2%
[c]Day 1	500–900 (5 points)	0.0023 ± 2.8%	0.038 ± 41.2%	0.997 ± 0.2%)
[c]Day 1	500–900 (5 points)	0.0024 ± 5.7%	0.077 ± 70.5%	0.996 ± 0.3%)
Day 6	50–400 (5 points)	0.0025 ± 3.6%	0.017 ± 69.2%	0.999 ± 0.1%
Day 6	500–900 (5 points)	0.0024 ± 2.4%	0.029 ± 21.6%	0.998 ± 0.2%
Day 6	50–900 (10 points)	0.0024 ± 2.7%	0.0077 ± 91.2%	0.999 ± 0.0%
Day 8	50–400 (5 points)	0.0025 ± 0.0%	0.013 ± 0.4%	0.999 ± 0.0%
Day 8	50–900 (10 points)	0.0024 ± 0.0%	0.022 ± 0.4%	0.999 ± 0.0%
[c]Day 10	50–400 (5 points)	0.0024 ± 2.3%	0.011 ± 32%	0.999 ± 0.0%
Day 10	50–900 (10 points)	0.0023 ± 2.7%	0.017 ± 64.4%	0.999 ± 0.0%
Day12	50–400 (5 points)	0.0025 ± 2.6%	0.003 ± 33.5%	0.998 ± 0.4%
Day12	500–900 (5 points)	0.0025 ± 4.6%	0.042 ± 100.5%	0.985 ± 0.4%
Day12	50–900 (10 points)	0.0025 ± 2.6%	0.010 ± 53.5%	0.999 ± 0.1%
[d]Day 30	25–90 (5 points)	0.0023 ± 1.6%	0.006 ± 57.5%	0.996 ± 0.3%
Day 30	100–400 (5 points)	0.0024 ± 3.1%	0.010 ± 7.5%	0.997 ± 0.1%
Day 30	25–900 (15 points)	0.0024 ± 1.8%	0.002 ± 119.0%	0.999 ± 0.0%
[d]Day 60	25–125 (5 points)	0.0022 ± 2.6%	0.005 ± 54.6%	0.994 ± 1.0%
Day 60	200–400 (5 points)	0.0022 ± 5.1%	0.020 ± 103%	0.994 ± 0.3%
Day 60	450–650 (5 points)	0.0024 ± 6.8%	0.06 ± 89.0%	0.993 ± 0.0%
Day 60	25–650 (15 points)	0.0022 ± 2.4%	0.01 ± 10.0%	0.999 ± 0.0%

[a]The concentration range of the caffeine standard solutions is reported in Tables 9.1, 9.3 and 9.5. [b]The absolute intercept values have been reported. [c]The concentration of NaCl was reduced to 10 mM (n = 15). [d]The separation capillary was replaced by a new capillary.

Precision was determined at two levels, i.e., repeatability (intra-day precision) and intermediate precision (inter-day precision).

5.3.1. Repeatability

Five different standard samples of caffeine at concentrations 500 to 900 μg mL-1 were prepared and analyzed in order to evaluate the precision of the multiple-injection CE analyses. Three replicate standard sample preparations were analyzed eight times at each concentration. The multiple-injection CE intra-day precision data obtained from the analysis of the caffeine samples are reported in Table 9.3. The reported data also include variations due to sample preparations. The results showed good precision for migration times and peak area ratios. The precision of the corrected peak areas was, as expected, significantly better than that of the un-corrected peak areas. The relative standard deviation values (RSD%) for the corrected peak areas at the used concentration levels ranged from 1.1% to 2.2%. The large variation in the un-

Table 9.3   Intra- day precision data for peak areas, corrected peak areas and migration times. Separation conditions as in Figure 9.3, (n = 8).

Injection number	*a* Peak number	Concentration ($\mu g \; mL^{-1}$)	$t_{mig(av)}$ (min) $\pm$ RSD%	*b* Peak area $\pm$ RSD%	*c* Peak area ratio $\pm$ RSD%
1	1	500	2.06 $\pm$ 12.6	36626.1 $\pm$ 19.9	1.169 $\pm$ 1.5
	2	900	2.62 $\pm$ 10.3	31318.6 $\pm$ 19.6	
2	3	600	4.21 $\pm$ 6.0	45155.0 $\pm$ 17.9	1.414 $\pm$ 2.2
	4	900	4.78 $\pm$ 5.0	31852.6 $\pm$ 16.3	
3	5	700	6.36 $\pm$ 3.9	46943.6 $\pm$ 25.5	1.649 $\pm$ 2.1
	6	900	6.91 $\pm$ 3.8	28453.7 $\pm$ 25.3	
4	7	800	8.50 $\pm$ 2.9	44013.2 $\pm$ 28.6	1.859 $\pm$ 1.1
	8	900	9.06 $\pm$ 2.8	23691.1 $\pm$ 28.7	
5	9	900	10.64 $\pm$ 2.3	54842.1 $\pm$ 22.7	2.112 $\pm$ 1.8
	10	900	11.19 $\pm$ 2.3	25935.9 $\pm$ 22.2	

a Peaks with odd numbers belong to caffeine. *b* Peak areas were not corrected with respect to their migration times. *c* Peak area ratio = (Caffeine peak area) : (acetaminophen peak area).

corrected peak areas could be explained by fluctuations in the injection volumes. However, the presence of the IM significantly improved the precision of the measurements.

9.5.3.2.   Intermediate Precision

The purpose of the assessment of intermediate precision was to estimate the variation between the determinations accomplished on different days. The operation conditions, such as the buffer composition, the separation temperature and the applied voltage, were not changed. Intermediate precision was assessed by injection of the same samples, being used for the evaluation of repeatability. The caffeine peaks at different concentrations, *i.e.*, 500 to 900 μg mL^{-1}, were integrated and the average peak areas and peak area ratios as well as RSD% values were calculated. The variations in the migration times were also determined. The obtained variations were shown to be less than 4.0% and 15% for the peak area ratios and migration times, respectively, Table 9.4. The major cause of variation in the migration times is the change in electroosmotic flow due to unstable surface conditions of the fused silica capillary wall or small variations in buffer pH and composition. However, the variations in the peak area ratios, being the most important factor for an accurate caffeine assay, were small.

Furthermore, the good agreement between the slope values of the standard curves constructed in different days, as summarized in Table 9.2, demonstrates repeatability and intermediate precision of the method. The variation in the slope values of the constructed lines was less than 10%.

Table 9.4 Intermediate precision data for peak areas, corrected peak areas and migration times. Separation conditions as those reported in Figure 9.3. The reported values were obtained for corrected peak areas and migration times for the same samples, being used for the intra-day precision investigation, on different days (n = 15).

Injection number	^{a}Peak number	Concentration caffeine and acetaminophen ($\mu g\ mL^{-1}$)	$t_{mig(av)}$ (min) $\pm$ RSD%	Peak area ratio $\pm$ RSD%
1	1	500	2.18 $\pm$ 13.3	1.173 $\pm$ 2.2
	*2	900	2.76 $\pm$ 10.9	
2	3	600	4.32$\pm$ 6.4	1.418 $\pm$ 3.2
	*4	900	4.91 $\pm$ 5.3	
3	5	700	6.45 $\pm$ 4.2	1.650 $\pm$ 3.1
	*6	900	7.04 $\pm$ 5.9	
4	7	800	8.53 $\pm$ 6.1	1.864 $\pm$ 1.9
	*8	900	9.17 $\pm$ 5.0	
5	9	900	10.86 $\pm$ 5.3	2.118 $\pm$ 2.8
	*10	900	11.30 $\pm$ 4.4	

^{a}Peaks 2, 4, 6, 8 and 10 belong to acetaminophen (the IM).

9.5.4 Accuracy

Accuracy of the analytical procedure was defined as the proximity between the reference values and those determined experimentally. It was given as the ratio between the determined (C_d) and nominal (C_n) concentrations of caffeine, *i.e.*, $C_d : C_n$%. The standard solutions of caffeine at concentrations ranging from 25 to 900 $\mu g\ mL^{-1}$ were analyzed and the determined concentrations were compared to the nominal concentrations. The results reported in Table 9.5, show that the accuracy of the measurements varied between 98 to 104% with RSD values less than 10%. The results indicated that the method was reliable for the assay of caffeine. It implies that the determined caffeine concentration may deviate from its true concentration by about $\pm$ 5%.

9.5.5 Robustness

In order to explore the robustness of the method, the separation of caffeine and acetaminophen was performed under conditions slightly different from those used in the original method. The robustness study was mainly performed in the single-injection mode instead of multiple-injection CE in order to facilitate interpretation of the electropherograms, *i.e.*, to evaluate t_{PE}, Δt_{mig}, and the number of injections (n_{inj}). The impact of induced changes on the migration times, migration time differences (Δt_{mig}) and separation selectivity ($\alpha = t_{mig2} / t_{mig1}$) were studied. The induced changes affected mainly the EOF and thereby the migration time of the solutes. The affinity of the analyte for the S-β-CD may also be influenced by the induced changes. The migration

Table 9.5   Accuracy of the caffeine determination at different concentrations
(n = 10). Operation conditions as those given in Figure 9.3.

Caffeine Concentration ($\mu g\ mL^{-1}$)	*Accuracy % ($\pm RSD$)*
25	98 ± 6.0%
50	99 ± 5.8%
75	104 ± 2.1%
100	101 ± 3.4%
125	100 ± 2.9%
200	100 ± 2.4%
250	99 ± 2.8%
300	100 ± 2.5%
350	100 ± 1.9%
400	101 ± 1.6%
450	102 ± 2.3%
500	100% ± 2.6%
600	101% ± 1.5%
650	102% ± 2.5%
700	100% ± 1.3%
800	99.4% ± 1.2%
900	100.2% ± 1.3%

times were shorter at higher temperature, *i.e.*, 25 °C, as a result of increased
EOF and decreased viscosity of the BGE. The migration times were longer at
pH 5.3 because of a decrease in the electroosmotic mobility.

The observed variations in the separation parameters did not complicate the
conversion of these modified methods to the multiple-injection mode. The
relatively small changes in the migration times and Δt_{mig} could be compensated
for by the adjustment of the t_{PE} in order to maintain the original number of
injections (*i.e.*, n_{inj} = 5), Table 9.6.

The results illustrate the necessity of performing an initial separation in
single injection mode upon any changes in the separation conditions, *e.g.*,
replacement of the capillary column (Lodén *et al* 2008). The primary analysis
in single injection mode can be considered as a system suitability test.

The robustness of the developed method was further demonstrated by
constructing a standard graph consisting of twenty one different caffeine
concentrations analyzed in different capillaries and days, Figure 9.5.

The results demonstrated that the separation system was robust towards the
induced changes. The slight changes in the separation parameters could be
compensated by adjusting the t_{PE}.

9.6   System Suitability and Acceptance Criteria

The migration time of caffeine in the first injected sample should be between 2
and 4 minutes. A significant drift in the migration times is a result of a change
in the BGE composition, *e.g.*, evaporation of acetonitrile, or in the inner
surface of the capillary causing variations in EOF. The later issue can be

Table 9.6   Robustness study was performed in the single injection mode. Operation conditions are given in the table otherwise as in Figure 9.2.

Separation conditions	[a]*Selectivity factor (α)*	$t_{(mig)}$ *caffeine (min)*	$Δt_{(mig)}$ *(min)*	[b]t_{PE} *(min)*	[b]*n*
Phosphate buffer (50 mM, pH 7.04), 10% ACN, 20 mM NaCl, 15 mg mL^{-1} S-B-CD, T 25 °C	1.06	10.8	1.5	2.0	5
Phosphate buffer (50 mM, pH 7.04), 10% ACN, 20 mM NaCl, 15 mg mL^{-1} S-B-CD, T 20 °C	1.06	11.70	1.5	2.2	5
Phosphate buffer (50 mM, pH 7.04), 10% ACN, 15 mg mL^{-1} S-B-CD, T 20 °C	1.08	9.36	1.4	1.8	5
Phosphate buffer (50 mM, pH 7.04), 5% ACN, 10 mM NaCl, 15 mg mL^{-1} S-B-CD, T 20 °C	1.08	11.45	1.7	2.2	5
Phosphate buffer (50 mM, pH 7.04), 10% ACN, 20 mM NaCl, 10 mg mL^{-1} S-B-CD, T 20 °C	1.04	11.50	1.2	2.2	5
Phosphate buffer (50 mM, pH 5.3), 10% ACN, 15 mg mL^{-1} S-B-CD, T 20 °C	1.07	14.44	2.0	2.5	5

[a]Selectivity factor was calculated by: $α = t_{mig2} / t_{mig1}$. [b]t_{PE} and n were calculated by equations (5) and (6), respectively.

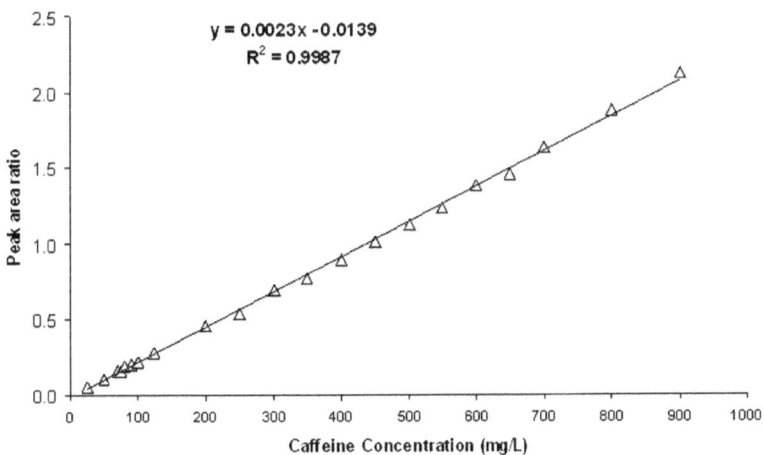

Figure 9.5   A standard graph plotting peak area ratio (Area$_{caffeine}$: Area$_{acetaminophen}$), *versus* caffeine concentration. Standard graph constructed through analysis of twenty standard solutions in the concentration range 25 to 900 mg L^{-1} caffeine, containing acetaminophen at 0.9 mg mL^{-1}. These samples were prepared and analyzed at different occasions within a two-month period.

partially solved by washing the capillary with sodium hydroxide or replacing the capillary tube. The ratio between the y-intercept and the peak area ratio at the midpoint of the standard curve should be less than 10% and correlation coefficient (R^2) should be larger than 0.990.

9.7 Determination of Caffeine in Dietary Products

The caffeine was extracted from the ground coffee and tea leaves with solid/liquid extraction to hot water. Solid/liquid extraction, sometimes called leaching, involves the removal of caffeine from ground coffee, as well as from tea leaves with hot water. During the solid/liquid extraction lower molecular weight compounds, such as caffeine, are dissolved in the hot water and are removed from the high molecular weight protein, cellulose and lipid materials.

The products were accurately weighed and suspended in an appropriate volume of hot water to give a concentration of approximately 100 mg product per mL water. After at least three minutes extraction one volume of the supernatant was mixed with the same volume of acetaminophen solution at a concentration of 1.8 mg mL^{-1}. It was sometimes necessary to further dilute the analyte sample with acetaminophen solution (0.9 mg mL^{-1}) to such an extent that no other peaks or interferences than the caffeine and acetaminophen peaks appeared in the electropherograms, see Figure 9.6. The results from the multiple-injection CE analysis of coffee products showed that the caffeine peak was accompanied with an unknown and partially resolved peak, Figure 9.4 and 9.6 (A). In order to reduce the influence of this peak on the caffeine assay the extract was further diluted prior to analysis.

The multiple-injection CE determinations have been performed through analyzing the sample and reference standards simultaneously (Lodén and Amini 2007). However, in this study, because of relatively small sample capacity (n) of the method, the reference standards and samples were analyzed separately in order to enhance sample throughput. Therefore, a calibration graph was constructed with five appropriate standard concentrations prior to analysis of the dietary samples. Four analyte samples were analyzed in each multiple-injection CE run, Figure 9.6. The purity and identity of caffeine and acetaminophen peaks were performed by UV spectral matches. The UV spectra obtained from the samples agreed with the standard UV spectra, Figure 9.4.

The results of the determination of caffeine content of different dietary products, such as tea, chocolate and coffee are summarized in Table 9.7. The variations in the caffeine content of the tea products (Table 9.7) may depend on the growing conditions, and processing methods. The reported values are, however, not representative for the total caffeine content of these products, since the samples were analyzed immediately after three minutes solid/liquid extraction. The results obtained from electrophoretic analysis of the dietary samples after 30 minutes extraction, revealed that only about 30% of the

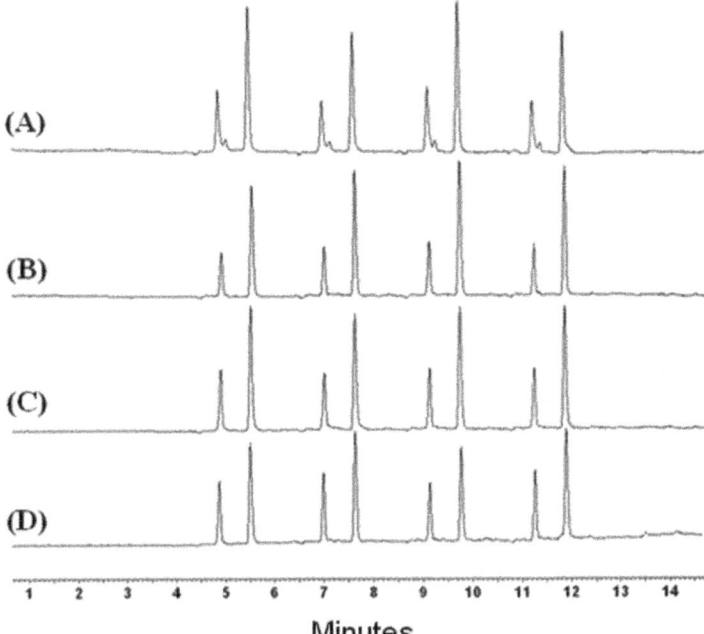

Figure 9.6   Multiple-injection CE mode I analysis of different coffee and tea products. Representative multiple-injection CE mode I analysis of a coffee product (A), tea product G (B), tea product F (C), tea product H (D). Operation conditions as those given in Figure 9.3, with the exception that only four injections were performed. The caffeine was extracted through hot water liquid/solid extraction. The small peak beside the large caffeine peak, in panel (A), was not identified. Operation conditions as those in Figures 9.2 and 9.3.

caffeine content was extracted from the leaves during the first three minute extraction, compare Tables 9.7 and 9.8.

In order to investigate the release profile of the caffeine, the caffeine concentration-time profile was constructed by continuously monitoring the caffeine concentration under 1-hour solid/liquid extraction period. As illustrated in Figure 9.7, an extension of the extraction time up to twenty minutes led to an increase of the caffeine concentration. However, after thirty minutes there was no extraction, indicating that all of the caffeine had already been extracted.

9.8   Conclusions

A simple, rapid, specific and reliable multiple-injection CE method for the determination of caffeine present in dietary products has been developed and validated. The method facilitates separation of caffeine from other compounds present in the dietary sample. The separations were performed using a

Table 9.7   Determination of caffeine in different dietary products by multiple-injection CE (8 determinations per product). These determinations were performed by calculating the caffeine concentration by using six different standard curves, each consisting of fifteen points. Each extract sample was analyzed in duplicate in the multiple-injection CE mode I by performing four injections. The samples were taken after 3.5 minutes solid/liquid extraction to hot water. Other separation conditions as those reported in Figure 9.3.

[a]Dietary product	Caffeine (RSD%) (mg g[-1])
Tea (A)	3.9 ± 4.1%
Tea (B)	5.9 ± 3.3%
Tea (C)	8.8 ± 2.9%
Tea (D)	8.8 ± 4.3%
Tea (E)	6.8 ± 4.3%
Tea (F)	7.4 ± 3.8%
Tea (G)	5.4 ± 3.7%
Tea (H)	9.2 ± 3.0%
Tea (I)	6.2 ± 3.8%
Ground Coffee	11.4 ± 3.4%
Nescafé®	32.3 ± 2.5%
Chocolate powder	3.3 ± 3.6%
Coca-Cola®	120 mg L[-1] ± 5.1%

background electrolyte consisting of 15 mg mL^{-1} sulfated β-cyclodextrin dissolved in 50 mM phosphate buffer at pH 7.0, containing 10% acetonitrile (v/v %) and 20 mM sodium chloride. Acetaminophen was used as an injection marker to ensure precision and accuracy of the determinations. The analyte samples from solid/liquid extraction were directly analyzed by the multiple-injection CE method. The method provided the possibility of analyzing four analyte samples simultaneously. The caffeine concentrations were determined

Table 9.8   Comparison between the caffeine concentration of four different tea products after 3 and 30 min extraction. The assays were performed by the multiple-injection CE (eight determinations were done per product). Separation conditions as those reported in Figure 9.3.

Tea product	[a]Caffeine (RSD%) (mg g[-1] tea)	[b]Caffeine (RSD%) (mg g[-1] tea)
Tea product (B)	6.0 ± 3.3%	17 ± 3.0%
Tea product (C)	9.0 ± 2.9%	22 ± 4.0%
Tea product (G)	5.4 ± 3.7%	21 ± 2.4%
Tea product (H)	9.2 ± 3.0%	24 ± 2.9%

[a]Three minutes extraction. [b]Thirty minutes extraction.

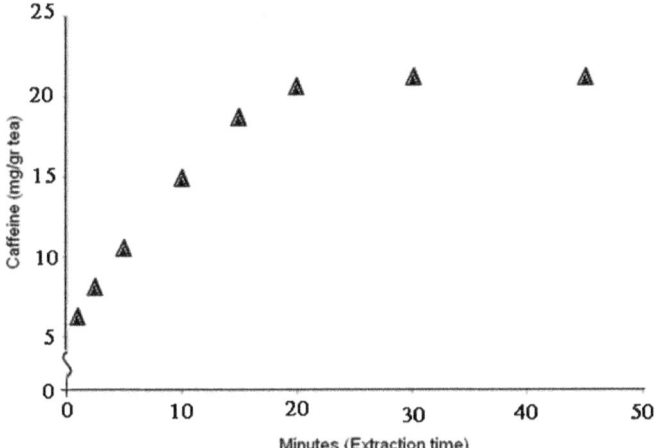

Figure 9.7 Concentration of caffeine extract as a function of the extraction time. Caffeine concentration determined at different solid/liquid extraction time periods. Separation conditions as those given in Figure 9.2. The analyzed dietary product: tea product G.

using a standard curve consisting of at least five points. The concentration of the reference standards was selected on the basis of the extract concentration. The caffeine content of thirteen different dietary products, including nine tea and two coffee products, was determined. The results showed that the concentration of caffeine increased continuously with time during the solid/liquid extraction to reach its maximum concentration after about 20 minutes. It indicates that all the caffeine content may be extracted during this time period.

Summary Points

- This chapter describes the quantitative analysis of caffeine in dietary products, such as tea, coffee and chocolate, by multiple-injection capillary electrophoresis.
- Multiple-injection capillary electrophoresis improves the precision of the assays since the samples to be determined are analyzed under the same conditions.
- With the method developed it was possible to analyse up to five samples within a single electrophoretic run.
- A calibration curve was constructed by analysing simultaneously five standard solutions of caffeine at different concentrations within a single run. Acetaminophen as an injection marker was added to the standard and to the analyte samples to compensate for injection volume fluctuations.
- Caffeine was extracted from ground coffee, chocolate, as well as tea leaves with hot water through solid/liquid extraction.

- The concentration of caffeine in the dietary samples was determined by a calibration curve.
- The caffeine concentration increased with solid/liquid extraction time and the maximum caffeine concentration was reached after about 20 min.
- Chocolate was much lower in caffeine than coffee and tea.

Key Facts

- Capillary electrophoresis in open tubes was first described by Hjertén in 1967 (Hjertén 1967).
- Zone electrophoresis in narrow bore fused-silica capillaries was introduced by Jorgenson and Lukacs in 1981 (Jorgenson 1981; Jorgenson and Lukacs 1981).
- Multiple-injection capillary electrophoresis is a special mode of capillary electrophoresis which enables analysis of more than one sample in a single run.
- An injection marker is added to the analyte samples to compensate for injection volume fluctuations and thereby improving the accuracy and precision of the assays.
- Different modes of multiple-injection capillary electrophoresis are available. Mode I is suitable for components exhibiting small migration mobility differences. This mode was applied to the assay of caffeine in the examined products.

Key Terms

- t_{mig}. Migration time, *i.e.*, the time period it takes for an analyte to migrate from the injection site to the detection site.
- *S-β-CD*. Sulfated β-cyclodextrin is a heterogeneous mixture of β-cyclodextrin molecules with sulfate esters at the C-6 primary and C-2 secondary hydroxyl groups. Sulfated β-cyclodextrin is added to the running buffer to separate caffeine from other components present in the analyte sample.
- t_{PE}. In multiple-injection capillary electrophoresis, each sample is subjected to partial electrophoresis for a short period of time (t_{PE}) immediately after hydrodynamic injection in order to be isolated from the previously injected samples. A correct adjustment of this time is important to avoid inter-plug interferences.
- n_{inj}. Total number of injections performed in multiple-injection capillary electrophoresis.
- n_s. Sample capacity is defined as the highest number of injections which can be performed in multiple-injection capillary electrophoresis ($n_s \geq n_{inj}$).

List of Abbreviations

ACN	Acetonitrile
BGE	Background electrolyte
CE	Capillary electrophoresis
DAD	Diode array detector
Δt_{mig}	Migration time difference
Δn	Difference between n_s and n_{inj}
EOF	Electroosmotic flow
IM	Injection marker
LOD	Limit of detection
LOQ	Limit of quantification
μ_{app}	Apparent electrophoretic mobility
μ_{eff}	Effective electrophoretic mobility
μ_{eo}	Electroosmotic mobility
μ_{ep}	Electrophoretic mobility
Multiple-injection CE	Multiple-injection capillary electrophoresis
n_s	Sample capacity
n_{inj}	Total number of injections
RSD	Relative standard deviation
S-β-CD	Sulfated β-cyclodextrin
SDS	Sodium dodecyl sulfate
t_{mig}	Migration time
t_{PE}	Time of partial electrophoresis
UV	Ultra-violet

References

Altria, K.D. 1993. Essential peak area normalisation for quantitative impurity content determination by capillary electrophoresis. Chromatographia. 35: 177–182.

Amini, A. 2001. Recent development in chiral capillary electrophoresis and applications of this technique to pharmaceutical and biomedical analysis. Electrophoresis. 22: 3107–3130.

Amini, A., Barcley, V., Rundlöf, T., Jönsson, S., Karlsson, A. and Arvidsson, T., 2006. Determination of ephedrine, pseudo-ephedrine and caffeine in a dietary product by capillary electrophoresis. Chromatographia. 63: 143–148.

Amini, A., Lodén, H., Pettersson, C. and Arvidsson, T., 2008. Principles for different modes of multiple injection CZE. Electrophoresis. 29: 3952–3958.

Amini, A., Rundlöf, T., Grön Rydberg, M.-B. and Arvidsson, T., 2004. Characterization of sulfated β-cyclodextrins and determination of enantiomeric purity of (1*R*,2*S*)-ephedrine by capillary zone electrophoresis. Journal of Separation Science. 27: 1102–1108.

Ashton, C. H. 1987. Caffeine and health. British Medical Journal. 295:1293–1294.

Astrup, A., Toubro, S., Cannon, S., Leif Breum, P. H. and Madsen J., 1990. Caffeine: a double-blind, placebo-controlled study of its thermogenic, metabolic, and cardiovascular effects in healthy volunteers. American Journal of Clinical Nutrition. 51:759–67.

Aucamp, J. P., Hara, Y. and Apostolidez, Z., 2000. Simultaneous analysis of tea catechins, caffeine, gallic acid, theanine and ascorbic acid by micellar electrokinetic capillary chromatography. Journal of Chromatography A. 876: 235–242.

Cianchino, V., Acosta, G., Ortega, C., Martinez, L.D. and Gomez, M.R., 2008. Analysis of potential adulteration in herbal medicines and dietary supplements for the weight control by capillary electrophoresis. Food Chemistry. 108: 1075–1081.

Collier, P.D., Mallows, R. and Thomas, P.E., 1972. Interactions between theaflavins, flavanols and caffeine. Phytochemistry. 11: 867–873.

Conte, E. D., Barry, E. F. and Rubinstein H., 1996. Determination of caffeine in beverages by capillary electrophoresis. Journal of Chemical Education. 73:1169–70.

Curatolo, P. W. and Robertson, D., 1983. The health consequences of caffeine. Annals of Internal Medicine. 98:641–53.

Emre, D. and Özaltın, N., 2007. Simultaneous determination of paracetamol, caffeine and propyphenazone in ternary mixtures by micellar electrokinetic capillary chromatography. Journal of Chromatography B. 847: 126–132

Geldart, S. E. and Brown P. R., 1999. Separation of purine and pyrimidine bases by capillary zone electrophoresis with carbonate buffers. Journal of Chromatography A. 831: 123–129.

Heller, C. 1997. Analysis of nucleic acids by capillary electrophoresis. Vieweg Publishing, Wiesbaden

Hjertén, S. 1967. Free zone electrophoresis. Chromatographic Reviews. 9: 122–219.

Horie, H. and Kohata, K., 1998. Application of capillary electrophoresis to tea quality estimation. Journal of Chromatography A. 802: 219–223.

Jimidar, M., Hamoir, T. P., Foriers, A. and Massart, D. L., 1993. Comparison of capillary zone electrophoresis with HPLC for the determination of additives in the foodstuffs. Journal of Chromatography. 636: 179–186.

Jorgenson. J. W.. 1981. Zone electrophoresis in open-tubular glass capillaries. Analytical Chemistry. 53: 1298–1302.

Jorgenson, J. W. Lukacs, K. A.,1981. Free- zone electrophoresis in glass capillaries. Clinical Chemistry. 27:1551–1553.

Juliano, L. M. and Griffiths, R. R., 2004. A critical review of caffeine withdrawal: empirical validation of symptoms and signs, incidence, severity, and associated features. Psychopharmacology. 176:1–29

Jumppanen, J., Sirén, H. and Riekkola, M.-L., 1993. Screening of diuretics in urine and blood serum by capillary zone electrophoresis. Journal of Chromatography A. 652: 441–450.

Krogh, M., Brekke, S., Tønnesen, F. and Rasmussen, K. E., 1994. Analysis of drug seizures of heroin and amphetamine by capillary electrophoresis. Journal of Chromatography A. 674: 235–240.

Larger, P. J., Jones, A. D. and Dacombe, C., 1998. Separation of tea polyphenols using micellar electrokinetic chromatography with diode array detection. Journal of Chromatography A. 799: 309–320.

Lee, C. 2000. Antioxidant ability of caffeine and its metabolites based on the study of oxygen radical absorbing capacity and inhibition of LDL peroxidation. Clinica Chimica Acta. 295: 141–154.

Lilley, K. A. and Wheat, T. E., 1996. Drug identification in biological matrices using capillary electrophoresis and chemometric software. Journal of Chromatography A. 683: 67–76.

Lindeberg, J. 1996. Capillary electrophorsis in food analysis. Food Chemistry. 55:73–94.

Lodén, H. and Amini, A., 2007. Quantification of buserelin in a pharmaceutical product by multiple-injection CZE. Electrophoresis. 28: 1548–1556.

Lodén, H., Pettersson, C., Arvidsson T. and Amini, A., 2009. Quantitative determination of salbutamol in tablets by multiple-injection capillary zone electrophoresis. Journal of Chromatography A. 1207: 181–185.

Mills, M. J., Maltas, J. and Lough, W. J., 1997. Assessment of injection volume limits when using on-column focusing with microbore liquid chromatography. Journal of Chromatography A. 759: 1–11.

Morneau, A., Pialli, V., Nigam, S., Winnik, F. M. and Ziolo, R.F., 1999. Analysis of ferrofluids by capillary electrophoresis .Colloids and Surfaces, A: Physicochemical and Engineering Aspects. 154: 295–301.

Muijselaar, P. M., Claessens, H. A. and Cramers, C. A., 1997. Migration behaviour of micelle counterions in micellar electrokinetic chromatography: influence on micelle mobility, efficiency and selectivity. Journal of Chromatography A. 764: 127–133.

Pedersen-Bjergaard, S., Gabel-Jensen, C. and Hansen, S. H., 2000. Selectivity in microemulsion electrokinetic chromatography. Journal of Chromatography A. 897: 375–381.

Thanh, T. C., Timothy, A., Huq, T. A., Kantes, H. L., Crane, J. N. and Strein, T. G., 1997. Determination of creatinine and other uremic toxins in human blood sera with micellar electrokinetic capillary electrophoresis. Journal of Chromatography B. 690: 35–42.

Thompson, C. O., Trenerry, V. C. and Kemmery, B., 1995. Micellar electrokinetic capillary chromatographic determination of artificial sweeteners in low-Joule soft drinks and other foods. Journal of Chromatography A. 694: 507–514.

Thormann, W., Minger, A., Molteni, S., Caslavska, J. and Gebauer, P., 1992. Determination of substituted purines in body fluids by micellar electrokinetic capillary chromatography with direct sample injection. Journal of Chromatography. 593: 275–288.

Vogt, C. and Conradi, S., 1997. Determination of caffeine and other purine compounds in food and pharmaceuticals by micellar electrokinetic chromatography. Journal of Chemical Education. 74: 1126–1130.

Walker, J. C., Zaugg, S. E. and Walker, E. B., 1997. Analysis of beverage by capillary electrophoresis. Journal of Chromatography A. 781: 481–485.

Zhang, Q.-L., Lian, H.-Z., Wang, W.-H. and Chen, H.-Y., 2005. Separation of caffeine and theophylline in poly(dimethylsiloxane) microchannel electrophoresis with electrochemical detection. Journal of Chromatography A. 1098 : 172–176.

Zhao, Y. and Lunte, C. E. 1997. Determination of caffeine and its metabolites by micellar electrokinetic capillary electrophoresis. Journal of Chromatography B. 688: 265–274.

CHAPTER 10

Determination of Caffeine in Various Coffee Types by Capillary Electrophoresis Through the Anionic Complex with 3,4-Dimethoxycinnamate

THIAGO NOGUEIRA[1] AND CLAUDIMIR LUCIO DO LAGO*[1]

[1] Departamento de Química Fundamental - Instituto de Química, Universidade de São Paulo, Av. Prof. Lineu Prestes, 748 - CEP 05508-000, São Paulo – SP, Brazil
*E-mail: claudemi@iq.usp.br

10.1 Introduction

Electrophoresis has been used for decades as a separation technique of ionic species. Although it can be done on preparative scale, the analytical technique is of particular interest, because of the wide applicability, which has been demonstrated for a great number of analytes and matrices. Due to the inherent Joule heating developed during the electrophoretic running, the efficiency of the separation technique was limited until the small bore silica capillary become available, because the better efficiency in heating dissipation makes possible the use of a high electric field, which improves the number of plates.

Food and Nutritional Components in Focus No. 2
Caffeine: Chemistry, Analysis, Function and Effects
Edited by Victor R Preedy
© The Royal Society of Chemistry 2012
Published by the Royal Society of Chemistry, www.rsc.org

Besides high efficiency of separation, low consumption of sample, and low consumption of reagents, capillary electrophoresis (CE) gained popularity as it may be applied to more and more different classes of substance; among them are the neutral species. Micellar Electrokinetic Chromatography (MEKC) is the most popular approach in this case. However, when we think of the meaning of a neutral species and how we can move it under an electric field in a viscous medium—the basic idea behind electrophoresis—we will realise that it is a very common approach. For instance, when a CE practitioner has to think about the separation of a mixture of carboxylic acids, a straightforward idea comes to mind: to work in a pH around or above the pK_a of these acids to make sure that these species will acquire a partial or full negative charge, allowing the migration process under the electric field. On the other hand, sugars are considered neutral species and the use of CE sounds unlikely, because "CE is for ionic species". However, virtually all sugars become anions above pH around 12. Thus, sugars as well as carboxylic acids are candidate species to be separated by CE, because they may be converted to ions at an appropriate pH value. By the way, the pH control is the most famous way to obtain separations for a great number of substances, from amino acids to proteins and from inorganic anions to sugars or phenols.

Although powerful in electrophoresis, the pH control is not the key to separate caffeine from other neutral constituents of the matrix, because caffeine remains as a neutral species for a wide range of pH. Differently from what happens for theobromine and theophylline—two other important xanthines—caffeine has neither a good nucleophilic nor an electrophilic group available in its structure. Thus, a different approach must be used in this case.

Capillary electrophoresis has been previously used in quantification of caffeine. One approach is to use the migration process to remove ionic interferents from the sample. Basically, caffeine is carried by the electroosmotic flow (EOF) from the injection side of the capillary until the detector, while ionic species migrate way from the sample region, leaving caffeine as the unique detectable species there. Of course, such an approach has strong limitations, because only under specific conditions can one be assured that only caffeine will behave like that.

A better approach is MEKC, where the interaction of caffeine with ionic micelles is used to allow caffeine to migrate under an electric field. Sodium dodecyl sulphate (SDS) forms negative micelles and has been proposed to obtain separation of caffeine from other neutral drugs (Hyotylainen *et al* 1996). A microemulsion formed with SDS and heptane was also demonstrated to be useful for this task (Huang *et al* 2005). In both cases, one should expect that the partition will occur by interaction with the surface of the micelle, because although it is neutral, caffeine is polar and should not have great affinity for the apolar inner part of the micelles.

The spectrophotometric detection of caffeine and other xanthines is possible due to the absorption in the UV region. Since SDS and other common components of the background electrolyte (BGE) do not absorb in the same region, the UV detection in CE is possible. However, another growing approach for detection in CE does not give the same good compatibility: capacitively coupled contactless conductivity detection (C^4D). Although previously used as a detection scheme for MEKC, we are not able to reach the desirable limit of detection of caffeine in different matrices by using C^4D. The foundations for this kind of detection will be discussed later, but in a few words the point is the occurrence of a noisy baseline caused by the high conductivity BGE and the small difference caused by caffeine in the mobility of the micelles.

The seeking of another way to promote the electrophoretic mobility of caffeine resulted in the π-complex formation with chlorogenic acids or their analogues. Such a complex occurs naturally in plants that synthesise this complex as a way to store caffeine. Along the next sections, the CE-C^4D is discussed as well as the experimental conditions and results obtained in this new approach to separate neutral species in electrophoresis.

10.2   Capillary Electrophoresis with Capacitively Coupled Contactless Conductivity Detection

There are many good text books and review articles dealing with capillary electrophoresis (Altria 1996; Camilleri 1997) with different detection systems, including C^4D (Brito-Neto *et al* 2005a; 2005b; Kuban and Hauser 2004a; 2004b). Of course, the purpose of this chapter is not to cover in detail all the aspects of this technique, but to show to the reader the most relevant features to improve the understanding about how it is possible to determine the concentration of caffeine, a neutral species, by capillary electrophoresis, a well-known technique for ions (Table 10.1).

Table 10.1   Key features of CE-C^4D. This table lists some fundamental aspects about capillary electrophoresis and particularly how detection occurs.

1. Separation by electrophoresis is not only for separation of naturally charged species. There are plenty of chemical reactions that can be used to change neutral species into ions previously or during the electrophoretic run.
2. While migrating under the application of an electric field, a species changes the chemical composition of that region and, consequently, the electric field there. Thus, an analyte changes the local concentration of the co-ion and counter ion.
3. The perturbation of the concentration profile of all ionic species is the base for conductivity detection as well as UV indirect detection. In this last case, the co-ion or the counterion is an absorbing species, and negative peaks are formed when analytes reach the detection point.
4. Contactless conductivity detection is an advantageous approach, because it prevents corrosion of the electrodes, eases the electrode positioning, and precludes the contact of the detector electronics with the inner side of the capillary, which can be at as high as tens of kilovolts.

Mobility is a fundamental concept in electrophoresis. When submitted to an electric field (E), a charged particle develops a constant velocity (v) in a viscous medium, and mobility (μ) is the proportionality constant:

$$v = \mu E \tag{1}$$

Electrophoresis as a separation technique is possible, because different species have different mobilities in different media. Although the Stokes' relation was developed for large molecules and particles, it is still reasonable to derive a model from it to describe the mobility of small ions:

$$\mu = \frac{ze}{6\pi\eta r} \tag{2}$$

where z and r, respectively, are charge and hydrodynamic radius of the ion, η is the viscosity of the medium and e is the elemental charge. This equation shows the importance of z-to-r ratio to the mobility and then to the velocity of the different ions of a mixture.

Another important concept is the effective mobility, which is the mobility of a species that is engaged in chemical equilibriums that may change the size and–or the charge of the ion. If the direct and indirect reactions are fast enough when compared to the time scale of an electrophoresis experiment, the species behaves as a new species with intermediary charge and size. The effective mobility may be modelled by:

$$\mu_{eff} = \sum_i \alpha_i \mu_i \tag{3}$$

where α_i is the molar fraction for each species i involved in the equilibrium. The effective mobility is the reason why, for example, one can separate isomers like o-, m-, and p-toludine, which are species with the same mobility when protonated, but with different pK_a. The right choice of the pH of the BGE allows these isomers to develop different effective mobilities, because of the different values of molar fractions of the protonated isomers.

When observed in a capillary electrophoresis experiment, the analytes seem to behave, most of the time, as species with a different mobility. The cause is the electroosmotic flow, which is generated by the permanent charge at the inner wall of the capillary. A satisfactory explanation about the EOF is beyond the scope of this text. To put in simple words, the result is a continuous pumping process of the solution from one to the other reservoir during which time the electric field is kept constant. This pumping process carries neutral species from injection point to the detector and may contribute positively or negatively to the migration of the charged species. Thus, when EOF is present, species migrate with an apparent mobility (μ_{app}), which is given by:

$$\mu_{app} = \mu_{eff} + \mu_{EOF} \tag{4}$$

where μ_{EOF} is a positive or negative value, depending on the chemistry composition of the of the inner surface of the capillary and the solution.

Based on equations (10.1) to (10.4), one can note that the analytes develop different velocities, which allows the separation along the path from the injection point to the detector. At this point, different properties can be explored in order to allow the identification and quantification of the components. UV-visible light absorption seems to be a straightforward approach for analytes with a significant molar absorptivity in this region. Conductivity detection, on the other hand, may cause some confusion, because it depends on the composition of the solution as a whole, which is modified along the electrophoresis process.

There are some articles dealing with the phenomena and the instrumentation for conductivity detection and particularly with C^4D (Brito-Neto *et al* 2005a; 2005b; Katzmayr *et al* 1999; Kuban and Hauser 2004a; 2004b). Again, a deep understanding about the matter is out of scope of this text. Thus, we select some fundamental concepts for better understanding of the next sections.

Equation (10.1) shows that the velocity of an ion depends on the electric field present in the region where this ion is migrating to. An interesting fact about the migration process of the different ions from the BGE and from the sample is that it changes the chemical composition and, consequently, the conductivity of that region and the electric field. In other words, the electric field causes the migration process, which by its turn modifies the electric field. This is the reason why a simple explanation about the concentration profiles of all ionic species along the capillary is not an easy task. Simulations using programs, such as Simul developed by Gas' group (Hruska *et al* 2006), gives a better insight about the complex events that rule the chemical composition, which is also the base for a right understanding of the conductivity detection. For this moment, a simple model is enough.

Considering monovalent and full-dissociated ions, the sensitivity of a conductivity detector may be described by (Brito-Neto *et al* 2005a; Katzmayr *et al* 1999):

$$\frac{d\kappa}{dc_a} = \frac{(\mu_A - \mu_S)(\mu_A - \mu_O)}{\mu_A} F \tag{5}$$

where $d\kappa/dc_a$ is the variation of conductivity with the analyte concentration, μ_A, μ_S, and μ_O are the mobilities of the analyte, its co-ion, and its counterion at the BGE, respectively, and F is the Faraday's constant.

Equation (10.5) gives some important directions for planning the use of conductivity detection. One can note that the higher the difference of mobility between the analyte and the co-ion, the higher the sensitivity. In this case, the difference may be positive or negative, which is responsive by the formation of positive or negative peaks, respectively. Another issue to be considered is the conductivity of the BGE, because the higher the conductivity, the higher the noise, which obviously impacts the signal-to-noise ratio and consequently, the limit of detection.

The basic difference between a classic conductivity measurement and C^4D is that the second approach is not based on Faradaic contact of the electrodes with the solution to be evaluated. There is a dielectric layer—more specifically, the silica capillary wall—that prevents this contact, which is beneficial because electrode corrosion is also prevented and the high electric potential inside the capillary does not destroy the electronics of the conductivity meter.

10.3 The Formation of π-Complexes of Caffeine and Chlorogenic Acids and Analogues

Caffeine is produced by many plants as a pesticide, as a substance that discourages insect feeding and also as an inhibitor for the germination of competitors (Baumann and Gabriel 1984; Nathanson 1984). Caffeine can permeate the membrane of the cells of the plant that provides its synthesis, and remains inside the cell as a complex with the chlorogenic acid (Waldhauser and Baumann 1996). A chlorogenic acid is an ester of *trans*-cinnamic acid and quinic acid, 5-*O*-caffeoylquinic acid being the most common (Clifford 1999).

The first observation of the caffeine-chlorogenate complex has been described at the beginning of last century (Gorter 1908). Six decades after that, the behavior of solutions of caffeine and chlorogenic acids as well as with some analogues was studied (Horman and Viani 1972), and they concluded that the structure is an 1 : 1 π-complex, which is formed between the flat molecule of caffeine and the flat region of the chlorogenic acid, *i.e.*, the cinnamic acid portion of this molecule. In this study, based on NMR spectra, the authors calculated the association constant of caffeine and some chlorogenic acid analogues, which inspired the approach used to develop a capillary electrophoresis method published recently (Nogueira and do Lago 2007).

10.4 The Ionic π-Complex with Cinnamic Acid Derivatives as A Mobile Version of Caffeine and the Resulting Electrophoretic Method

Taking into account the possibility of forming a π-complex with caffeine as shown on the previous section, one can easily envision an electrophoretic method based on the formation of an anionic or cationic structure, which can be manipulated by applying a convenient electric field. Obviously the size of such a structure is bigger than caffeine alone, but the effective charge makes the mobility of the complex different from zero (Equation (10.2)).

The thermodynamic data suggest that more than one of the complexes studied could be used (Horman and Viani 1972). However, no kinetic data were available. Taking into account the concept of effective mobility (Equation

(10.3)), one can note that there are two limiting conditions that can be accepted in the method development:

(1) The new species formed by complexation is not labile. In such a case, one could prepare a mixture of the sample with the reagent to form the ionic complex.

(2) The formation and decomposition reactions of the complex are fast enough when compared with the electrophoretic experiment, which consumes some minutes to be concluded. In this case, Equation (10.3) allows predicting that the effective mobility of the complex can be controlled by the concentration of the complexing agent used in the composition of the BGE.

A preliminary study suggested that the second condition is obeyed. Thus, the next step is the choice of the cinnamate, which presents an appropriate stability constant and attends some practical requirements.

Although chlorogenic acid is the original complex found in plants, one can argue that it is not a good candidate to be included in the BGE of a new electrophoresis method, because it is too expensive for routine analysis. Of course, the reagent consumption is an advantageous feature of CE, and chlorogenic acid could be used despite its cost. However, the study by Horman and Viani clearly shows that the quinic moiety is not essential to the complex formation. For instance, the association constants for chlorogenate and caffeate are, respectively, 16.9 and 12.2 kg mol^{-1} (Horman and Viani 1972). Thus, caffeic acid seems to be natural choice, because it is the cinnamate moiety in the chlorogenic acid structure.

The experiments with caffeic acid were carried out with a homemade CE-C^{4}D equipment (da Silva and do Lago 1998; da Silva *et al* 2002) with a 60 cm long fused silica capillary and 75 μm of inner diameter. The voltage applied was +25 kV, which results in an electric field of 417 V cm^{-1}. This is the average electric field that allowed the species to migrate from the injection point to the detector positioned 50 cm downstream. The samples were hydrodynamically injected at 10 cm H$_2$O by 30 s.

Figure 10.1 shows an electropherogram obtained for a synthetic sample of caffeine (500 mg L^{-1}) using caffeate as the complexing agent. The BGE was composed by 30 mmol L^{-1} caffeic acid and pH adjusted to 8.0 with Tris. The carboxylic group of caffeic acid has pK$_a$ 4.44, while the phenolic groups have pK$_a$ values of 7.6 and 11.85 (Jovanovic *et al* 1994). Thus, the carboxylic group should be completely dissociated at pH 8.0, while one of the phenolic groups has a partial negative charge. Thus, the complexing agent should have a net negative charge about 1.5.

The remarkable transient observed on the electropherogram from Figure 10.1 (between 5 and 6 min) is related to the EOF transportation of the region occupied originally by the sample at the end of the capillary. The cationic species migrate faster and so should be recorded before this time. On the other hand, anionic species have migration times greater than it. Thus, the

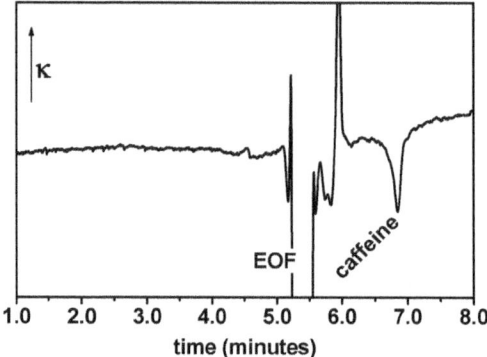

Figure 10.1   Electropherogram of a standard solution of caffeine using caffeic acid in
the BGE. Electropherogram obtained for a synthetic sample of caffeine
(500 mg L^{-1}). Background electrolyte composed by 30 mmol L^{-1}
caffeic acid and pH adjusted to 8.0 with Tris. Figure adapted (Nogueira
and do Lago 2007).

negative peak attributed to the caffeate-caffeine π-complex agrees with the
predicted behaviour for this species, because its migration time demonstrates
that it is an anionic species and the negative peak suggests that this complex is
bigger than caffeate alone.

Despite the effective mobility of the caffeate-caffeine complex, caffeic acid
share with chlorogenic acid a significant disadvantage: both of them are
catechols, which means that they are prone to oxidation by oxygen that is
naturally absorbed from the air. Thus, the complexing agent was substituted
by 3,4-dimethoxycinnamic acid, which is less expensive, more stable, and has
an even higher association constant.

Figure 10.2 shows the electropherogram when caffeic acid is substituted by
3,4-dimethoxycinnamic acid in the BGE. In this case, for sake of reference, the
sample contained also theobromine (200 mg L^{-1}) and theophylline (100 mg
L^{-1}) besides caffeine (200 mg L^{-1}). In this case, the pH of the BGE was
adjusted to 8.5 and the acid concentration was reduced from 30 to 20 mmol
L^{-1}. Both of these changes augment the EOF mobility and consequently
reduce the total time of analysis. The migration time for caffeine was reduced
from 6.9 to 3.9 min.

10.5   Validation of the Method

Taking the solution containing 20 mmol L^{-1} 3,4-dimethoxycinnamic acid and
pH adjusted to 8.5 with Tris as the BGE, a great number of experiments were
carried out to determine some important figures of merit.

The relative standard deviation (RSD) for the migration time was 0.18%,
which suggests that the caffeine peak can be identified in real samples within
0.4 s of variation of the migration time.

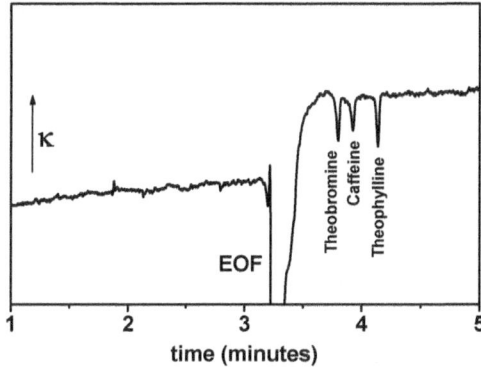

Figure 10.2   Electropherogram of a standard solution of caffeine, theobromine, and theophylline using 3,4-dimethoxycinnamic acid in the BGE. Electropherogram obtained for a synthetic sample containing theobromine (200 mg L^{-1}), theophylline (100 mg L^{-1}), and caffeine (200 mg L^{-1}). Background electrolyte composed by 20 mmol L^{-1} 3,4-dimethoxycinnamic acid and pH adjusted to 8.5 with Tris. Figure adapted (Nogueira and do Lago 2007).

The RSD for peak area was 1.1%, and an intermediate precision evaluation using real coffee samples resulted in a global RSD of 3.7%. The 5-points analytical curve from 100 to 500 mg L^{-1} had r = 0.9998. The limit of detection based on signal-to-noise ratio (SNR = 3) was 33 mg L^{-1} for caffeine in the injected sample, which corresponds to 0.8 and 0.3 mg g^{-1} of dry matter of instant coffee and ground coffee, respectively. The different values are due to the different preparation of the beverage: 1.0 g of ground coffee or 0.4 g of instant coffee for 10 mL of water.

The recovery percentage and comparison with high performance liquid chromatography (HPLC) were used to evaluate the accuracy of the new method. Instant coffee (4 samples) and ground coffee (2 samples) were spiked with two levels of caffeine concentration: 50 and 100 mg L^{-1}. The recovery was between 92 and 109% for the two concentration levels and for coffees of different origins.

The same 6 samples of coffee of different origins were analysed by both CE and HPLC. The HPLC experiments were based on the method developed by Trugo and co-workers (Trugo *et al* 1983), using C18 column, water : methanol 60 : 40 (v/v) as mobile phase, and UV detection at 272 nm. The concentrations obtained were around 15 mg g^{-1} for ground coffee and 30 mg g^{-1} of instant coffee, and a good agreement between the two methods was obtained. The RSDs were systematically smaller for HPLC with UV detection. However, the CE runs were done in half of the time and obviously generated a lesser amount of residue.

10.6 Decaffeinated Coffee and the Need For a Better Limit of Detection

The industrial processes for removing caffeine from coffee products are not quantitative, and a minimum level is acceptable depending on the country. For instance, the Brazilian norms (Anvisa Resolution RDC number 277 from September 22, 2005) state that maximum limits for caffeine are 0.1% and 0.3% (w/w) for decaffeinated ground and instant coffee, respectively. The above reported limit of detection is enough attend this norm. However, for good quantitation, a preconcentration step is needed, because otherwise only a positive or negative answer could be given since the limits stated in the norm and the limit of detection are too close.

The preconcentration was obtained by liquid/liquid extraction using dichloromethane. Initially, the pH of the coffee solution was adjusted to 10 using NaOH 1 mol L^{-1} to dissociate organic acids also available in coffee and diminish the preconcentration of them in the organic phase. The initial 10 mL sample is shaken three times with 20 mL of dichloromethane. After evaporation of the organic solvent, the residual solid is solubilized in 1 mL of deionized water, which results in a 10 : 1 preconcentration factor.

The recovery tests were between 89 and 97%, and the electropherograms (Figure 10.3) show that the preconcentration did not cause enrichment of co-migrating species. In fact, the whole profile is quite similar to the one obtained for regular coffee.

Although the improvement of sensitivity allows quantitation of caffeine in decaffeinated products, there are two drawbacks: (1) the preconcentration step is time consuming and (2) it needs dichloromethane, which is eliminated by evaporation at the end of the procedure.

This subject was revisited after the release of a new version of the C^4D (Francisco and do Lago 2009). This new version uses a compact electronics and allows obtaining electropherograms with a better signal-to-noise ratio. The limit of detection was reduced from 33 mg L^{-1} to 5.4 mg L^{-1}, which is enough to eliminate the preconcentration step. However, there is yet an advantage in using liquid/liquid extraction: the better reproducibility in migration time. The great number of substances present in coffee contributes negatively in two different ways: (1) the ionic species of the beverage contribute to the conductivity of the sample, which impairs the initial mobilization of the analyte; and (2) cationic species (remarkably proteins and cationic amphiphilic species) can adsorb on the inner wall of the capillary and change the EOF, which impacts the apparent mobilities of the species. The pH elevation and solubilization steps in dichloromethane and water diminish the content of such species, which is beneficial to the reproducibility of the migration time. Nevertheless, considering the trade-off between the migration time reproducibility and the improvement of the limit of detection without dichloromethane, we believe that this second approach is the best one.

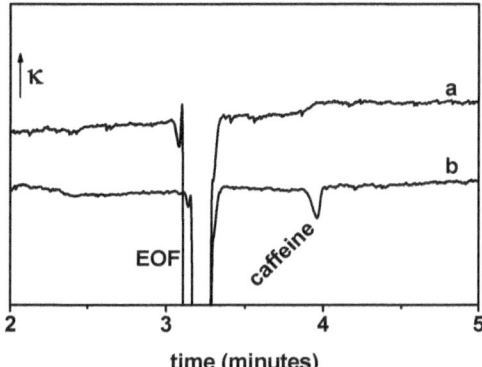

Figure 10.3 Electropherograms of decaffeinated instant coffee with and without preconcentration. Electropherograms obtained for a sample of decaffeinated instant coffee without (a) and with (b) preconcentration step (enrichment factor of 10 times). The separation was carried out with BGE composed 20 mmol L^{-1} 3,4-dimethoxycinnamic acid and pH adjusted to 8.5 with Tris. Figure adapted (Nogueira and do Lago 2007).

10.7 Conclusion

The formation of a π-complex seems to be a promising approach to the mobilization not only for caffeine and other xanthines, but for other classes of compounds that have planarity and conjugation of double bounds along the structure. This feature of the molecular structure is closely related to the absorption in the UV-visible region, which suggests that spectrophotometric detection is not a natural option for detection, because of the high baseline value due to the cinnamic group, with small absorbance variation due to the complex formation. However, as the C^4D becomes more and more popular, the analytical method here described and similar proposals become competitive with other methods, such as HPLC. In this particular case, precision is not the strongest point for CE, but rather the small consumption of sample and reagents, and the small amount of waste and shorter analysis time.

Summary Points

- This chapter is focused on a capillary electrophoresis method for caffeine quantitation in coffee products.
- Caffeine is a neutral species, but it can form π-complexes with some cinnamic acid analogues and then assume a net charge.
- The resulting π-complex allowed the separation of caffeine from other ionic and non-ionic species, including other xanthines.
- Capacitively coupled conductivity detection was the key to the development of the method, because it allows distinguishing species with low difference of molar absorptivity, which is typically needed in UV detection.

- The limit of detection of the methods is compatible with the requirement of the Brazilian norm about decaffeinated products.

Key Facts

- Pi-complexes are formed because of the planar structure of two molecules and the interaction between π-electrons as well as polar and hydrogen bonds.
- Caffeine is stored inside the plant cells as a complex with chlorogenic acid. When released, caffeine can permeate through the membrane.
- The π-complexes of caffeine and cinnamate analogues are bigger than caffeine alone, which reduces the diffusion coefficient. However, as a charged species, it can be now conducted by an electric field.

Definitions of Words and Terms

Electroosmotic flow (EOF): Flow of solution originated when the surface has a zeta potential difference of zero and a significant electric field is applied. In the electrophoresis context, it is originated by the interface between the solution and the inner wall of the capillary.

Micellar Electrokinetic Chromatography (MEKC): Electrophoretic technique in which ionic micelles are used to mobilize neutral analytes.

Background electrolyte (BGE): The electrolyte solution that is used to fill the capillary and the reservoirs in an electrophoresis experiment. Frequently, this solution is also a buffer.

Co-ion: The BGE ion with the same charge of the analyte.

Counterion: The BGE ion with the opposite charge of the analyte.

Joule heating: Heating caused when an electrical current flows through a body with resistivity greater than zero. In capillary electrophoresis, although the heating power is typically below one watt, the volume of solution is so small that Joule heating can cause evaporation of the solution inside the capillary.

Mobility: The proportional constant that describes the velocity of a charged particle in a viscous medium as a function of the electric field applied on it.

Ionic mobility: *natural mobility*, or *intrinsic mobility*, The mobility of an ion completely free and fully dissociated in a medium.

Effective mobility: The average mobility of ionic species involved in a set of fast equilibriums.

Apparent mobility: The result of the combination of effective mobility and the EOF mobility.

EOF mobility: The proportionality constant that describes the velocity of the EOF as a function of the electric field applied on the capillary.

Hydrodynamic radius: It is the apparent radius of a sphere that encompasses the central molecule or ion and its solvation shell.

List of Abbreviations

BGE	background electrolyte
C^4D	capacitively coupled contactless conductivity detection
CE	capillary electrophoresis
EOF	electroosmotic flow
HPLC	high performance liquid chromatography
MEKC	micellar electrokinetic chromatography
NMR	nuclear magnetic resonance
RSD	relative standard deviation
SDS	sodium dodecyl sulphate
Tris	tris(hydroxymethyl)aminomethane
UV	ultraviolet

References

Altria, K. D. 1996. Capillary Electrophoresis Guidebook: Principles, Operation, and Applications. Humana Press Inc., New Jersey, 349 pp.

Baumann, T. W. and Gabriel, H., 1984. Metabolism and excretion of caffeine during germination of *coffea-arabica* L. Plant and Cell Physiology. 25(8): 1431–1436.

Brito-Neto, J. G. A., da Silva, J. A. F., Blanes, L. and do Lago, C. L., 2005a. Understanding capacitively coupled contactless conductivity detection in capillary and microchip electrophoresis. Part 1. Fundamentals. Electroanalysis. 17(13): 1198–1206.

Brito-Neto, J. G. A., da Silva, J. A. F., Blanes, L. and do Lago, C. L., 2005b. Understanding capacitively coupled contactless conductivity detection in capillary and microchip electrophoresis. Part 2. Peak shape, stray capacitance, noise, and actual electronics. Electroanalysis. 17(13): 1207–1214.

Camilleri, P. 1997. Capillary electrophoresis : theory and practice. CRC Press, Boca Raton, 552 pp.

Clifford, M. N. 1999. Chlorogenic acids and other cinnamates - nature, occurrence and dietary burden. Journal of the Science of Food and Agriculture. 79(3): 362–372.

da Silva, J. A. F. and do Lago, C. L., 1998. An oscillometric detector for capillary electrophoresis. Analytical Chemistry . 70(20): 4339–4343.

da Silva, J. A. F., Guzman, N. and do Lago, C. L., 2002. Contactless conductivity detection for capillary electrophoresis - Hardware improvements and optimization of the input-signal amplitude and frequency. Journal of Chromatography A. 942(1–2): 249–258.

Francisco, K. J. M. and do Lago, C. L., 2009. A compact and high-resolution version of a capacitively coupled contactless conductivity detector. Electrophoresis. 30(19): 3458–3464.

Gorter, K. 1908. Contributions to the knowledge of coffee. Justus Liebigs Annalen der Chemie. 358(1/3): 327–348.

Horman, I. and Viani, R., 1972. Nature and conformation of caffeine-chlorogenate complex of coffee. Journal of Food Science. 37(6): 925–&.

Hruska, V., Jaros, M. and Gas, B., 2006. Simul 5 - free dynamic simulator of electrophoresis. Electrophoresis. 27(5–6): 984–991.

Huang, H. Y., Lien, W. C. and Chiu, C. W., 2005. Comparison of microemulsion electrokinetic chromatography and micellar electrokinetic chromatography methods for the analysis of phenolic compounds. Journal of Separation Science. 28(9–10): 973–981.

Hyotylainen, T., Siren, H. and Riekkola, M. L., 1996. Determination of morphine analogues, caffeine and amphetamine in biological fluids by capillary electrophoresis with the marker technique. Journal of Chromatography A. 735(1–2): 439–447.

Jovanovic, S. V., Steenken, S., Tosic, M., Marjanovic, B. and Simic, M. G., 1994. Flavonoids as antioxidants. Journal of the American Chemical Society. 116(11): 4846–4851.

Katzmayr, M. U., Klampfl, C. W. and Buchberger, W., 1999. Optimization of conductivity detection of low-molecular-mass anions in capillary zone electrophoresis. Journal of Chromatography A. 850(1–2): 355–362.

Kuban, P. and Hauser, P. C., 2004a. Fundamental aspects of contactless conductivity detection for capillary electrophoresis. Part I: Frequency behavior and cell geometry. Electrophoresis. 25(20): 3387–3397.

Kuban, P. and Hauser, P. C., 2004b. Fundamental aspects of contactless conductivity detection for capillary electrophoresis. Part II: Signal-to-noise ratio and stray capacitance. Electrophoresis. 25(20): 3398–3405.

Nathanson, J. A. 1984. Caffeine and related methylxanthines - possible naturally-occurring pesticides. Science. 226(4671): 184–187.

Nogueira, T. and do Lago, C. L., 2007. Determination of caffeine in coffee products by dynamic complexation with 3,4-dimethoxycinnamate and separation by CZE. Electrophoresis. 28(19): 3570–3574.

Trugo, L.C., Macrae, R. and Dick, J., 1983. Determination of purine alkaloids and trigonelline in instant coffee and other beverages using high-performance liquid-chromatography. Journal of the Science of Food and Agriculture. 34(3): 300–306.

Waldhauser, S. S. M. and Baumann, T. W., 1996. Compartmentation of caffeine and related purine alkaloids depends exclusively on the physical chemistry of their vacuolar complex formation with chlorogenic acids. Phytochemistry. 42(4): 985–996.

CHAPTER 11

Analysis of Caffeine and Related Compounds by Automated Flow Methods

PARASKEVAS D. TZANAVARAS*[1] AND
CONSTANTINOS K. ZACHARIS[1,2]

[1] Laboratory of Analytical Chemistry, Department of Chemistry, Aristotelian University of Thessaloniki, GR-54124, Greece; [2] Department of Food Technology, School of Food Technology and Nutrition, Alexander Technological Educational Institute (ATEI) of Thessaloniki, GR-57400, Greece
*E-mail: ptzanava@chem.auth.gr

11.1 Introduction

Caffeine, (1,3,7-trimethylxanthine), is a member of the methylxanthine compounds (theophylline, theobromine, paraxanthine, *etc.*) and it is classified as purine alkaloid.

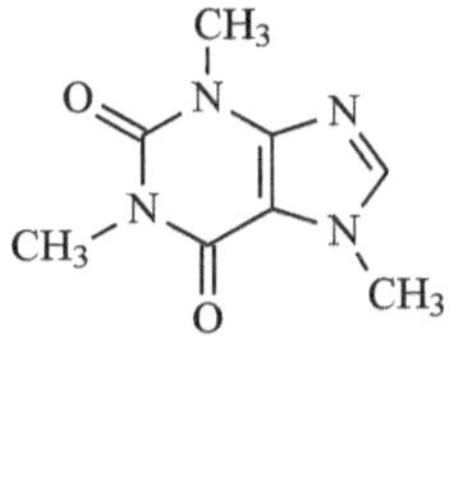

Scheme 11.1

Food and Nutritional Components in Focus No. 2
Caffeine: Chemistry, Analysis, Function and Effects
Edited by Victor R Preedy
© The Royal Society of Chemistry 2012
Published by the Royal Society of Chemistry, www.rsc.org

Historically, caffeine was first isolated from coffee by a German chemist Friedlieb Ferdinand Runge in the early 1820s. One year later, French researchers (Robiquet, Pelletier, Caventou) who were not aware of Runge's work also isolated caffeine from coffee beans. From an etymology point of view, the word "caffeine" derives from the French word "café".

Caffeine is a compound that occurs naturally in over 60 different types of plants *e.g.* coffee, tea, cacao pod, guarana, *etc.* However, there are several types of common food, beverages and medications around the world that contain caffeine, including coffee (brewed–drip, instant, espresso, *etc.*), tea (brewed, instant, cocoa, hot chocolate, *etc.*), chocolate bar (milk, dark) and caffeine-containing pharmaceutical formulations (analgesics, stimulants, weight-loss products, *etc.*). For example, in some of these consuming products, caffeine levels vary from 50–100 mg 100 mL^{-1} (brewed coffee) and 20–73 mg 100 mL^{-1} (instant coffee and tea). Cola-type beverages and chocolate bars usually contain 5–20 mg 100 mL^{-1} or g^{-1}, respectively. Finally, pharmaceutical formulations include caffeine in the range of 15–200 mg table depending of the type (Dlugosz and Bracken 1992).

In North America and in some European countries (Sweden, Finland, Switzerland), coffee and tea are the major sources of caffeine in the adult diet, while caffeinated soft drinks and chocolate are the major sources of caffeine in the diet of children (Nawrot *et al* 2003). In the developed countries, it is considered that the adults consume on average 200 mg caffeine on a daily basis from coffee, tea, and beverages. However, in Northern European countries the daily intake often exceeds this to 400 mg. Studies have showed that 30 mg or less of caffeine can alter self-reports of mood and affect behavior and 100 mg per day can lead to physical dependence and withdrawal symptoms upon abstinence (Griffths *et al* 2003).

Consumers and the scientific community have expressed concern about the potential for caffeine to produce adverse effects on human health. The consumption of caffeine in moderate amounts (*e.g.* 400 mg per day) increases the energy availability, alertness, wakefulness, and feelings of "energy" and decreases fatigue. On the other hand, higher amounts may lead to "caffeine toxicity" which is accompanied with a variety of clinical symptoms including nervousness, irritability, insomnia, arrhythmia, tachycardia, diuresis, gastrointestinal disturbances, *etc.* (Nawrot *et al* 2003 and Glade 2010). In the human body, caffeine is rapidly absorbed *via* the gastrointestinal path. The highest concentration of caffeine in blood is usually observed between 15 to 120 min . Caffeine intake of 5–8 mg kg^{-1} body weight results in concentrations of 8–10 mg L^{-1} in plasma. Looking to its metabolic pathway in the human body, *ca.* 80% of caffeine is metabolized to paraxanthine while 16% is converted to theophylline and theobromine (Higdon and Frei 2006).

11.2 Automated Flow Injection Techniques

In the following sections three basic automated flow-based analytical techniques are discussed briefly; Flow Injection Analysis (FI), Sequential Injection Analysis (SI) and Multicommutated flow systems.

11.2.1 Flow Injection Analysis

The concept of Flow Injection Analysis (FI) was originally proposed by Ruzicka and Hansen in 1975 as an alternative approach to perform chemical analysis under flow conditions (Ruzicka and Hansen 1975). Nowadays, FI is a well-established and mature analytical technique with more than 12 000 FI papers having been published in scientific journals. A simplified definition of flow injection analysis might be, "FI is an automated, continuous flow approach to perform chemical analysis". It is based on injecting a small, well-defined volume of sample, into a continuously flowing carrier stream to which appropriate auxiliary reagent streams can be added.

The main feature of FI that differentiates this technique from previous Segmented Flow Analyzers (SFA) and of course from batch methodologies, is its kinetic character. Due to precise timing and geometry, FI does not rely on complete mixing of sample and reagent(s) (physical homogenization), nor to completion–equilibrium of chemical reactions (chemical homogenization). Thus, in FI it is feasible to base the principles of an analytical assay on the

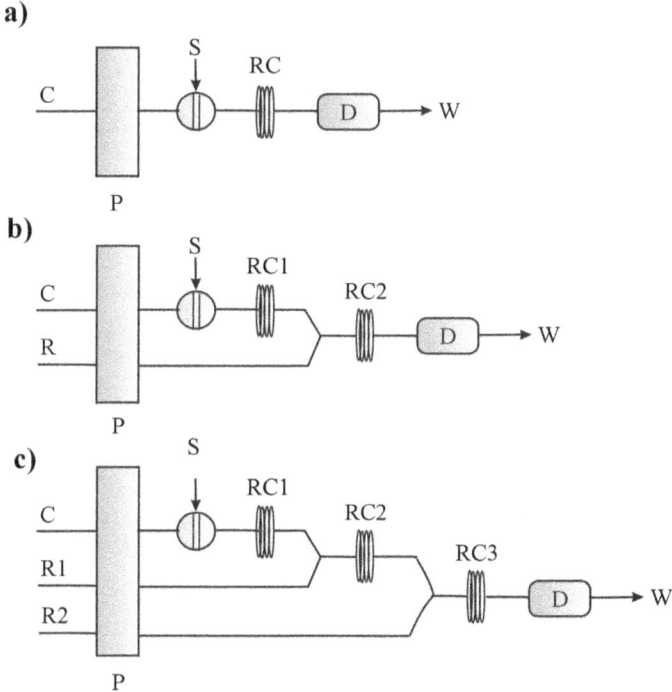

Figure 11.1   Representative FI manifolds of different configurations. Single- (a), two- (b) and three-channeled (c) FI manifolds for automation of various chemical systems. Samples (S) are injected in the carrier stream (C) and merge downstream with suitable reagents (R/R1/R2). On-line reaction takes place on passage through reaction coils (RC/RC1/RC2/RC3). P = pump; D = detector; W = waste.

measurement of unstable intermediate compounds, which might exhibit particularly interesting analytical characteristics; to exploit extremely fast detection principles relying on bio- or chemi-luminescence; to develop assays based on kinetic discrimination-based schemes by taking advantage of the reaction rates of the individually occurring reactions in methods comprising a series of reactions; to perform so-called *stopped-flow* measurements, where a sample–reagent segment is stopped within the detector, and where the signal recorded directly can be related to the concentration.

FI is generally considered as a simple and inexpensive technique. It employs instrumentation and hardware that is standard equipment in most analytical chemistry laboratories, such as peristaltic pumps and low-pressure injection valves. Compared to batch methods it offers increased sampling rate, lower reagent consumption, better precision and high versatility.

Figure 11.1 depicts the most typical manifold configurations in FI. In single-stream manifolds (Figure 11.1 (a))—that are the most simplified setups—the sample is injected in the carrier stream and flows directly towards the detector. If a chemical reaction is necessary the reagent is contained in the carrier stream. Otherwise (inert carrier) detection is based on the properties of the molecule of interest. Multi-streams manifolds (*e.g.* Figure 11.1 (b) and 11.1 (c)) are suitable for more complex "chemistries". In such configurations the carrier is typically inert and it simply propels the sample zone towards confluence points were mixing with reagent and–or buffer streams takes place.

11.2.2   Sequential Injection Analysis

Sequential-injection analysis (SI) is considered to be the second generation of flow injection techniques and was initially developed by Ruzicka and Marshall in 1990's (Ruzicka and Marshall 1990). The primary goal of the researchers was to develop a novel technique that would be more suitable than FI for process analytical applications and operation on a 24-h basis. In other words, to eliminate some of the disadvantages of FI: (i) different manifold configuration is necessary to implement each analytical method employing in certain cases multi-channel manifolds; (ii) continuous flowing of solutions through the manifold leads to increased consumption wastes generation; (iii) optimization of the geometrical and chemical parameters for each assay is tedious and time-consuming because alterations to the configuration of the flow manifold must be made.

On the other hand, as can be seen in a typical SI setup in Figure 11.2, the heart of an SI manifold is a multiposition selection valve. Fluids are manipulated within the manifold by means of a bi-directional pump. A holding coil is placed between the pump and the common port of the multiposition selection valve. The selection ports of the valve are reservoirs, detectors, pumps, reactors, separators, special cells, other manifolds *etc.* After aspiration of a discrete volume (zone) of sample into the holding coil *via* the sample line, the sample can be subjected to very complex physical and chemical

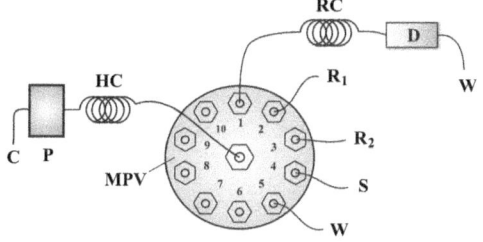

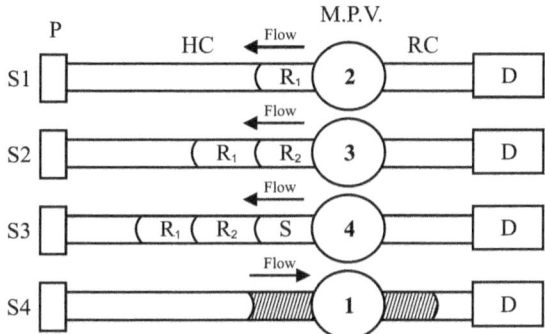

Figure 11.2 Typical SI setup and sequence steps for a three-zone chemistry. The "heart" of every SI setup is the multiposition valve (MPV). Samples (S) and reagents (R1, R2) are aspirated as discrete zones in the holding coil (HC). Reactions take place at the overlapped regions upon flow reversal towards the detector (D). C = carrier; P = bi-directional pump; RC = reaction coil; W = waste; S1–S4 = analysis steps.

pre-treatment in different ways within the SI manifold. SI offers great potential for sample handling because it is a bidirectional, stopped-flow sample-handling technique enabling the sample to be serially processed in the different modules connected to the selection valve by means of repetitive aspiration and delivery steps. The advantages of SI over FI are the following: (i) SI makes use of a simpler manifold that can be employed for a larger range of analytical methods without (or minimal) alterations In its physical configuration; (ii) in SI, discrete volumes of sample and reagents are aspirated and their consumption is drastically reduced; (iii) the bidirectional and stopped-flow operation of SI provides great scope for pre-treatment of the sample.

11.2.3 Multicommutation-based Flow Injection Techniques

The characteristic components of the flow network of a multicommutation-based manifold are solenoid valves. Injection of defined zones of samples and reagents in the flow are carried out through these solenoid valves by strict computer-controlled function (ON/OFF) and time (Rocha *et al* 2002). Both FI and SI methods and analytical protocols can be applied using multicommuta-

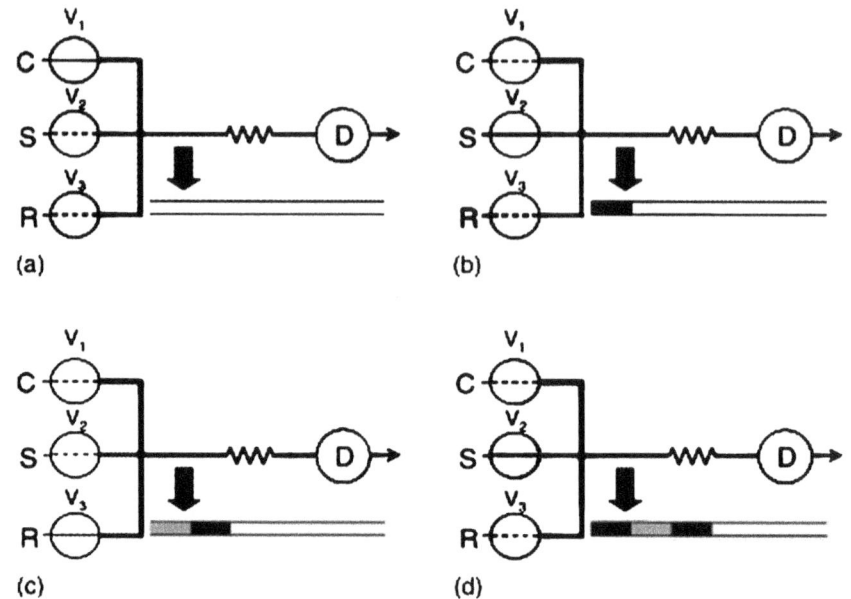

Figure 11.3   Representative multicommutation flow manifold. In multicommutated
flow systems all actions are carried out by the precisely time-based
controlled operation of solenoid valves that are strategically positioned
in the flow network: (a) flushing with carrier (C) through valve V1; (b)
injection of first sample zone (S) through valve V2; (c) injection of
reagent (R) through valve V3; (d) injection of second sample zone (S)
through valve V2. (With permission from Elsevier, Rocha *et al* 2002).

tion flow setups. A typical multicommutation flow configuration including
injection of sample (b) and (d) and reagent (c) in a "sandwich" format is
depicted in Figure 11.3 (a)–(d). Multicommutation enables increased repeatability, easier sample handling, and reduced reagent consumption, providing
an invaluable tool to develop environmentally friendly methods, which
provides benefits in terms of reagent waste generation or sample–reagent
consumption (Garciía-Reyes *et al* 2006).

11.3   Determination of Caffeine and Related Compounds by Coupling of SI/FI to Separation Techniques

Historically, one of the major challenges of flow injection based techniques is
the ability to simultaneously analyze mixtures of more than one analyte in
complex matrices. Despite the numerous ingenious approaches towards this
direction based on *e.g.* kinetic discrimination, sample splitting and so on
(Hlabangana *et al* 2006), the most straightforward and viable solution appears
to be the direct coupling of flow injection techniques to separation ones
(HPLC, GC, CE). The hyphenated analytical schemes combine the separation

efficiency and increased selectivity of the latter techniques and the automation potentials of on-line flow injection protocols. Figure 11.4 (A) depicts a typical setup of coupling SI and HPLC.

In 2004 our research group proposed, for the first time, an automated SI-HPLC method for the determination of caffeine in biological material and beverages (soft drinks and coffee) (Theodoridis *et al* 2004). The SI part was used to automate an SPE procedure for sample preparation based on the enhanced selectivity of a molecularly imprinted polymer material (MIP). All necessary steps of the SPE protocol including activation of the column, loading of the sample, washing to remove endogenous compounds and elution of the target analyte (caffeine) were carried out efficiently and reproducibly by SI. The eluent was directed to the injection loop of the HPLC instrument for

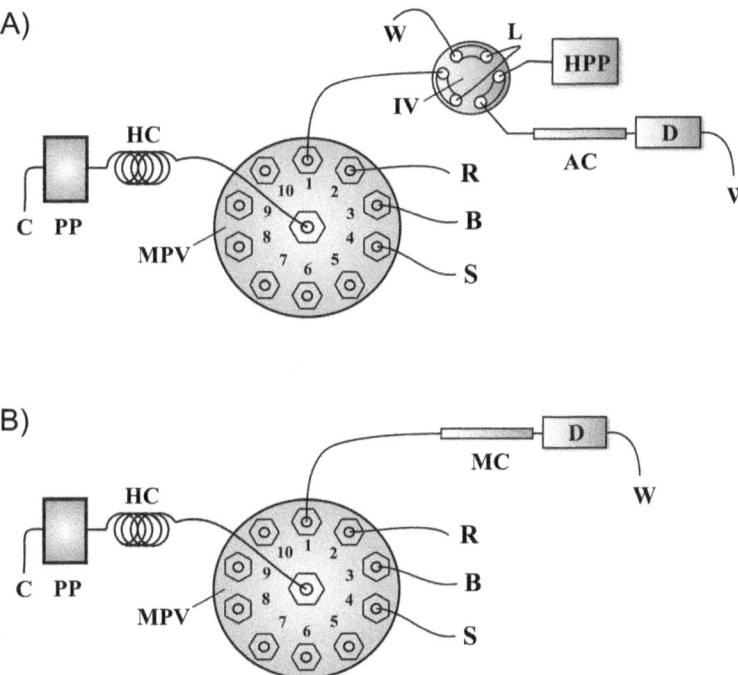

Figure 11.4   (A) Representative SI-HPLC setup and (B) typical SIC configuration. In SI-HPLC (A) coupling of the techniques is carried out by simply connecting a port of the low-pressure multiposition valve (MPV) of SI to the appropriate port (typically port 5 or 6) of the high pressure injection valve of HPLC (IV). A critical parameter is the "tuning" of the desired section of the SI zones to the loop (L) of the HPLC injection valve; in SIC (B) a short monolithic column (MC) is positioned in the flow network between the low-pressure MPV and the detector (D). C = carrier; PP = low pressure pump; HC = holding coil; R,B = reagents, mobile phases, buffers *etc.*; S = sample; AC = analytical column; HPP = high pressure HPLC pump; W = waste.

subsequent analysis and UV detection at 274 nm. Selectivity experiments using mixtures of caffeine and other major xanthines (theophylline and theobromine) confirmed the specificity of the MIP against the target analyte.

One of the major breakthroughs in flow injection techniques was the incorporation of monolithic stationary phases in low–moderate pressure flow systems. A new technique was developed under the term "Sequential Injection Chromatography – SIC" (Kika 2009) and enabled the chromatographic separation of mixtures of analytes at flow injection conditions. This new technique took advantage of the unique properties of monolithic reversed phase materials to operate at extremely lower pressures compared to particulate HPLC columns. A depiction of a representative SIC manifold is shown in Figure 11.4 (B). In this concept, Satinsky and co-workers developed a SIC method for the simultaneous separation and determination of paracetamol, caffeine and acetylsalicylic acid in pharmaceutical preparations. Using a short monolithic column (25 × 4.6 mm i.d., FlashRod, Chromolith, Merck) and isocratic elution at a flow rate of 0.6 mL min^{-1}, the separation of the three analytes and the internal standard (benzoic acid) was completed in less than 6 min (Satinsly *et al* 2004). The same group from Czech Republic, continuing their research on SIC proposed quite recently a two-column SIC setup as an interesting alternative to gradient elution based separations (Chocholous *et al* 2010). The idea behind this concept was to achieve isocratic separation of compounds of different polarities. A mixture of paracetamol, caffeine, salicylic acid (internal standard) and propyphenazone was selected for this purpose. The first three compounds were eluted through a first monolithic column (25 × 4.6 mm i.d.) by a "weak" mobile phase containing 10% acetonitrile, while the more hydrophobic propyphenazone was eluted in a second run through a shorter monolithic column (10 × 4.6 mm i.d.) by a mobile phase containing 30% acetonitrile.

Recently, a research group from Japan introduced the term *"quasi-FI"* in order to describe the incorporation of a short reversed phase particulate column (GL-Sciences Inertsil ODS-3, 3μm, 10mm × 1.5 mm) to a low-pressure HPLC system for the determination of caffeine in tea samples (Yamauchi *et al* 2008). The carrier–mobile phase consisted of MeOH : Water : Acetic acid (15 : 84 : 1 v/v), while peak height—as in typical FI applications—was used for the evaluation of the signals. The proposed configuration had limited selectivity and separation efficiency and analysis of real tea samples was feasible only after treatment of the samples through a specially prepared polyvinylpolypyrrolidone (PVPP) cartridge. A representative series of chromatograms of tea samples with and without treatment is depicted in Figure 11.5. The precision of the proposed system was better than 1% (within-day and day-to-day). The efficiency of the clean-up step was also demonstrated by recovery experiments that were in the range of 97.3–103.9% for caffeine concentration up to 20 mg g^{-1}.

An interesting alternative on the simultaneous determination of caffeine, theophylline and theobromine in real samples (cola beverages, tea leaves and

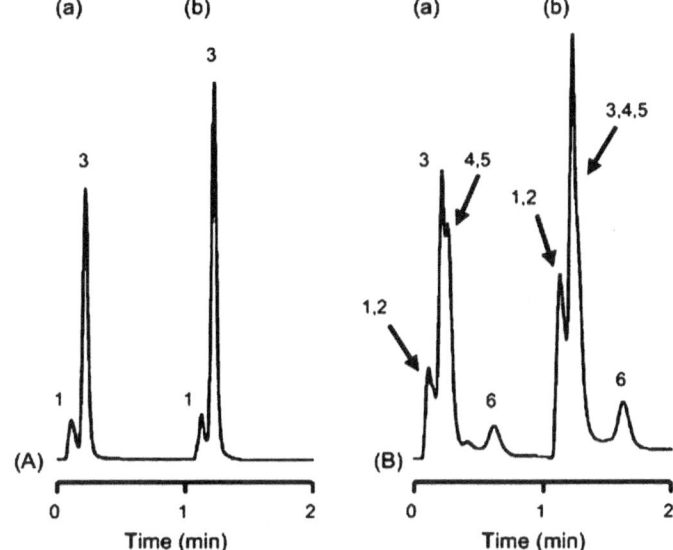

Figure 11.5 Determination of caffeine in tea samples by the quasi-FI method. Typical chromatograms of (A) PVPP (50 mg) cartridge-treated and (B) non-treated samples of (a) oolong and (b) black tea by the quasi-FIA system; (i) Due to the low separation efficiency of the short column pretreatment with PVPP is critical and (ii) caffeine (3) and theobromine (1) are adequately resolved for quantitative analysis. Other peaks: 2, (−)-epigallocatechin; 4, (−)-epigallocatechin gallate; 5, (−)-epicatechin; 6, (−)-epicatechin gallate. (With permission from Elsevier, Yamauchi *et al* 2008).

drink, coffee) has been proposed by coupling of FI to capillary electrophoresis (CE) (Pu and Fang 1999). CE offers the critical advantages of minimum waste generation, aqueous running buffers and high separation efficiency of both organic and inorganic compounds. On the other hand, typical injection modes in CE (electrokinetic or hydrodynamic) are performed in a batch way limiting the potentials of automated sample preparation. A viable solution was the on-line coupling of FI to CE (Kuban and Karlberg 2009). Pu and Fang separated the three xanthines by FI-CE with a simple background electrolyte of borate buffer (30 mmol L^{-1}, pH = 10) and precision of better than 6% in all cases. The recoveries from the analysis of real samples ranged between 91.6 and 105.8%.

11.4 Determination of Caffeine and Related Compounds by Flow Optosensors

A major drawback of application of direct UV-Vis measurements in analytical chemistry is the lack of sufficient sensitivity and selectivity. A viable solution was developed in the mid-seventies by the proposal of the concept of Solid Phase Spectroscopy. The principle of this new type of sensors is that the native absorbance of the analytes is measured after absorption of the latter on a suitable solid support immobilized in the cell of the detector. Such configurations have been automated extensively by flow injection techniques. The resulting flow optosensors combine the advantages of enhanced selectivity and sensitivity with the high precision and sampling rate of flow injection techniques. Additionally, based on the different retention properties of the analytes of interest, there are potentials for multi-analyte determinations. Typical solid phase supports range from hydrophobic reversed phase and ion-exchange beads to tailor-made molecularly imprinter polymers.

The simultaneous determination of pairs of xanthines, caffeine–theophylline and caffeine–theobromine, using a flow optosensor, has been proposed by Llorent-Martínez and co-workers (Llorent-Martínez *et al* 2005). Reversed phase silica gel beads served as the sensing zone in the flow cell, while all analytes were monitored at 272 nm. In order to achieve separation of the above-mentioned xanthine-pairs prior to entering the sensing zone, the authors placed a mini-column with the same material between the injection valve and the detector. Using an aqueous carrier stream, caffeine was retained in the mini-column, while theophylline (or theobromine) was detected in the flow-cell. Elution of caffeine was carried out by replacing the carrier with a solution containing 25% MeOH. The detection limits were at the 0.1 mg L^{-1} level and the precision at 3%. The analytical features were sufficient for the simultaneous determination of caffeine–theophylline in pharmaceuticals and caffeine–theobromine in soft drinks and food samples.

The same research group adopted a similar approach for the separation–determination of mixtures of caffeine and accompanying pharmaceutical compounds: caffeine, salicylamide, propyphenazone (Gilbert-López *et al* 2007); caffeine, paracetamol, propyphenazone (Vidal *et al* 2003); caffeine, paracetamol (Ortega-Barrales *et al* 2002). In all cases, polar analytes *e.g.* paracetamol and salicylamide were not retained on the reversed phase pre-column, while more hydrophobic compounds were eluted by suitable organic–water mixtures. Critical analytical figures of merit included: (i) detection limits in the range of 0.21–0.61 mg L^{-1} (Gilbert-López *et al* 2007), 0.65–12 mg L^{-1} (Vidal *et al* 2003) and 0.5–0.75 mg L^{-1} (Ortega-Barrales *et al* 2002); (ii) relative standard deviations between 2.97–4.86% (Gilbert-López *et al* 2007), 2.7–3.8% (Vidal *et al* 2003) and 0.5–3.1% (Ortega-Barrales *et al* 2002); (iii) sampling throughput between 11–20 h^{-1} (Vidal *et al* 2003) and 15–20 h^{-1} (Ortega-Barrales *et al* 2002).

11.5   Determination of Caffeine and Related Compounds by On-line IR Detection

The coupling of IR detectors to continuous flow systems has proven to be an advantageous alternative to conventional detection systems that are typical in most FI methods (UV-Vis, FL, CL *etc.*). IR detection offers the potential of simultaneous determinations of more than one analyte, while on-line measurements under flow conditions have proven useful on the basis of automation, reproducibility and cost-effectiveness.

Since the first reports of the coupling of IR and FT-IR instruments to flow systems (Morgan *et al* 1985; Olesick *et al* 1986), there are several published reports on the analysis of caffeine and related compounds in various matrices. Most of the publications on this topic come from the group of M. De La Guardia in Spain and include the simultaneous determination of caffeine–acetylsalicylic acid (Garrigues *et al* 1993) and of caffeine–propyphenazone

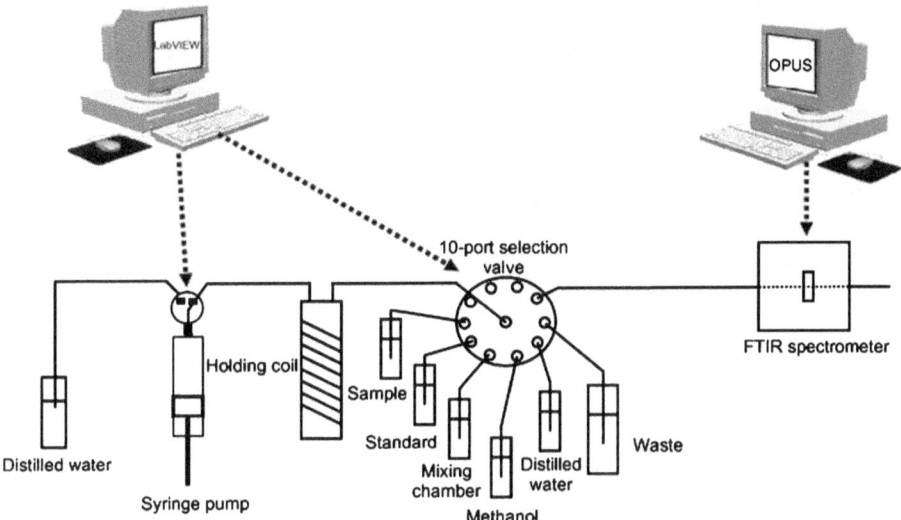

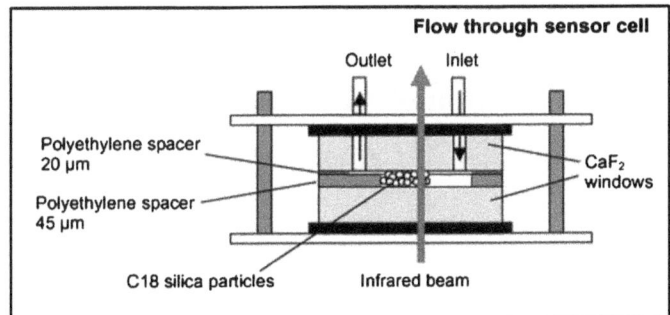

Figure 11.6   Setup of a SI system and a mid-IR flow through cell containing C18 silica beads. Precise handling of samples and standards is accomplished by a typical SI system; the flow IR sensor is constructed by the trapping of reversed phase C18 beads in the flow cell by the use of suitable polyethylene spacers (With permission from Elsevier, Armenta and Lendl 2009).

(Bouhsain *et al* 1997) in pharmaceuticals and of caffeine in soft drinks (Daghbouche *et al* 1997) and coffee samples (Bouhsain *et al* 1999). Organic solvents such as dichloromethane and chloroform were employed as carrier and sample diluents in all the above-mentioned studies. Some unique and worth-mentioning features of the methods include: (i) the incorporation of an on-line distillation apparatus in the flow-line that recycles the organic solvent and reduces the environmental impact (Bouhsain *et al* 1997); (ii) the automated pretreatment of soft-drinks by on-line solid phase extraction prior to FT-IR detection of caffeine (Daghbouche *et al* 1997) and (iii) the on-line extraction of caffeine from slurry coffee samples in a closed-flow manifold (Bouhsain *et al* 1997). The sampling rates varied between 6 h^{-1} (Bouhsain *et al* 1999) and 120 h^{-1} (Bouhsain *et al* 1997) and the limits of detection between 9 mg L^{-1} (Bouhsain *et al* 1999) and 1500 mg L^{-1} (Bouhsain *et al* 1997).

Despite the obvious usefulness of IR detectors for qualitative purposes and structure elucidation, their main bottleneck is the limited sensitivity. Two recent approaches attempted to overcome this problem by the construction of IR sensors based on the incorporation of solid sorbents in the detection cell (Alcudia-Leon *et al* 2008; Armenta and Lendl 2009). Both methods were automated by sequential injection analysis (SI) and were both applied to the determination of caffeine in soft drinks. Alcudia-Leon *et al* incorporated a typical hydrophobic SPE sorbent material (LiChrolut EN) in an Attenuated Total Reflection (ATR) flow cell *via* a home-made PTFE accessory for connection to the flow system, enabling the sensitive and selective analysis of caffeine at a throughput of 4.3 h^{-1} and a LOD of 7 mg L^{-1}. Better analytical figures of merit (10 h^{-1}, LOD = 1.8 mg L^{-1}) were achieved by Armenta and Lendl (2009) by entrapping C18 sorbent (mesh size of 25 μm) in the flow cell of a mid-IR detector. As can be seen in Figure 11.6, the hydrophobic silica beads were retained in the flow cell by the use of specially designed polyethylene spacers.

11.6 Various Flow Injection Methods for the Determination of Caffeine and Related Compounds

11.6.1 Determination of Caffeine by FI Coupled to Amperometric Detection

Electrochemical-based methods, although quite sensitive and cost-effective, suffer from one major drawback that limits their applicability to routine, real samples analysis: the fouling of the electrodes by the endogenous components of the samples that makes the frequent mechanical and–or electrochemical cleaning of the active surface necessary. Incorporating amperometric flow cells in FI systems might provide a viable solution to the above-mentioned problem due to the limited contact time between the electrode and the sample.

A simple and fast FI amperometric method for the determination of caffeine in soft drinks was developed by Lima and co-workers (Lima *et al* 1998).

Untreated samples were injected in a two-channel FI manifold and propelled towards a wall-jet flow cell. The continuous flow character of the method minimized the need for regeneration of the glassy carbon electrode to once a day. Caffeine could be determined at a sampling rate of 120 h^{-1} and a precision of better than 4%.

11.6.2   On-line Extraction of Caffeine from Solid and Slurry Samples

The group of Valcarcel from Spain proposed an FI method for the determination of caffeine in solid and slurry samples after on-line solvent extraction using a supported liquid membrane (SLM) and UV detection (Luque-Perez *et al* 1999). The concept of SLM is an interesting alternative to traditional batch and FI solvent extraction protocols in terms of organic solvents consumption and extraction efficiency. In brief, a suitable organic solvent is immobilized within the pores of a hydrophobic membrane (*e.g.* PTFE). The membrane facilitates the transfer of the analytes from the aqueous sample through the immobilized solvent to an acceptor solution. Extraction of caffeine from solid and slurry samples was accomplished on-line using *n*-undecane : hexylether entrapped in the pores of a PTFE membrane (Figure 11.7). An acidic aqueous solution was proved as the most suitable acceptor. UV detection at 274 nm offered a wide determination range between 0.5 and 15 g L^{-1} caffeine, at a satisfactory sampling throughput of 7–8 h^{-1}.

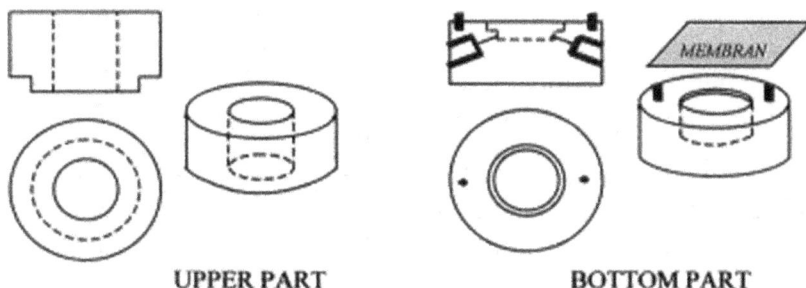

UPPER PART   **BOTTOM PART**

Figure 11.7   Membrane unit for the extraction of caffeine from coffee samples using supported liquid membranes. The extractant—typically a non-miscible organic solvent—is immobilized to the hydrophobic membrane. The aqueous sample flows on the top of the membrane and the analytes are extracted *via* the immobilized organic phase to a suitable acceptor that is positioned below the membrane. (With permission from Elsevier, Luque-Pérez *et al* 1999).

11.6.3 Enzymatic FI Method for the Determination of Theophylline

A useful group of analytical methodologies that were really "boosted" by the development of flow injection techniques was the analytical applications of enzymatic reactions. The advantage of the specificity of the latter chemical systems was suppressed by the necessity of strictly controlled conditions, mainly in terms of timing due to their kinetic nature–character. Flow injection-based methods solved this problem and lead to the publication of numerous methods using both soluble and immobilized enzymes.

A typical example is the FI determination of theophylline in human serum based on its inhibitory effect on immobilized alkaline phosphatase (Sánchez-Cabezudo *et al* 1995). The enzyme was immobilized on controlled pore glass (CPG) and packed on glass tubes of fixed dimensions. The immobilization procedure enables unaltered activity for at least two months. The selective and sensitive determination of theophylline was based on its inhibitory effect on the hydrolysis of 4-methylumbelliferone-phosphate by the enzyme. Detection was carried out fluorimetrically (365–455 nm). From a flow configuration point of view, the analyte and the substrate were injected simultaneously to two different carrier streams and overlapped on passage through a confluence point prior to entering the column containing the immobilized alkaline phosphatase. The analytical figures of merit were quite satisfactory for clinical applications, since the determination range was 10–200 μmol L^{-1}, the precision better than 4% and the sampling rate 40 h^{-1}.

Summary Points

- This chapter reviews flow-based automated methods for the determination of caffeine and related compounds (e.g. xanthines etc).
- Caffeine is one of the most highly consumed compounds world-wide.
- There is need for the development of fast and reliable analytical methods for the determination of caffeine and related compounds in various matrices.
- Automated flow-based techniques (e.g. flow / sequential injection analysis) offer increased sampling rate, precision and accuracy for the analysis of caffeine in real samples.
- Sequential injection chromatography (SIC) using low-pressure monolithic stationary phases expanded the potentials of flow injection-based techniques to multianalyte determinations.
- Flow-through optosensors, on-line infrared detection and the coupling of FI/SI to separation techniques allow the simultaneous automated determination of caffeine and accompanying compounds in pharmaceutical formulations and biological material.
- Direct amperometric detection of caffeine in a flow injection manifold using a glassy carbon electrode allows the analysis of soft drinks without pretreatment at a sampling rate of 120 h^{-1}.

- Incorporation of supported liquid membranes (SLM) in a flow configuration enables the on-line extraction and subsequent detection of caffeine from solid and slurry coffee samples.
- Selective automated extraction of caffeine from urine samples in the presence of other similarly structured xanthines (theophylline and theobromine) can be achieved by the incorporation of a molecularly imprinted polymer (MIP) sorbent in a sequential injection system coupled to HPLC.

Key Facts

1. Key Facts on Flow / Sequential Injection Analysis

- FI is the first generation of automated flow based techniques and was introduced in the seventies.
- In FI the reagents are flowing continuously towards the detector, offering high sampling rate but relatively high consumption.
- FI affords automation of various physical-chemical processes e.g. reactions, solvent extraction, gas-diffusion, distillation etc.
- SI is considered as the second generation of automated flow based techniques and was introduced in 1990.
- In SI advanced flow manipulation through a multi-position valve enables the use of precise volumes of reagents and samples at the micro-litre level.
- SI is particularly useful for stopped-flow applications due to the computer-controlled operation.

2. Key Facts on Sequential Injection Chromatography

- The bottleneck of FI/SI techniques is the difficulty in multi-analyte determination.
- Monolithic stationary phases are new chromatographic materials that operate at extremely low backpressures.
- The backpressure of short monolithic columns (5 – 50 mm) is compatible to FI/SI instrumentation.
- The incorporation of monolithic columns in SI systems in the mid-2000 was described as a new technique called "Sequential Injection Chromatography-SIC".
- SIC applications "boosted" automated flow-based methods to more complicated matrices.
- SIC offers the advantage of automated sample preparation (derivatization, extraction, dilution etc) and low-pressure chromatographic separation in a single configuration.
- SIC systems are cost effective and often portable, allowing field analysis.

Definitions of Words and Terms

Automation: In chemical analysis, automation is generally the replacement of manual (batch) stages of a process by mechanical devices. Typical examples include autosamplers, robotic analyzers, flow-through systems etc. Upon proper performance and operation validation, automated systems generally offer increased robustness and ruggedness, reduced cost, increased precision and potentials for process monitoring on 24-h basis.

Flow Injection Analysis: Is the first generation of flow-based automated techniques and was proposed in the seventies by Ruzicka and Hansen. A pre-defined volume of sample is injected in a continuously flowing stream of a suitable carrier/reagent. The analyte of interest is subjected to an on-line physical/chemical process prior to detection on passage through a suitable flow-through cell.

Infrared (Fourier Transform) detection: Spectroscopic detection is carried out at wavelengths > 700 nm. It provides valuable information, both qualitative and quantitative. The infrared spectrum, particularly the $600–1500$ cm^{-1} region, can be considered as the fingerprint of a given molecule. Further processing of the collected IR data through Fourier-Transform algorithms (FT-IR) enhance dramatically the analytical power of the technique.

Molecularly Imprinted Polymer: They are novel materials that are synthesized by polymerization in the presence of the analyte of interest. The analyte is "imprinted" in the structure of the polymer and upon removal the remaining cavities enable specific recognition of the target compound (loch-and-key mechanism). These materials have been extensively applied in analytical science in order to enhance the selectivity of the sample preparation protocols.

Monolithic stationary phases in HPLC: These novel materials started being commercially available in 2000. The principle of the monolithic column is completely different from traditional particulate-based analogues. They consist of a single piece of organic or inorganic (silica-based) polymer prepared *in situ*. The typical structure is bimodal, consisting of macropores (the main network of the monolithic material that allows the flowing of the mobile phase) and mesopores (where chromatographic separation takes place). The two main features of monolithic stationary phases are the extremely low pressure drop and the fast mass transfer kinetics. These features offer the advantages of efficient separations at elevated flow rates (up to 9 mL min^{-1} or higher).

Multicommutation: Is an alternative flow-based automated approach that can "mimic" both traditional FI and SI. It is based on the design of a strictly programmable flow network that is composed of solenoid valves. By controlling the time and the ON/OFF operation of these valves precise volumes of reagents and samples can be inserted in flow lines and undergo the desired manipulation.

Sequential Injection Analysis: Is considered to be the 2nd generation of flow injection techniques. Specific volumes of reagents and samples are sequentially aspirated through the ports of a multiposition valve and stacked in a suitable carrier stream. Reactions take place on the overlapped sections of the zones in a practically single-channelled configuration.

Sequential Injection Chromatography: This new term was introduced in the mid-2000 in order to describe the incorporation of short monolithic stationary phases for liquid chromatography to SI systems. The new technique is a hybrid between SI and HPLC and combines the automation potentials of SI (on-line derivatization, extraction, dilution etc) and the separation efficiency of liquid chromatography.

Solid Phase Spectrophotometry: The concept of solid phase spectrophotometry (or solid phase spectroscopy) was introduced in mid-1970 as an alternative approach for increasing the sensitivity and selectivity of classical solution-based UV-Vis methods. Measurement of the analyte(s) of interest – either directly or after a suitable reaction – is carried out after absorption on a suitable solid support that is located in the flow-cell of a UV-Vis detector.

Supported Liquid Membranes: An alternative approach for performing liquid-liquid extraction under flow conditions. The extractant, typically a non-miscible organic solvent, instead of flowing continuously is immobilized within the pores of a suitable hydrophobic membrane (e.g. made of PTFE). The analytes of interest are extracted upon contact with the membrane from an aqueous donor stream (sample) through the immobilized solvent to a suitable acceptor solution that undergoes detection.

List of Abbreviations

ATR	Attenuated Total Reflection
CE	Capillary Electrophoresis
CL	Chemiluminescence
FI	Flow Injection Analysis
FL	Fluorescence
FT-IR	Fourier Transform Infrared
GC	Gas Chromatography
HPLC	High Performance Liquid Chromatography
IR	Infrared Spectroscopy
LOD	Limit of Detection
MeOH	Methanol
MIP	Molecularly Imprinted Polymer
PTFE	Polytetrafluoroethylene (Teflon)
PVPP	Polyvinylpolypyrrolidone
SFA	Segmented Flow Analysis
SI	Sequential Injection Analysis
SIC	Sequential Injection Chromatography
SLM	Supported Liquid Membrane
SPE	Solid Phase Extraction
UV-Vis	Ultraviolet – Visible spectrophotometry

References

Armenta, S. and Lendl, B. 2009. Flow through FTIR sensor based on solid phase spectroscopy (SPS) on conventional octadecyl (C18) silica. Vibrational Spectroscopy. 51: 60–64.

Bouhsain, Z., Garrigues, J. M., Garrigues, S. and De La Guardia, M., 1999. Flow injection Fourier transform infrared determination of caffeine in coffee. Vibrational Spectroscopy. 21: 143–150.

Bouhsain, Z., Garrigues, S. and De La Guardia, M., 1997. Clean method for the simultaneous determination of propyphenazone and caffeine in pharmaceuticals by flow injection Fourier transform infrared spectrometry. Analyst. 122: 441–445.

Chocholouš, P., Šatínský, D., Sklenářová, H. and Solich, P., 2010. Two-column sequential injection chromatography-new approach for fast and effective analysis and its comparison with gradient elution chromatography. Analytica Chimica Acta. 668: 61–66.

Daghbouche, Y., Garrigues, S., Teresa Vidal, M. and De La Guardia, M., 1997. Flow injection Fourier transform infrared determination of caffeine in soft drinks. Analytical Chemistry. 69: 1086–1091.

Dlugosz, L. and Bracken, M. B., 1992. Reproductive effects of caffeine: A review and theoretical analysis. Epidemiologic Reviews. 14: 83–100.

Garciía-Reyes, J. F., Llorent-Martiínez, E. J., Ortega-Barrales, P. and Molina-Diíaz, A., 2006. The potential of combining solid-phase optosensing and multicommutation principles for routine analyses of pharmaceuticals. Talanta. 68: 1482–1488.

Garrigues, S., Gaállignani, M. and De La Guardia, M., 1993. Simultaneous determination of acetylsalicylic acid and caffeine in pharmaceuticals by flow injection with Fourier transform infrared detection. Talanta. 40: 1799–1807.

Gilbert-López, B., Llorent-Martínez, E. J., Ortega-Barrales, P. and Molina-Díaz, A., 2007. Development of a multicommuted flow-through optosensor for the determination of a ternary pharmaceutical mixture. Journal of Pharmaceutical and Biomedical Analysis. 43: 515–521.

Glade, M. J. 2010. Caffeine – not just a stimulant. Nutrition. 26: 932–938.

Griffiths, R. R., Juliano, L. M. and Chausmer, A. L., 2003. Caffeine pharmacology and clinical effects. In: Graham A.W., Schultz T. K., Mayo-Smith M. F., Ries R. K. and Wilford, B. B. (ed.) Principles of Addiction Medicine. Chevy Chase, MD: American Society of Addiction, Third Edition, pp. 193–224.

Higdon, J. V. and Frei, B., 2006. Coffee and health: a review of recent human research. Critical Reviews in Food Science and Nutrition. 46:101–123.

Hlabangana, L., Hernández-Cassou, S. and Saurina, J., 2006. Multicomponent determination of drug using flow-injection analysis. Current Pharmaceutical Analysis. 2: 127–140.

Kika, F.S. 2009. Low pressure separations using automated flow and sequential injection analysis coupled to monolithic columns. Journal of Chromatographic Science. 47: 648–655.

Kuban, P. and Karlberg, B., 2009. Flow/sequential injection sample treatment coupled to capillary electrophoresis. A review. Analytica Chimica Acta. 648: 129–145.

Lima, J. L. F. C., Delerue-Matos, C., Nouws, H. P. A. and Vaz, M. C. V. F., 1998. An automatic determination of caffeine in soft drinks using flow injection system with amperometric detection. Food Additives and Contaminants. 15: 265–269.

Llorent-Martínez, E. J., García-Reyes, J. F., Ortega-Barrales, P. and Molina-Díaz, A., 2005. Solid-phase ultraviolet sensing system for determination of methylxanthines. Analytical and Bioanalytical Chemistry. 382: 158–163.

Luque-Pérez, E., Ríos, A., Valcárcel, M., Danielsson, L.-G. and Ingman, F., 1999. Spectrophotometric flow injection determination of caffeine in solid and slurry coffee and tea samples using supported liquid membranes. Laboratory Automation and Information Management. 34: 131–142.

Morgan, D. K., Danielson, N. D. and Katon, J. E., 1985. Aqueous flow injection analysis with Fourier transform infrared detection. Analytical Letters. 18: 1979–1998.

Nawrot, P., Jordan, S., Eastwood, J., Rotstein, J., Hugenholtz, A. and Feeley, M., 2003. Effects of caffeine on human health. Food Additives and Contaminants. 20: 1–30.

Olesick, S. V., French, S. B. and Novotny, M., 1986. Reaction monitoring in supercritical fluids by flow injection analysis with Fourier transform infrared spectrometric detection. Analytical Chemistry. 58: 2256–2258.

Ortega-Barrales, P., Padilla-Weigand, R. and Molina-Díaz, A., 2002. Simultaneous determination of paracetamol and caffeine by flow injection-solid phase spectrometry using C18 silica gel as a sensing support. Analytical Sciences. 18: 1241–1246.

Pu, Q.-S. and Fang, Z.-L., 1999. Combination of flow injection with capillary electrophoresis. Part 6. A bias-free sample introduction system based on electroosmotic-flow traction. Analytica Chimica Acta. 398: 65–74.

Rocha, F. R. P., Reis, B. F., Zagatto, E. A. G., Lima, J. L. F. C., Lapa, R. A. S. and Santos, J. L. M., 2002. Multicommutation in flow analysis: Concepts, applications and trends. Analytica Chimica Acta. 468: 119–131.

Ruzicka, J. and Hansen, E. H., 1975. Flow injection analyses. Part I. A new concept of fast continuous flow analysis. Analytica Chimica Acta. 78: 145–157.

Ruzicka, J. and Marshall, G.D., 1990. Sequential injection: A new concept for chemical sensors, process analysis and laboratory assays. Analytica Chimica Acta. 237: 329–343.

Sánchez-Cabezudo, M., Fernández-Romero, J. M. and Luque De Castro, M. D., 1995. Fluorimetric-flow injection determination of theophylline based on its inhibitory effect on immobilized alkaline phosphatase. Analytica Chimica Acta. 308: 159–163.

Sătínský, D., Neto, I., Solich, P., Sklenárõvá, H., Conceicão, M., Montenegro, B. S. M. and Araújo, A. N., 2004. Sequential injection chromatographic determination of paracetamol, caffeine, and acetylsalicylic acid in pharmaceutical tablets. Journal of Separation Science. 27: 529–536.

Theodoridis, G., Zacharis, C. K., Tzanavaras, P. D., Themelis, D. G. and Economou, A., 2004. Automated sample preparation based on the sequential injection principle: Solid-phase extraction on a molecularly imprinted polymer coupled on-line to high-performance liquid chromatography. Journal of Chromatography A. 1030: 69–76.

Vidal, A. D., Barrales, P. O. and Díaz, A. M., 2003. Simultaneous determination of paracetamol, caffeine and propyphenazone in pharmaceuticals by means of a single flow-through UV multiparameter sensor. Mikrochimica Acta. 141: 157–163.

Yamauchi, Y., Nakamura, A., Kohno, I., Hatanaka, K., Kitai, M. and Tanimoto, T., 2008. Quasi-flow injection analysis for rapid determination of caffeine in tea using the sample pre-treatment method with a cartridge column filled with polyvinylpolypyrrolidone. Journal of Chromatography A. 1177: 190–194.

CHAPTER 12

Analysis of Caffeine by Immunoassay

JOSÉ JOÃO CARVALHO AND RUDOLF J. SCHNEIDER*

BAM Federal Institute for Materials Research and Testing, Working Group Immunochemical Methods, Richard-Willstaetter-Str. 11, D-12489 Berlin, Germany
*E-mail: rudolf.schneider@bam.de

12.1 Development of Immunoassays for Caffeine

The development of caffeine immunoassays has been driven by the clinical need to determine caffeine levels in human body fluids. A radioimmunoassay was developed in the mid 1970s to measure caffeine in plasma and saliva in concentrations higher than 25 µg L^{-1} (Cook et al 1976). The main barrier for its widespread dissemination was the required use of radiolabeled (tritiated) compounds.

During the 1980s, the first enzyme label was produced and made commercially available (from Syva): an automated enzyme multiplied immunoassay technique (EMIT) showing a cross-reactivity of 28% to paraxanthine – caffeine's main metabolite – and working in the range 0.5–155 mg L^{-1} (Zysset et al 1984). Despite the high cross-reactivity, the assay seemed suitable for clinical purposes and the results correlated well with those obtained by high-performance liquid chromatography (HPLC) in the mg L^{-1} range tested (Miceli et al 1984; Zysset et al 1984). Some other assays have been developed since but all of them aimed at measuring biological fluids (Helgeson et al 1983; Pearson et al 1984; Fickling et al 1990; Wong et al 1995).

Food and Nutritional Components in Focus No. 2
Caffeine: Chemistry, Analysis, Function and Effects
Edited by Victor R Preedy
© The Royal Society of Chemistry 2012
Published by the Royal Society of Chemistry, www.rsc.org

A new antibody fragment for caffeine, produced in a llama, was described in 2006 – a camelised antibody. The authors proposed a caffeine binder (antibody fragment) that could work at temperatures as high as 70°C and issued a patent for a fluid-based device (dipstick) to measure caffeine directly in hot beverages (Ladenson *et al* 2006), but describing only its application in the mg L^{-1} range and no cross-reactivity figures for paraxanthine were provided.

Table 12.1   Key facts of caffeine occurrence and analysis by immunoassay.

Caffeine, Where Is It Found?
  1. Caffeine (1,3,7-trimethylxanthine) is probably the most widely used psychoactive substance in the world today (Rogers and Dernoncourt 1998). It can be found in a variety of sources, the most well-known being coffee, tea, cocoa, soft drinks, energy drinks, pills and chocolate (Barone and Roberts 1996; Fredholm et al 1999; Mandel 2002).
  2. Caffeine can also be found in modern beers and shampoo (Fredholm et al 1999).
  3. Caffeine was found in seawater (Weigel et al 2004), surface water (Buerge et al 2003), drinking water (Focazio et al 2008) and rain water (Seitzinger et al 2003).
  4. Caffeine is an important constituent of some human pharmaceuticals, either alone or in combination with other active substances. It is used as a respiratory stimulant for treating apnea, as a cerebral and cardiac stimulant and also as a diuretic. Due to its enhancing effect, it is part of several combined formulations used for cough, cold and headache (Buerge et al 2003).
  5. Caffeine clearance is a key indicator in clinical chemistry, mainly to assess hepatic function in humans. Following caffeine ingestion, its faster or slower concentration decay in biological fluids, like saliva for example, is helpful to diagnose liver dysfunction (Tarantino et al 2006).
  6. Due to its widespread use by humans, its concentration in environmental waters has been suggested as an indicator of anthropogenic input into the environment. After being metabolised in the human body, residual caffeine is excreted via urine and eventually reaches surface waters (Buerge et al 2003).

Why Quantify Caffeine?
  7. In the beverages industry it is important to control caffeine concentration during the production and in the final product to assure consumer protection. This control is even more relevant when caffeine is added to the beverages in the form of plant extracts – where caffeine concentration can vary widely from one extract batch to another.
  8. Analytical laboratories, either controlling the quality of the final industrial product or for ensuring consumer protection, have the need to quantify caffeine concentrations in a high number of samples within the shortest time possible.

Why Immunoassay?
  9. Immunoassays are useful analytical tools to process large batches of samples within a short period of time. They have been mostly used for clinical purposes but within the past two decades they have also been applied in environmental and food sciences (Schneider 2003).
 10. Their major advantage is, besides the high throughput, that immunoassays usually do not require the use of organic solvents. The described immunoassay for caffeine requires neither the use of organic solvents nor any sample pre-treatment procedure, making it an interesting tool for laboratories looking for green chemistry principles.
 11. Online and real-time caffeine measurements during beverage production and bottling can be achieved by adapting the microtitre plate based immunoassay to special needs.

Recently, a caffeine enzyme-linked immunoassay (ELISA), using horse-radish peroxidase as signaling enzyme was reported (Carvalho *et al* 2010) after conjugating the enzyme with a caffeine derivative (Carvalho *et al* 2007). The assay, using a mouse monoclonal antibody, proved capable of measuring (with high precision and accuracy) caffeine concentrations in very different matrices such as environmental waters, beverages, shampoos and human saliva (Carvalho *et al* 2010; 2011).

In this chapter focus is given to the immunoassay application for determining caffeine concentration in beverages. An experimental protocol, describing in detail the assay procedure, is provided.

12.2 Enzyme Immunoassay for Caffeine

12.2.1 How It Works

Immunoassays are based on the binding properties of an antibody which has been selected to recognize a certain antigen. This antibody can be used as an analytical tool to quantify that antigen—caffeine in this case—using different formats (set-ups). The direct competitive format, in a 96-well microtitre plate support, is the one herein described and it is represented in detail in Figure 12.1.

A generic calibration curve is illustrated by Figure 12.2 to show a distinct characteristic of immunoassays when compared with instrumental detection methods, such as UV–Vis or mass spectrometry: in immunoassay the lowest signals are obtained for the highest concentrations of analyte in the samples and the highest signals for the blanks. Additionally, because the binding phenomenon antibody-antigen is non-linear, the resulting calibration functions are alike. Such less-conventional dose–response relationships should be discouraged from their use for quantitation purposes as comparable precision and accuracy can be obtained compared to linear calibration with instrumental methods. Within this chapter, evidence is provided for the equivalence between results of the immunoassay and an LC-MS-MS method. A method to linearize the calibration model is also offered.

12.2.2 A Detailed Experimental Protocol

12.2.2.1 General

A detailed protocol is given assuming the immunoassay is to be performed by a qualified operator. Bracketed trademarks and companies, the ones used in our lab, are merely indicative but are included here to help in case problems appear while running the assay. Critical, however, is the selection of the anti-caffeine monoclonal antibody. Several murine antibodies are available in the market, all of them stating to have been produced by using as immunogen a conjugate of 7-ethoxycarboxypentyl-1,3-dimethylxanthine coupled to a protein (keyhole

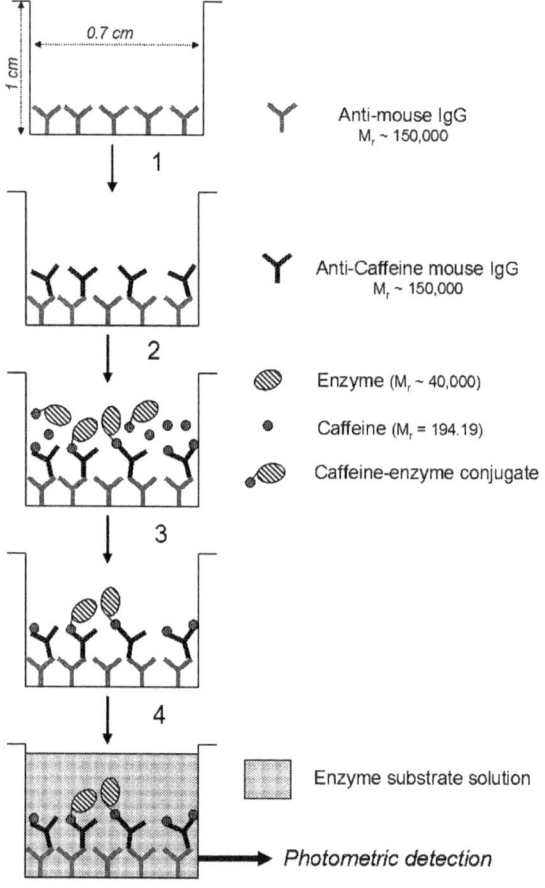

Figure 12.1   The caffeine immunoassay procedure. A generic immunoassay workflow
using the direct competitive format and caffeine as model analyte. The
square-like shape represents a single well in a 96-well microtitre plate
(volume max. 360 μL per well). The plate is incubated with the coating
antibody, an anti-mouse antibody (from a sheep), typically over night
under constant shaking; 1) After washing away the coating antibody
solution, the plate containing the coating antibody adsorbed on the
wells' surface is incubated with the primary antibody: the anti-caffeine
IgG (from mouse); 2) The coating antibody immobilises the anti-caffeine
antibody, to which the analyte (caffeine) will bind, in a quantitative
manner. To remaining binding sites of the anti-caffeine antibody pool,
the caffeine enzyme conjugate will bind. As a result, the lower the
caffeine concentration in the samples, the higher the number of binding
sites available for the enzyme conjugate and consequently the higher the
signal; 3) The sample as well as unbound enzyme conjugate is washed
away; 4) The enzyme substrate is added into the wells and after a certain
development time an acidic solution is added to stop the enzymatic
reaction. Thereafter, the plate is photometrically read with the highest
signals registered for wells containing the highest amount of bound enzyme
conjugate, *i.e.* the lowest analyte (caffeine) concentration, and *vice-versa*.

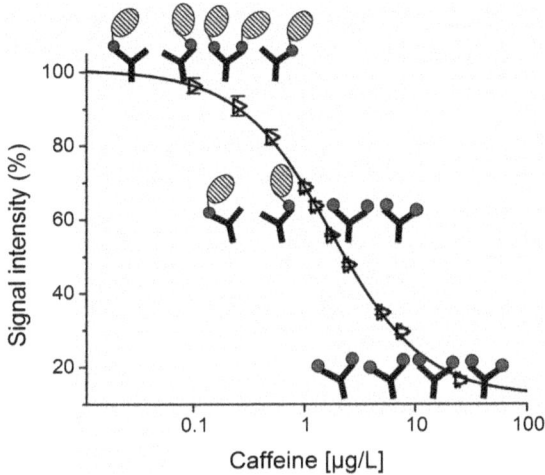

Figure 12.2 A generic calibration curve with direct competitive immunoassays. A calibration curve showing a sigmoidal curve following a 4-parametric logistic model fitted to the calibration standards (open triangles). The highest signal intensity is obtained for the blank samples, as no analyte (antigen) is available to bind to the antibodies and therefore they will all bind the enzyme conjugate which catalyses a specific substrate reaction. Contrarily, high quantities of the analyte will "block" all antibody binding sites making them unavailable for the enzyme conjugate and, consequently, the lowest signals will be registered.

limpet hemocyanin or bovine serum albumin), which sometimes does not seem to have been the case. Such antibodies will not bind to the enzyme conjugate produced and therefore the assay will not work. To ensure the right one is being purchased, an antibody presenting the lowest cross-reactivity to paraxanthine should be chosen. Typical values encountered in the suppliers' datasheets are 0.1–1%. Antibodies and enzyme conjugate dilutions must be optimized depending on the purchased products. In the present protocol the (antibody and enzyme conjugate) dilutions mentioned were optimized for the purchased product, indicated within brackets.

12.2.2.2 Equipment and Reagents

- UV–Vis microtitre plate reader (SpectraMax Plus384, Molecular Devices);
- Automatic 96-channel plate washer (ELx405 Select, BioTek Instruments);
- Plate shaker (Titramax 101, Heidolph) set at 750 rpm;
- Transparent microtitre plates with 96 flat-bottom wells possessing high protein binding capacity (MaxiSorp™, Nunc–Thermo Scientific);
- Coating antibody: polyclonal anti-mouse IgG (from sheep, R12569, 2.2 mg mL^{-1}, Acris Antibodies);

- Anti-caffeine monoclonal antibody, from mouse, produced using as immunogen 7-ethoxycarboxypentyl-1,3-dimethylxanthine conjugated to a protein (mouse IgG 2b, C0110-06, clone 1.BB.877, 1.37 mg mL^{-1}, USBiological);
- Enzyme conjugate: horseradish peroxidase (HRP) coupled with 7-ethoxycarboxypentyl-1,3-dimethylxanthine (CAF-HRP conjugate, BAM CAF-T-01, BAM, Berlin, Germany);
- Buffer salts: Fluka "ultra" quality grade; 3,3',5,5'-Tetramethylbenzidine (TMB): research grade; Tween 20: pure; Hydrogen peroxide 30%: Fluka Trace select® grade; caffeine: reference standard, Sigma.

12.2.2.3 Buffers and Solutions

- Assay buffer: Tris-NaCl buffer (10 mM tris(hydroxylmethyl)amino-methane, 150 mM NaCl, pH 8.5);
- PBS buffer (10 mM sodium dihydrogen phosphate; 70 mM disodium hydrogen phosphate, 145 mM NaCl, pH 7.6);
- Washing buffer (0.05% (v/v) Tween^{TM} 20 in PBS buffer pH 7.6;
- Substrate buffer (220 mM potassium dihydrogen citrate, 0.5 mM sorbic acid potassium salt, pH 4.0);
- TMB solution (40 mM tetramethylbenzidine and 8 mM tetrabutylammonium borohydride in N,N-dimethylacetamide);
- Substrate solution (540 μL TMB solution and 3 mM hydrogen peroxide in 21.5 mL of substrate buffer). Must be prepared not earlier than 30 min before its use;
- Sulfuric acid 1M solution.

12.2.2.4 Protocol

Buffers, standards and samples must be at RT before starting the assay, else edge-effects can deleteriously affect the well-to-well repeatability. Covering the plates with Parafilm® is recommended during the incubation steps, especially for overnight (o/n) incubations, in order to minimize evaporation.

- Coating antibody (anti-mouse IgG), diluted 1 : 1000 in assay buffer, incubated o/n at RT on the microtitre plate shaker at 750 rpm (200 μL per well);
- Plate washing (3x) with washing buffer;
- Anti-caffeine monoclonal antibody, diluted 1 : 100 000 in assay buffer, incubated for 1 h 30 min at RT on the microtitre plate shaker at 750 rpm (200 μL per well);
- Plate washing (3x) with washing buffer;
- Standards and samples are pipetted into the wells (100 μL per well) and topped by the labeling enzyme conjugate (100 μL per well) which was

previously diluted 1 : 50 000 in assay buffer. The plate is incubated for 30 min under the same conditions as mentioned for the antibody;

- Plate washing (3x) with washing buffer;
- Substrate solution is added into the plate (200 µL per well) and incubated for 30 min. A blue color gradation should be observed within the plate: dark blue for the blanks and fading according to caffeine concentration increments;
- Sulfuric acid 1 M solution added (100 µL per well) producing a yellow-colored solution in the wells;
- Spectrophotometric read-out at two fixed wavelengths: 450 nm and 620 nm;
- Plotting caffeine concentration *versus* OD signal difference 450 nm–620 nm and fitting the sigmoidal curve to the calibrators using a 4-parameter logistic model (Rodbard 1974).

12.2.3 Quantifying Caffeine in Beverages, Tablets and Shampoo

12.2.3.1 Sample Pre-treatment

Samples containing CO_2 must be degasified before analysis. Briefly, 5 mL of the sample can be transferred into a 10 mL glass tube and degassed by immersing a metal tube dispensing argon into the sample for some minutes. Coffee and tea can be used without any pre-treatment but they must be diluted in ultrapure water so their caffeine concentration will fall within the immunoassay calibration range. Caffeine in caffeine-containing tablets has to be brought into solution prior to analysis. A typical protocol encompasses a 3 hour dissolution of the tablets in 1 liter of acidified (pH $\sim$ 3) ultrapure water, under stirring (Carvalho *et al* 2010).

The shampoo samples should be dissolved in ultrapure water prior to analysis. For shampoos containing about 5 g L^{-1} caffeine, a dilution of 0.5 g shampoo in a liter of ultrapure water is suggested.

Because the assay is measuring caffeine concentration in the µg L^{-1} range, samples containing caffeine in the mg L^{-1} or even g L^{-1} range have to be diluted. Caffeine-free beverages like Coca-Cola® caffeine-free and some intensely colored fruit infusions, *e.g.* raspberry tea, can also be measured after a 10-fold dilution. Thus, a caffeine concentration under 0.010 mg L^{-1} (10 x LOQ) can be attributed to such samples (Carvalho *et al* 2010).

12.2.3.2 Determining Caffeine Concentration

Depending on the accuracy and the precision desired, or legally required, the number of standards to be used, as well as the number of replicates should be established by the analytical laboratory or industrial plant. For the results discussed within this chapter, standards ranging from 1 to 100 µg L^{-1} caffeine in ultrapure water were used (1, 1.5, 2.5, 5, 10, 15, 25, 50, 75 and 100 µg L^{-1}), assayed in triplicate, the same as for the samples. A typical calibration curve is depicted in Figure 12.3, fitted by the prevalent calibration model for immunoassays—a logistic 4-parameter sigmoidal curve (A)—but also including its logarithmic transformation to create a linear calibration model. Binding assays are characterized by non-linear calibration models which seem to be, somehow, less accepted than linear ones by analytical staff and, therefore, an attempt to linearize the dose-signal relationship is also provided (B).

The logistic model, although it might look complex at first glance, just needs some extra minutes of attention. The sigmoidal equation, represented in Figure 12.3 (A), is user-friendly (Carvalho *et al* 2010) and has been comprehensively explained in the scientific literature (Schneider *et al* 2005; Findlay and Dillard 2007). A liner transformation is nevertheless easily reachable by the simple logarithmic transformation of the obtained concentrations and signals.

The proposed working range, depicted in Figure 12.3, can be further extended according to the analytical needs. By decreasing the antibody solution concentration, *i.e.* diluting it more, the working range can be lowered, allowing quantification of caffeine in the ng L^{-1} range (Carvalho *et al* 2010). And the opposite, increasing the antibody concentration, will shift the working range upwards as more antibody binding sites will then be available for the analyte. Extending the working range either ways requires, however, drawing additional attention to some validation and quality assurance issues: a) Quantifying caffeine in the ng L^{-1} is associated with lower precision figures ($\pm$ 20%) when compared with the proposed µg L^{-1} range (Carvalho *et al* 2010) and, if sample dilution factors have to be increased for samples containing caffeine in the mg L^{-1} range, then such high dilutions can deleteriously affect the final result accuracy; b) Measuring in the upper µg L^{-1} or mg L^{-1} range requires that sample matrix effects are taken into account as well as sample with color or containing peroxidase-active compounds.

12.2.4 Immunoassay Cross-Reactivity – Selectivity Towards Caffeine

The immunoassay selectivity for caffeine with respect to related xanthines was tested by assaying an aqueous dilution series of theophylline, theobromine, paraxanthine, xanthine, caffeine-*trimethyl*-^{13}C, and caffeine itself. The cross-reactivity was calculated as the ratio of mass concentrations at the inflection points (midpoints, in the 4-parameter logistic curve ca. half maximum inhibitory concentration, IC50) of the corresponding calibration curves and expressed in percent relative to caffeine (Equation 1) as described elsewhere (Schneider *et al* 2005; Bahlmann *et al* 2009).

$$CR = \frac{C_{Caffeine}}{C_{Test}} \times 100\% \qquad (1)$$

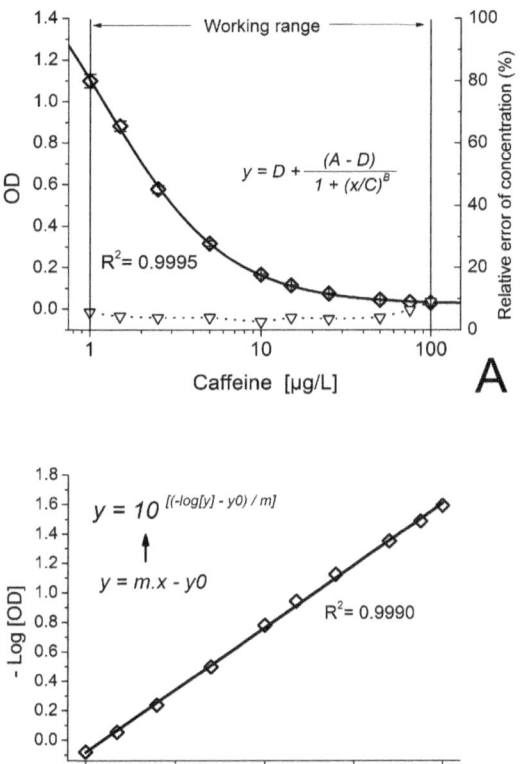

$$y = D + \frac{(A - D)}{1 + (x/C)^B}$$

$R^2 = 0.9995$

Caffeine [µg/L]   **A**

$y = 10^{[(-log[y] - y0) / m]}$

$y = m.x - y0$

$R^2 = 0.9990$

Caffeine [µg/L]   **B**

Figure 12.3 A calibration curve for determining caffeine concentrations by immunoassay: logistic model (A) and the respective linear transformation (B). A) The immunoassay calibration curve used to quantify caffeine in beverages, shampoo and tablets after dissolution ($R^2 = 0.9995$). Each calibration standard was determined in triplicate and the obtained average value represented by an open rhomb associated with the respective standard deviation—the error bars—and it reads on the left ordinate axis (OD, Optical Density). The abscissa axis is represented as a logarithmic scale, as typical in immunoassays. The assay working range, which is dependent on the antibody and enzyme conjugate concentrations used, starts at 1 µg L^{-1} and spans until 100 µg L^{-1}. It can still be further expanded to lower or higher concentration ranges by changing the reactants' concentration as detailed in the text. The right axis represents the relative error of each concentration calculated from the precision profile for binding assays (Ekins 1981); the obtained points are represented by the open down triangle. B) Linear transformation of the same dataset as represented in A, achieved by plotting –Log [OD] *vs.* Log [caffeine]. The obtained calibration can be thus described by the general model y = m * x + y$_0$. The depicted functions' parameters are (A): A = 1.983; B = 1.208; C = 1.190; D = 0.023; R^2 = 0.9995; (B) m = 0.846; y$_0$ = −0.0837; R^2 = 0.9990.

Table 12.2   Key facts for the division "Immunoanalytics".

- Division "Immunoanalytics" is a division within the Department of Analytical Chemistry; Reference Materials of BAM Federal Institute for Materials Research and Testing; Head of Division: Dr. Rudolf J. Schneider.
- The group is dedicated to the development of immunoanalytical methods, especially immunoassays.
- The group develops applications for enzyme-linked immunosorbent assay (ELISA) in the fields of food analysis, environmental monitoring and *in vitro* diagnostics with a focus on quality assurance and uncertainty assessment for these methods.
- We developed an ELISA for caffeine based on a commercial monoclonal antibody. Enzyme conjugates will be available in the BAM webshop http://www.webshop.bam.de/.

Table 12.3   Cross-reactivity of the immunoassay with caffeine related compounds.

Substance	CAS number	Cross-reactivity
Caffeine (1,3,7-trimethylxanthine)	58-08-02	100%
Caffeine-trimethyl-^{13}C	78072-66-9	93%
Theophylline (1,3-dimethylxanthine)	58-55-9	12%
Theobromine (3,7-dimethylxanthine)	83-67-0	0.13%
Paraxanthine (1,7-dimethylxanthine)	611-59-6	0.08%
Xanthine	69-89-6	< 0.002%

CR describes the cross-reactivity in percent; C_{caffeine} is a parameter of the 4PL giving the caffeine concentration at the inflection point and C_{Test} refers to the concentration of the cross-reacting compound at its inflection point. The results are summarized in Table 12.3.

The assay shows very low cross-reactivity overall, theophylline being the only cross-reactant worth mentioning. Samples containing theophylline in equal or higher concentration than caffeine are an obvious limitation of the assay. For concentrations of theophylline 10-fold below caffeine, for example, an overestimation of only 1.2% of the final concentration is expected and when ratios of caffeine : theophylline are higher than 10, the assay can be used without major concern.

12.2.5   Comparing the Results with an LC-MS/MS Method

The samples were analyzed on the same day by immunoassay and by LC-MS/MS and the results obtained by both methods compared as presented in Figure 12.4. Moreover, plotting caffeine concentration by immunoassay *vs.* caffeine concentration by LC-MS/MS produces a regression line with a slope of 1.004 and a correlation coefficient of 0.9946.

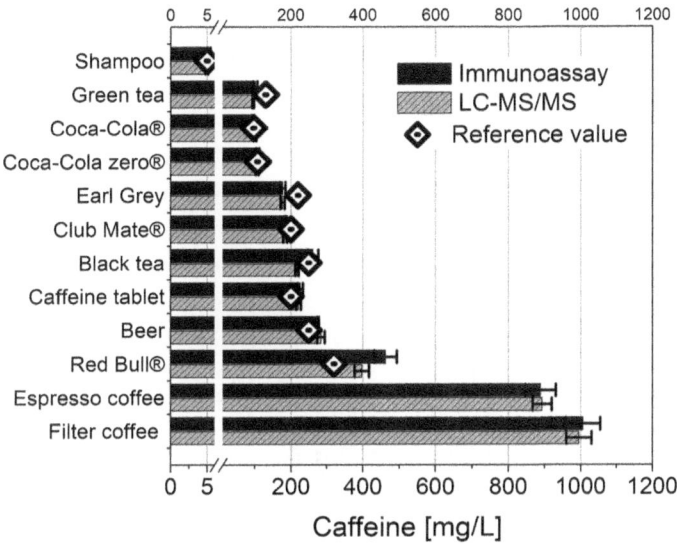

Figure 12.4   Caffeine concentration in beverages, shampoo and tablets determined using the immunoassay and LC-MS/MS, respectively. Concentrations of caffeine in several beverages, shampoo and a tablet. The grey-hatched bars represent results obtained by liquid chromatography-tandem mass spectrometry (LC-MS/MS) while the black-bold bars show results obtained by immunoassay. The error bars, to the right of the columns, correspond to the standard deviation of a triplicate analysis under repeatability conditions. The reference values were obtained from the literature (Pena *et al* 2005; Roehrs and Roth 2008; Carvalho *et al* 2010) and are represented by the black rhombus. For the coffee samples no reference value is provided as the caffeine concentration in such preparations varies broadly depending on the origin of the coffee beans, the preparation time and technique (*e.g.* brewing or dripping), the amount of coffee and water used, *etc.* The shampoo concentration, which cannot be clearly seen in the figure, is 5.1 $\pm$ 0.23 mg L^{-1} (immunoassay) and 4.9 mg L^{-1} $\pm$ 0.16 (LC-MS/MS).

Differences between the caffeine concentration obtained by immunoassay and LC-MS/MS are within a $\pm 7\%$ interval for most samples. Two exceptions were, however, observed for black tea and Red Bull, for which an overestimation of about 15% was registered when using the immunoassay. A possible explanation is the presence of a high concentration of some cross-reactant to the anti-caffeine antibody used. The high theobromine amounts in black teas (Hicks *et al* 1996; Lopez-Martinez *et al* 2003) may as well contribute to the obtained differences. As for Red Bull, further studies will be needed to identify possible cross-reactants in such a complex mixture as well as possible matrix effects when measuring with the immunoassay. The immunoassay results are comparable to the reference values encountered in the scientific literature (Pena *et al* 2005; Roehrs and Roth 2008; Carvalho *et al* 2010) giving confidence in the assay accuracy.

Table 12.4 Assay intra-day and inter-day precision.

Sample	Sample dilution [v/v]	Caffeine concentration measured [µg L^{-1}]	Caffeine concentration in original sample [mg L^{-1}]	Intra-day precision n = 5, same plate [% variation]	Inter-day precision n = 9, over 3 days [% variation]
Black tea	1 : 10 000	2.64	26.4	1.5	6.6
Red Bull® 1	1 : 100 000	4.86	486	2.9	4.6
Shampoo 1	1 : 1000	4.86	4.86	3.2	4.3
Coca-Cola® 1	1 : 10 000	9.81	98.1	1.0	2.9
Filter coffee 1	1 : 100 000	11.1	1110	11	7.8
Green tea 1	1 : 10 000	15.5	155	2.9	4.9
Club-Mate®	1 : 10 000	18.5	185	0.70	1.7
Red Bull® 2	1 : 10 000	46.2	462	3.6	6.7
Espresso coffee	1 : 10 000	88.9	889	1.2	4.7
Coca-Cola® 2	1 : 1000	94.8	94.8	3.7	4.5
Filter coffee 2	1 : 10 000	100.6	1006	4.8	8.2
Green tea 2	1 : 1000	101	101	5.7	5.7
Caffeine tablet	1 : 10 000	22.6	226	0.80	3.3
Shampoo 2	1 : 100	52.4	5.24	0.30	10

Intra-day and inter-day precision was assessed, as displayed in Table 12.4, by assaying the sample dilutions five times on the same day (intra-day) and repeating them over 3 days in triplicate analysis (inter-day). Variations below a 10% threshold were typically obtained.

The accuracy and precision figures obtained, oaths the immunoassay a bright perspective in industrial plants as process control and for quality control of the final product. Its low-cost and high-throughput are matchless within available analytical methods for assessing caffeine concentrations in liquid samples. Within a few hours batches of 66 samples can be analyzed in a single microtitre plate and, several plates can be run simultaneously. Furthermore, the cross-bordering application of the immunoassay allows for the quantitation of caffeine in several matrices other than beverages and consumer products.

12.3 Final Considerations

Caffeine is produced by nature, consumed by humans and sent back to nature through our home toilets. All along this cycle, caffeine concentration can be measured using a single analytical tool, the proposed immunoassay, after some adjustment regarding the calibration range selection and an appropriate sample pre-treatment. Concerning beverages, a pre-treatment is required only if samples contain CO_2, which can be easily removed by bubbling an inert gas like argon, for example. Having a calibration range adapted to the $\mu g\ L^{-1}$ range, the immunoassay can ensure a caffeine-free status to some beverages, at least below $1\ \mu g\ L^{-1}$ caffeine.

The high-throughput of the assay associated with the low volumes of sample required for testing (below 50 μL)—without compromising accuracy and precision—constitute a major improvement in cost-saving for process quality control in industrial plants and analytical laboratories.

Summary Points

- This chapter focuses on immunoassay technique to quantify caffeine in several aqueous sample types.
- Caffeine concentrations as low as 1 microgram per liter can be determined.
- The assay is robust, reliable, precise and accurate to verify caffeine content in several beverages and other consumer products such as caffeine tablets and shampoos.
- No sample pre-treatment is required besides degasification (with carbon dioxide containing beverages) and dilution in ultrapure water.
- The assay is based on a commercially available monoclonal antibody.
- A step-by-step description of the method is provided and it can be easily performed in analytical labs possessing conventional immunoassay equipment: 96-well microtitre plates, microtitre plate photometer ("reader"), microtitre plate shaker and microtitre plate washer.

• In less than 3 hours a batch of 66 samples, assayed in triplicate, can be analyzed in a single microtitre plate. Bearing in mind that at least two microtitre plates can be handled in parallel by a trained analyst, the immunoassay offers a remarkable throughput among available analytical methods for caffeine quantitation.

Definition of Words and Terms

Antibody: A biomolecule produced in response to an antigen. It will bind specifically to the antigen which induced its formation.

Antigen: A molecule that generates an immunological response and reacts with an antibody and the specific receptor on T and B cells of the immune system.

Carrier: An immunogenic molecule or part of a molecule that is recognized by T cells in an antibody response. It is usually used coupled to a hapten to induce immune response toward that hapten. Proteins like bovine serum albumin (BSA) and keyhole limpet hemocyanin (KLH) are among the most popular carriers for immunochemistry applications.

Conjugate: A reagent that is formed by covalently coupling two molecules together, such as a hapten coupled to an enzyme like horseradish peroxidase (HRP).

Cross-reactivity: The ability of an antibody, specific for one antigen, to also bind a slightly different antigen.

Epitope: The part of an antigen that contacts the antigen binding site of an antibody (paratope). Small molecules like caffeine (haptens) are believed to have a single epitope.

Fc fragment: The portion of an antibody that is responsible for binding to antibody receptor in cells (*in vivo*) and binding to the microtitre plate in immunoassays.

Hapten: A small molecule that can act as an epitope, but is incapable by itself of triggering an antibody response unless bound to a carrier.

Immunoassay: A binding-based analytical procedure that makes use of the antibodies' recognition properties to bind a specific analyte.

Immunogenic: Having the ability to stimulate an immune reaction and to produce specific antibodies against that immunogenic substance.

Immunoglobulins: The serum antibodies, including IgG, IgM, IgA, IgE and IgD.

Immunoglobulin G (IgG): Most abundant antibody in serum and the one predominantly used in immunoassays.

Monoclonal antibodies: Antibodies produced by a single clone of B cells in the immune system.

List of Abbreviations

Ab	Antibody
Ag	Antigen

BSA	Bovine serum albumin
CV	Coefficient of variation
CR	Cross-reactivity
DMF	N,N-Dimethylformamide
e.g.	Latin: exempli gratia (for example)
EIA	Enzyme immunoassay
ELISA	Enzyme-linked immunosorbent assay
EMIT	Enzyme multiplied immunoassay technique
HRP	Horseradish peroxidase
i.e.	Latin: id est. (that is)
IC50	Concentration that is required for 50% inhibition *in vitro*
Ig	Immunoglobulin
LC-MS/MS	Liquid chromatography-tandem-mass spectrometry
Mab	Monoclonal antibody
NSB	Non-specific binding
o/n	Over night
OD	Optical density
Pab	Polyclonal antibody
PBS	Phosphate-buffered saline
RIA	Radioimmunoassay
RT	Room temperature, 20 °C
SPE	Solid-phase extraction
TBS	Tris-buffered saline
TMB	3,3′,5,5′-tetramethylbenzidine
vs.	Latin: versus (against, turned)

References

Bahlmann, A., Weller, M. G., Panne, U. and Schneider, R. J., 2009. Monitoring carbamazepine in surface and wastewaters by an immunoassay based on a monoclonal antibody. Analytical and Bioanalytical Chemistry. 395: 1809–1820.

Barone, J. J. and Roberts, H. R., 1996. Caffeine consumption. Food and Chemical Toxicology. 34: 119–129.

Buerge, I. J., Poiger, T., Müller, M. D. and Buser, H. R., 2003. Caffeine, an anthropogenic marker for wastewater contamination of surface waters. Environmental Science & Technology. 37: 691–700.

Carvalho, J. J., Emmerling, F. and Schneider, R. J., 2007. 7-(5-Carboxypentyl)-1,3-dimethyl-xanthine monohydrate. Acta Crystallographica, Section E: Structure Reports Online. 63: O3718–U2211.

Carvalho, J. J., Weller, M. G., Panne, U. and Schneider, R. J., 2010. A highly sensitive caffeine immunoassay based on a monoclonal antibody. Analytical and Bioanalytical Chemistry. 396: 2617–2628.

Carvalho, J. J., 2011. Immunochemical and chromatographic methods for two anthropogenic markers of contamination in surface waters: caffeine and coprostanol. PhD Thesis, Humboldt-Universitaet zu Berlin, e-doc service: http://edoc.hu-berlin.de/docviews.

Cook, C. E., Tallent, C. R., Amerson, E. W., Myers, M. W., Kepler, J. A., Taylor, G. F. and Christensen, H. D., 1976. Caffeine in Plasma and Saliva by a Radioimmunoassay Procedure. Journal of Pharmacology and Experimental Therapeutics. 199: 679–686.

Ekins, R. P., 1981. The precision profile: its use in RIA assessment and design. The Ligand Quarterly. 4: 33–44.

Fickling, S. A., Hampton, S. M., Teale, D., Middleton, B. A. and Marks, V., 1990. Development of an enzyme-linked-immunosorbent-assay for caffeine. Journal of Immunological Methods. 129: 159–164.

Findlay, J. W. A. and Dillard, R. F., 2007. Appropriate calibration curve fitting in ligand binding assays. AAPS Journal. 9: E260–E267.

Focazio, M. J., Kolpin, D. W., Barnes, K. K., Furlong, E. T., Meyer, M. T., Zaugg, S. D., Barber, L. B. and Thurman, M. E., 2008. A national reconnaissance for pharmaceuticals and other organic wastewater contaminants in the United States - II) Untreated drinking water sources. Science of the Total Environment. 402: 201–216.

Fredholm, B. B., Battig, K., Holmen, J., Nehlig, A. and Zvartau, E. E., 1999. Actions of caffeine in the brain with special reference to factors that contribute to its widespread use. Pharmacological Reviews. 51: 83–133.

Helgeson, C., Hu, M., Chegwidden, K., Collins, C., Singh, P. and Jaklitsch, A., 1983. A homogeneous enzyme-immunoassay for caffeine in serum. Clinical Chemistry. 29: 1275–1275.

Hicks, M. B., Hsieh, Y. H. P. and Bell, L. N., 1996. Tea preparation and its influence on methylxanthine concentration. Food Research International. 29: 325–330.

Ladenson, J. H., Ladenson, R. C., Landt, Y. and Crimmins, D. L., 2006. International Patent WO 2006/036882 A2. Method for determining and lowering caffeine concentration in fluids. Assignee to Washington University, United States of America. Issue date: 06.04.2006. World Intellectual Property Organization (WIPO).

Lopez-Martinez, L., Lopez-de-Alba, P. L., Garcia-Campos, R. and de Leon-Rodriguez, L. M., 2003. Simultaneous determination of methylxanthines in coffees and teas by UV-Vis spectrophotometry and partial least squares. Analytica Chimica Acta. 493: 83–94.

Mandel, H. G., 2002. Update on caffeine consumption, disposition and action. Food and Chemical Toxicology. 40: 1231–1234.

Miceli, J. N., Aravind, M. K. and Ferrell, W. J., 1984. Analysis of caffeine: comparison of the manual enzyme multiplied immunoassay (EMIT), automated EMIT, and high-performance liquid chromatography procedures. Therapeutic Drug Monitoring. 6: 344–347.

Pearson, S., Smith, J. M. and Marks, V., 1984. Measurement of plasma caffeine concentrations by substrate labeled fluoroimmunoassay. Annals of Clinical Biochemistry. 21: 208–212.

Pena, A., Lino, C. and Silveira, M. I. N., 2005. Survey of caffeine levels in retail beverages in Portugal. Food Additives & Contaminants, Part A: Chemistry, Analysis, Control, Exposure & Risk Assessment. 22: 91–96.

Rodbard, D., 1974. Statistical quality-control and routine data-processing for radioimmunoassays and immunoradiometric assays. Clinical Chemistry. 20: 1255–1270.

Roehrs, T. and Roth, T., 2008. Caffeine: sleep and daytime sleepiness. Sleep Medicine Reviews. 12: 153–162.

Rogers, P. J. and Dernoncourt, C., 1998. Regular caffeine consumption: a balance of adverse and beneficial effects for mood and psychomotor performance. Pharmacology Biochemistry and Behavior. 59: 1039–1045.

Schneider, C., Schöler, H. F. and Schneider, R. J., 2005. Direct sub-ppt detection of the endocrine disruptor ethinylestradiol in water with a chemiluminescence enzyme-linked immunosorbent assay. Analytica Chimica Acta. 551: 92–97.

Schneider, R. J., 2003. Environmental immunoassays. Analytical and Bioanalytical Chemistry. 375: 44–46.

Seitzinger, S. P., Styles, R. M., Lauck, R. and Mazurek, M. A., 2003. Atmospheric pressure mass spectrometry: a new analytical chemical characterization method for dissolved organic matter in rainwater. Environmental Science & Technology. 37: 131–137.

Tarantino, G., Conca, P., Capone, D., Gentile, A., Polichetti, G. and Basile, V., 2006. Reliability of total overnight salivary caffeine assessment (TOSCA) for liver function evaluation in compensated cirrhotic patients. European Journal of Clinical Pharmacology. 62: 605–612.

Weigel, S., Berger, U., Jensen, E., Kallenborn, R., Thoresen, H. and Hühnerfuss, H., 2004. Determination of selected pharmaceuticals and caffeine in sewage and seawater from Tromsø/Norway with emphasis on ibuprofen and its metabolites. Chemosphere. 56: 583–592.

Wong, P., Leylandjones, B. and Wainer, I. W., 1995. A competitive enzyme-linked-immunosorbent-assay for the determination of N-acetyltransferase (Nat2) Phenotypes. Journal of Pharmaceutical and Biomedical Analysis. 13: 1079–1086.

Zysset, T., Wahlländer, A. and Preisig, R., 1984. Evaluation of caffeine plasma levels by an automated enzyme-immunoassay (EMIT) in comparison with a high-performance liquid chromatographic method. Therapeutic Drug Monitoring. 6: 348–354.

CHAPTER 13

Quantification of Self-Reported Caffeine Use

MERIDETH ADDICOTT

Department of Psychiatry and Behavioral Sciences, Duke University Medical Center, 2301 Erwin Road, Durham NC 27705, USA
E-mail: Merideth.addicott@duke.edu

13.1 Introduction

Caffeine is best known for its mild stimulant effects on the nervous system, but it acts on other peripheral tissues as well. The habitual use of caffeine has been extensively investigated as a potential health concern. Studies in this field include the effects of caffeine on blood pressure, bone loss, pregnancy, cerebral blood flow, sleep, vigilance, cognitive performance, reaction time, and mood (for review, see Fredholm *et al* 1999). All of this research on the chronic and acute effects of caffeine relies on the accurate estimation of caffeine consumption. Given its presence in many popular foods and beverages (*e.g.*, tea, coffee, soft drinks, chocolate), caffeine is a common component of a Western diet but consumption can vary widely among individuals. There is also variability within individuals since people do not consume the same caffeinated foods and beverages, or the same amount of those items, every day. Another source of variability is the caffeine concentrations of coffee and tea, which depend on brewing strength and manufacturer (Bracken *et al* 2002). Despite these challenges, caffeine use estimates are routinely calculated from self-reported information, and these estimates are easily validated with physiological concentrations.

Food and Nutritional Components in Focus No. 2
Caffeine: Chemistry, Analysis, Function and Effects
Edited by Victor R Preedy
© The Royal Society of Chemistry 2012
Published by the Royal Society of Chemistry, www.rsc.org

This chapter will discuss the utility of self-reported caffeine use as well as other methods used to measure caffeine use. Issues involving the use of standardized caffeine content information and the validation of self-reports with physiological caffeine concentrations will also be addressed.

13.2 Estimation of Dietary Caffeine Consumption

The long-term effects of habitual caffeine use on health and behavior are commonly studied by identifying and comparing naturally low caffeine consumers with naturally high caffeine consumers. The most economical and efficient characterization of normal caffeine use consists of an individual's self-reported behavior. Self-reports of caffeine intake may be either retrospective or prospective questionnaires or interviews. Self-reported drug use is sometimes suspected of being unreliable, since individuals may have reason to be deceptive about illicit drug use. However, self-reports of illicit drugs have been shown to be valid and reliable, and since caffeine use is an accepted part of Western culture, self-reports are not likely to be biased by social desirability concerns.

Caffeine use estimates are most accurate when they are adjusted for the approximate caffeine concentrations for each type of food, beverage, or medication, as well as the amount and frequency of use. Therefore, a thorough caffeine use questionnaire or interview should query each source of caffeine (*e.g.*, coffee, chocolate, tea, soft drinks, energy drinks, medications, supplements, *etc.*), the brand names, the number of occasions each source is consumed per day, number of beverages or amount ingested per occasion, the volume per beverage, the brewing method (*e.g.*, drip-brewed, French press, instant, espresso), the brewing duration, whether the chocolate was milk or dark, whether the coffee and tea was regular or decaf, and whether the tea was green or black and iced or hot. Dilution with milk or ice should also be noted and factored into the volume calculation. Previously we have used a caffeine consumption questionnaire (CCQ) described in Landrum (1992) and modified to include other sources of caffeine such as energy drinks and medications, see Table 13.1 for an example. The CCQ is divided into four columns for different time periods (*e.g.*, morning, afternoon, evening) with separate rows for each type of caffeinated food, beverage, and medication, as well as the brewing method for coffee. This was used both as a retrospective interview guide and as a prospective diary (Addicott *et al* 2009).

13.2.1 Self-report Methods

Retrospective self-reports of caffeine use are usually questionnaires or interviews of past or typical caffeine use. The retrospective report is more time-efficient than the prospective method and is more commonly used. In addition, a skilled interviewer may be capable of soliciting a more thorough and systematic history than a paper-and-pencil questionnaire. Most impor-

Table 13.1 Example of a caffeine quantification questionnaire. Adapted from Landrum (1992)

	Morning 6am–12pm		Afternoon 12pm–6pm		Evening 6pm–12pm	
	Volume of beverage	Number of servings	Volume of beverage	Number of servings	Volume of beverage	Number of servings
COFFEE						
Drip-brewed						
Percolated						
French press						
Espresso						
Instant						
Decaf						
Restaurant:						
TEA						
Black						
Green						
Hot						
Iced						
Decaf						
Brand:						
SOFT DRINKS						
Brand name(s):						
Regular						
Diet						
ENERGY DRINKS						
Red Bull						
Rock Star						
Monster						
5-hour Energy						
AMP Energy						

Table 13.1 (*Continued*)

	Morning 6am–12pm		Afternoon 12pm–6pm		Evening 6pm–12pm	
	Amount or serving size	Number of servings	Amount or serving size	Number of servings	Amount or serving size	Number of servings
CHOCOLATE						
Milk						
Dark						
Cocoa						
Beverage						
	Dose	Number of pills	Dose	Number of pills	Dose	Number of pills
DRUGS						
Midol						
Exedrin						
Anacin						
Goody's Headache Pwdr						
NoDoz						
Vivarin						
Dexatrim						
Other:						

tantly, a retrospective report may be the only means of determining onset of use, development of daily use patterns, and changes in past caffeine use. The main limitation of this method is that it relies on an individual's memory, and may be prone to estimation errors or over-generalization of typical caffeine use. A more precise measure of current daily caffeine use is the prospective self-report. This is a diary of caffeine use, usually recorded as it is consumed. Since caffeine intake varies from day to day within individuals, subjects should complete a diary for several days in order to calculate the average daily intake. To obtain a weekly average, it is important to include both work days and weekends, as caffeine consumption may decrease on days when cognitive demands are lower. A limitation of the diary method is that it may be inadvertently recorded during a period of atypical caffeine use, and therefore not be representative of an individual's normal use. The diary may also be inconvenient for subjects to fill out, resulting in poor compliance. In addition, since it is being recorded outside the experimental environment, there is no way to ensure subjects are recording caffeine use as it occurs or filling in the form at the end of the diary period. Another disadvantage is the potential for reactive self-monitoring, when the monitoring of one's own behavior alters the observed behavior. See Table 13.2 for a summary.

In general, studies have shown agreement between retrospective and prospective measures (Rapoport *et al* 1984; Lelo *et al* 1986). Addicott *et al* (2009) reported a moderate correlation ($r = 0.77$) between average daily caffeine use estimated from a retrospective interview of typical weekly caffeine use and a prospective diary of caffeine use across 7 days.

Estimated use from both measures also correlated with salivary caffeine concentrations [CAF] obtained during normal caffeine use ($r = 0.61$ and 0.68,

Table 13.2 Comparison between retrospective and prospective self-report methods. Summary of differences between the two methods described in the main article.

Retrospective Self-report	Prospective Self-report
• Measures past or typical use	• Measures current daily use
• Questionnaire or interview method	• Diary method
• Fast (<10 min)	• Slow (1–7 days or more)
• Estimate of caffeine use	• Precise record of time and date of caffeine use
• Can provide information regarding the history of caffeine use, onset of use, changes in use patterns	• Can provide information regarding current, ongoing caffeine use
• May correlate poorly with physiological [CAF]	• Correlates well with physiological [CAF]
Limitations	
• Can be prone to memory errors	• May be recorded during period of atypical use
• Susceptible to over-generalization	• No experimenter supervision
	• May be affected by reactive self-monitoring

respectively; samples obtained between 7am–2pm). The strongest correlation was between salivary [CAF] and a diary of caffeine use from that same day (r = 0.86). However, subjects who reported consuming more than 600 mg per day in the retrospective report, reported significantly less caffeine use in the prospective diary. High caffeine users may be more prone to memory or estimation errors on retrospective reports, or their caffeine use may be more variable and thus poorly represented by a brief diary (Addicott *et al* 2009).

Alternative options to self-reports are observer reports of a subject's caffeine use. This has shown agreement with self-reports (James *et al* 1988), and has been used to measure children's caffeine use by using parents as observers (Rapoport *et al* 1984). However, observers may only have a limited amount of time in which to record the subject's behavior. Lastly, ecological momentary assessment is designed to collect measures close in time to the experience. Typically, a PDA or other mobile device is provided to the subject to collect mood and behavior data throughout the day. Subjects may be prompted at random intervals to enter information, or be instructed to enter information regarding behaviors as they occur. Ecological momentary assessment has been used to study drug use, such as tobacco and alcohol, but, to our knowledge, has not yet been used to study caffeine consumption.

13.3 Standardized Caffeine Content

The actual caffeine contents of coffee, tea, and chocolate can vary widely due to differences in brewing strength and manufacturing. In one study, caffeine content in caffeinated coffees ranged from 58 to 259 mg per dose, and decaffeinated coffees had as much as 17.7 mg per dose. Also notable was the wide range of caffeine concentrations (259–564 mg per dose) in the same coffee beverage obtained from the same outlet restaurant on six consecutive days (McCusker *et al* 2003). Despite this variability, investigators rely on standardized caffeine content information to quickly estimate caffeine consumption from self-reported data. Adjusting caffeine estimates by the volume and the type of beverage correlates more strongly with actual caffeine concentrations from beverage samples than an earlier method of estimation based on the absolute number of caffeinated beverages consumed (Lelo *et al* 1986). A frequently cited source of standardized caffeine content is Bunker and McWilliams' *Caffeine Content of Common Beverages* (1979). Bunker and McWilliams sampled caffeine concentrations in coffee brewed by different methods and reported the relative concentrations to be: freeze dried < percolator < drip-brewed. Black and green tea, which had similar caffeine contents overall, had caffeine concentrations that increased by brewing duration (1 < 3 < 5 min). The average values they reported were: 97 mg 100 mL^{-1} in drip-brewed coffee, 20 mg 100 mL^{-1} in black tea (1-minute brew time), and 6 mg 100 mL^{-1} in cocoa (Bunker and McWilliams 1979). Although these absolute values may not be accurate today, the differences in relative values demonstrate the importance of taking into account the beverage

Table 13.3   Comparison of estimated caffeine concentrations by source. Concentrations adapted from Bunker and McWilliams (1979); Barone and Roberts (1996); Bracken *et al* (2002); and the Center for Science in the Public Interest (CSPI) website, accessed April 2011.

	Bunker and McWilliams (1979)	*Barone and Roberts (1996)*	*Bracken* et al. *(2002)*	*CSPI (2011)*
Coffee (drip-brewed)	29.6 mg oz^{-1}	17 mg oz^{-1}	9.8 mg oz^{-1}	16.6 mg oz^{-1}
Tea (brewed 3 min)	Black: 9.3 mg oz^{-1} Green: 5.7 mg oz^{-1}	6 mg oz^{-1}	4.2 mg oz^{-1}	6.6 mg oz^{-1}
Cocoa	1.8 mg oz^{-1}	1 mg oz^{-1}		1.1 mg oz^{-1}

preparation for estimation of caffeine use. Another popular source of standardized contents is Barone and Roberts' *Caffeine Consumption* (1996). Here, published caffeine contents from many different sources were reviewed and the authors recommend 85 mg 5 oz^{-1} of caffeine in ground roasted coffee, 30 mg 5 oz^{-1} in tea, and 4 mg 5 oz^{-1} in cocoa (Barone and Roberts 1996). See Table 13.3 for a comparison. In contrast to coffee, tea and cocoa, other caffeinated foods and beverages such as soft drinks, energy drinks, bottled ice tea, medications, and dietary supplements have fixed caffeine contents per serving, which is usually listed on the packaging or is available from the manufacturer.

Estimating caffeine use is useful within a study to separate and compare low caffeine consumers with high caffeine consumers. For instance, we used standardized caffeine content values of 135 mg 8 oz^{-1} of drip-brewed coffee, 50 mg 8 oz^{-1} black tea, 10 mg 1.5 oz^{-1} of milk chocolate and 31 mg 1.5 oz^{-1} of dark chocolate obtained from the Center for Science in the Public Interest web site (www.cspinet.org; accessed July 1997). These totals were used to classify subjects into low (<200 mg per day), moderate ($200–600$ mg per day), and high (>600 mg per day) caffeine use groups and the caffeine in mg per day was correlated with salivary caffeine concentrations ($r = 0.86$) (Addicott *et al* 2009). The limitation of using standardized caffeine content information is that these are not absolute values, and comparisons across studies are limited if different standardized content information was used. Subjects classified as low caffeine users in one study may be classified as moderate users if caffeine use is quantified differently.

13.3.1   Beverage Sampling

Some studies have asked for beverage samples from their participants, in order to assay the actual caffeine concentrations. In one study, 186 subjects provided coffee and tea samples on 2 occasions (Bracken *et al* 2002). The authors reported a wide range of caffeine content across coffee brands (M = 330 µg mL^{-1}, SD = 177, range 3 to 742), and even within a brand between brewing sessions performed by the same subject (*e.g.*, Maxwell House, 95 and 113 μg mL^{-1}). Caffeine concentration in tea brewed more than 3 min ranged between 24 to 359 μg mL^{-1}, (M = 142 μg mL^{-1}, SD = 137) (Bracken *et al* 2002). Bracken *et al* (2002) also compared their caffeine concentration values to those reported in Bunker and McWilliams (1979). Bracken reported substantially lower average caffeine concentrations in tea and coffee. In contrast, an earlier study asked high caffeine consumers (*n* = 12; ≥8 cups of tea or coffee per day) to provide samples of each unique caffeinated beverage consumed for a week (James *et al* 1988). The authors reported that the caffeine values they assayed were in agreement with standardized caffeine content, and the beverage sampling did not produce a stronger correlation than the estimates based on standardized caffeine content (Barone and Roberts 1984). Considering the cost and the time taken to assay so many samples, James *et al* (1988) concluded that beverage sampling is not warranted, but the Bracken *et al* (2002) article illustrates the importance of choosing standardized caffeine content information from the most recent data available.

13.3.2   Controlled-dose Studies

The variability of dietary caffeine consumption can be disadvantageous when studying the development of tolerance to caffeine. When knowing the exact dose of caffeine consumed each day is vital to a study, an alternative to tracking normal dietary caffeine consumption with self-reports is to administer controlled-doses over a period of time while participants abstain from dietary caffeine use. For instance, one study conducted a crossover trial in which subjects were maintained on placebo for one week, then 300 mg per day for a week, then 600 mg per day for another week to investigate tolerance to the hypertensive effects of caffeine (Lovallo *et al* 2004). Another study used an adjusted dose of 1.75 mg kg^{-1} administered 3 times per day for 7 days to investigate tolerance to the mood and psychomotor effects of caffeine (James *et al* 2005). The advantages of the controlled-dose method are that investigators know the exact amount and duration of caffeine use, and can compare the effects of different doses within the same subject sample. A disadvantage is that caffeine must be administered long enough for tolerance to develop, usually for a period of several days to a week, and compliance must be monitored. This increases the length of time participants must remain in the study, which could compromise feasibility and increase attrition. There are also ethical considerations about maintaining individuals on high daily doses of caffeine due to the risk of adverse health consequences (Nawrot *et al* 2003).

13.4   Self-reported Caffeine Use Validation with Physiological Caffeine Concentrations

Recent self-reported caffeine use is often validated with physiological caffeine concentrations [CAF]. However, there are significant inter-individual differences in the rate of caffeine metabolism, and the time of sample collection and the recency of caffeine use are also important considerations when interpreting physiological [CAF].

Caffeine absorbs rapidly from the gastrointestinal tract and peak concentrations are reached after approximately 30 minutes in plasma (Blanchard and Sawers 1983) and between 40–70 minutes in saliva (Liguori *et al* 1997). The majority of caffeine is metabolized in the liver by enzyme P450 to create paraxanthine (84%), theobromine (12%), and theophylline (4%) (Kalow and Tang 1993). These metabolites are further broken down in the liver before being excreted in urine. Less than 3% of caffeine is eliminated in urine unchanged (Kalow and Tang 1993).

The average half-life of caffeine metabolism for a 250 mg dose has been reported as 6.2 ± 1.6 hours (Patwardhan *et al* 1980). The half-life of caffeine is lengthened in women using oral contraceptive steroids and during pregnancy, and the half-life is reduced by tobacco use. For example, Campbell *et al* (1987) reported an average half-life of ($M \pm SE$) 6.8 ± 0.8 h for healthy controls, 9.2 ± 0.7 h for women using oral contraceptives, and 4.0 ± 0.6 h for smokers (Campbell *et al* 1987). Alcohol intake of 50 g per day (equal to 5 standard drinks) was also shown to prolong caffeine half-life by 72% (George *et al* 1986). No gender differences have been reported but there are differences according to age; fetuses and infants have less P450 activity and have much longer caffeine half-lives than adults (Fredholm *et al* 1999).

Caffeine concentrations can be assayed from blood, saliva, or urine samples. Blood samples may be considered the gold standard, but saliva samples are more comfortable for the participant and are commonly used. Salivary [CAF] are approximately 70% of serum concentrations and the correlation between serum and salivary [CAF] has been reported to be $r = 0.99$ from samples taken 105 minutes after caffeine administration (Biederbick *et al* 1997). However, saliva samples collected earlier had higher concentrations than serum, possibly due to gingival contamination (caffeine citrate was administered in a liquid solution) (Biederbick *et al* 1997). Therefore, when timing sample collection, it is important to consider that recent coffee, tea, or soft drink consumption could contaminate the saliva sample. Caffeine intake by athletes at sporting events is monitored by urinary [CAF], with an upper limit of 12 mg L^{-1} allowed. There is a strong correlation between urine and plasma [CAF] ($r = 0.93$) and the plasma : urine concentration ratios have been reported to range from 1.10 to 1.74 (Birkett and Miners 1991). See Table 13.4 for summary of caffeine metabolism.

The strength of the relationship between [CAF] and self-reported caffeine use varies depending on when the physiological sample was obtained. One study found no relationship between self-reported caffeine use and plasma [CAF] when blood samples were obtained in the morning following overnight caffeine abstention (Kennedy *et al* 1991). In another study, subjects' average self-reported caffeine consumption was correlated with salivary [CAF] and paraxanthine concentrations [PX] obtained at approximately 5pm. There were modest, but

Table 13.4 Summary about caffeine metabolism. Information summarized from the main article.

- Caffeine absorption peaks between 30 and 70 minutes following ingestion.
- In the human body, caffeine has a half-life of approximately 4–7 hours. The rate of metabolism can be affected by pregnancy, oral steroid contraceptives, tobacco use, and alcohol use.
- Caffeine concentrations can be assayed from blood, saliva, or urine.
- Caffeine concentrations are lower in saliva and urine than in blood, but concentrations correlate strongly between samples.
- The strength of correlation between physiological caffeine concentrations and self-reported caffeine use can depend on how much time has elapsed since caffeine consumption and sample collection.

significant, correlations with [CAF] ($r = 0.31$) and [PX] ($r = 0.42$) (James *et al* 1989). The authors conclude that late afternoon [PX] is more representative of caffeine consumption. Typically, caffeine consumption is greatest in the morning and tapers off throughout the day, therefore most of the caffeine in coffee or tea drank in the morning has been metabolized into PX by 5pm. PX has a similar half-life as caffeine, but PX concentrations [PX] are less sensitive to recent caffeine use and the concentrations are less variable throughout the day (Lelo *et al* 1986). However, a third study compared self-reported caffeine intake among moderate and high consumers across a 24-hour period with plasma [CAF]. Here, a single plasma [CAF] sample drawn at 5pm correlated strongly with estimated caffeine use from self-reports ($r = 0.97$) and [CAF] was a better index of caffeine ingestion than [PX] sampled at any time of day (Lelo *et al* 1986). However, moderate and high consumers probably continued drinking caffeine into the afternoon, in which case not all of the caffeine would have been metabolized by 5pm. If physiological concentrations are to be used to verify self-reported caffeine use, these studies illustrate the importance of considering the time of sample collection and the recency of caffeine use.

13.5 Conclusion

In conclusion, self-reported caffeine use is an effective measure of past and current caffeine use. A thorough caffeine use history should query the source of caffeine (*i.e.*, coffee, tea, soda, *etc.*), the amount consumed per occasion, and the frequency of use. When actual caffeine content information is not available, caffeine use can be estimated using standardized caffeine content. Adjusting for the relative amounts of caffeine contained in different beverages or foods will produce a more accurate estimation of caffeine use. Self-reported caffeine use is measured with questionnaires or interviews and may be either retrospective histories of past or typical caffeine use, or prospective diaries in which caffeine use is recorded as it occurs. Retrospective measures are fast and efficient, but can suffer from memory or estimation errors. Prospective diaries may be more accurate, but could inadvertently record atypical use. In general, studies have shown agreement between these two methods. Physiological caffeine concentrations are often obtained to validate self-reported information. The caffeine concentrations in blood and saliva samples tend to correlate more closely with self-reported caffeine use when the samples are obtained within a few hours after caffeine use. If samples are obtained later, the rate of caffeine metabolism needs to be considered.

Summary Points

- Caffeine is a mild neurostimulant found in many foods and beverages that are common in a Western diet.
- Dietary caffeine use can be estimated from an interview or self-report questionnaire, these measures should query every potential source of dietary caffeine, and the frequency and volume of each item consumed.
- Caffeine use estimates can be based on retrospective data of past caffeine use or prospective data of current caffeine use; retrospective measures are faster but may suffer from memory errors, while prospective measures may be more accurate but could be recorded during a period of atypical caffeine use.
- Actual caffeine concentrations in coffee and tea vary greatly, and depend on brewing method and duration.
- When manufacturer information regarding caffeine content is unavailable, caffeine estimates can be based on average caffeine concentrations reported in coffee, tea, and chocolate.
- When knowing the exact dose of caffeine ingested is vital to a study, an alternative to estimating dietary caffeine consumption is to administer a controlled-dose of caffeine to study participants.
- Peak caffeine absorption occurs 30–70 minutes after ingestion and caffeine has a half-life of 4–7 hours.
- The rate of caffeine metabolism can be affected by pregnancy, oral steroid contraceptives, tobacco use, and alcohol consumption.
- Recent caffeine use can be validated with physiological caffeine concentrations in blood, saliva, or urine samples. Saliva and urine caffeine concentrations are lower than those in blood, but saliva and urine concentrations correlate strongly with blood levels.
- Given the rate of caffeine metabolism, the timing of biological specimen sampling needs careful consideration.
- Under some circumstances, paraxanthine, the primary metabolite of caffeine, could also be used as a biological marker of caffeine use.

Key Facts about Methylxanthines in Coffee, Tea, and Cocoa

- Decaffeinated teas and coffees are not caffeine-free.
- Green coffee beans are roasted to enhance their flavor. The duration and temperature of roasting determines the flavor and can change the volume

and weight of the beans, but does not affect the amount of caffeine in the beans.

- Most commercial coffees are one of two varieties: robusta or arabica. Compared to robusta, arabica plants are more difficult to grow and take longer to produce beans. Robusta beans are less expensive, contain more caffeine, and are thought to produce a less flavorful coffee. Instant coffee is typically made from Robusta beans.
- Espresso is brewed by forcing hot water under pressure through finely ground coffee. Espresso may have a higher concentration of caffeine than other coffee preparations, but 1–2 oz of espresso has less caffeine than 8 oz of drip-brewed coffee.
- The difference between white, green, oolong, and black teas is the level of fermentation (*e.g.*, the extent of enzymatic oxidation). White tea leaves are not fermented, green tea is non- or lightly fermented, oolong tea is partially fermented, and black tea is fully fermented.
- Absolute caffeine concentrations cannot be reliably predicted based on the type of tea (*e.g.*, green, oolong, black). However, the brew time, agitation of the tea during brewing, and other preparation methods can affect caffeine concentrations and may vary between tea types.
- Herbal teas, made from plants other than *Camellia sinensis* generally do not contain caffeine.
- Cocoa contains more theobromine than caffeine. The amount of cocoa solids in chocolate determines the concentration of methylxanthines (dark > milk > white chocolate).

Definition of Words and Terms

Caffeine: 1,3,7-trimethylxanthine, a naturally occurring xanthine alkaloid found in several plant species that acts as a stimulant in the nervous system.

Controlled-dose study: a research method in which participants are administered a fixed amount of a substance at specific intervals while abstaining from dietary sources of that substance.

Half-life: the amount of time required for a substance to be decreased by half due to a process such as metabolization.

Paraxanthine: 1,7-dimethylxanthine, a dimethyl derivative of xanthine and the primary metabolite of caffeine.

Prospective: concerning a future occurrence. In research, a study that tracks individuals' behavior or health over time.

Retrospective: concerning a past occurrence. In research, a study that uses information about behaviors or events that occurred in the past.

Self-reported caffeine use: a measure of caffeine intake based on an individual's first-hand account of his–her behavior.

Standardized caffeine content: A list of average caffeine concentrations used in estimating caffeine contents of food and beverages, when the actual caffeine contents of those food and beverages naturally vary and cannot be directly measured.

Theobromine: 3,7-dimethylxanthine, a naturally occurring dimethyl xanthine found in several plant species and a metabolite of caffeine.

Theophylline: 1,3-dimethylxanthine, a naturally occurring dimethyl xanthine found in several plant species and a metabolite of caffeine.

List of Abbreviations

[CAF]	caffeine concentration
CCQ	caffeine consumption questionnaire
[PX]	paraxanthine concentration
M	mean
SD	standard deviation
SE	standard error
μg	microgram
mg	milligram
g	gram
kg	kilogram
mL	milliliter
oz	ounce
r	Pearson product-moment correlation coefficient
n	sample size
min	minute
h	hour
e.g.	exempli gratia

References

Addicott, M. A., Yang, L. L., Peiffer, A. M. and Laurienti, P. J., 2009. Methodological considerations for the quantification of self-reported caffeine use. Psychopharmacology. 203: 571–578.

Barone, J. J. and Roberts, H. R., 1996. Caffeine consumption. Food and Chemical Toxicology. 34: 119–129.

Biederbick, W., Joseph, G., Rump, A., Theisohn, M. and Klaus, W., 1997. Caffeine in saliva after peroral intake: Early sample collection as a possible source of error. Therapeutic Drug Monitoring. 19: 521–524.

Birkett, D. J. and Miners, J. O., 1991. Caffeine renal clearance and urine caffeine concentrations during steady-state dosing - Implications for monitoring caffeine intake during sports-events. British Journal of Clinical Pharmacology. 31: 405–408.

Blanchard, J. and Sawers, S. J. A., 1983. The absolute bioavailability of caffeine in man. European Journal of Clinical Pharmacology. 24: 93–98.

Bracken, M. B., Triche, E., Grosso, L., Hellenbrand, K., Belanger, K. and Leaderer, B. P., 2002. Heterogeneity in assessing self-reports of caffeine exposure: Implications for studies of health effects. Epidemiology. 13: 165–171.

Bunker, M. L. and McWilliams, M., 1979. Caffeine content of common beverages. Journal of the American Dietetic Association. 74: 28–32.

Campbell, M. E., Spielberg, S. P. and Kalow, W., 1987. A urinary metabolite ratio that reflects systemic caffeine clearance. Clinical Pharmacology and Therapeutics. 42: 157–165.

Fredholm, B. B., Battig, K., Holmen, J., Nehlig, A. and Zvartau, E. E., 1999. Actions of caffeine in the brain with special reference to factors that contribute to its widespread use. Pharmacological Reviews. 51: 83–133.

George, J., Murphy, T., Roberts, R., Cooksley, W. G. E., Halliday, J. W. and Powell, L. W., 1986. Influence of alcohol and caffeine consumption on caffeine elimination. Clinical and Experimental Pharmacology and Physiology. 13: 731–736.

James, J. E., Bruce, M. S., Lader, M. H. and Scott, N. R., 1989. Self-report reliability and symptomatology of habitual caffeine consumption. British Journal of Clinical Pharmacology. 27: 507–514.

James, J. E., Gregg, M. E., Kane, M. and Harte, F., 2005. Dietary caffeine, performance and mood: enhancing and restorative effects after controlling for withdrawal reversal. Neuropsychobiology. 52: 1–10.

James, J. E., Paull, I., Cameron-Traub, E., Miners, J. O., Lelo, A. and Birkett, D. J., 1988. Biochemical validation of self-reported caffeine consumption during caffeine fading. Journal of Behavioral Medicine. 11: 15–30.

Kalow, W. and Tang, B. K., 1993. The use of caffeine for enzyme assays - a critical-appraisal. Clinical Pharmacology and Therapeutics. 53: 503–514.

Kennedy, J. S., Vonmoltke, L. L., Harmatz, J. S., Engelhardt, N. and Greenblatt, D. J., 1991. Validity of self-reports of caffeine use. Journal of Clinical Pharmacology. 31: 677–680.

Landrum, R. E. 1992. College students' use of caffeine and its relationship to personality. College Student Journal. 26: 151–155.

Lelo, A., Miners, J. O., Robson, R. and Birkett, D. J., 1986. Assessment of caffeine exposure - caffeine content of beverages , caffeine intake, and plasma-concentrations of methylxanthines. Clinical Pharmacology and Therapeutics. 39: 54–59.

Liguori, A., Hughes, J. R. and Grass, J. A., 1997. Absorption and subjective effects of caffeine from coffee, cola and capsules. Pharmacology Biochemistry and Behavior. 58: 721–726.

Lovallo, W. R., Wilson, M. F., Vincent, A. S., Sung, B. H., McKey, B. S. and Whitsett, T. L., 2004. Blood pressure response to caffeine shows incomplete tolerance after short-term regular consumption. Hypertension. 43: 760–765.

McCusker, R. R., Goldberger, B. A. and Cone, E. J., 2003. Technical Note: Caffeine content of specialty coffees. Journal of Analytical Toxicology. 27: 520–522.

Nawrot, P., Jordan, S., Eastwood, J., Rotstein, J., Hugenholtz, A. and Feeley, M., 2003. Effects of caffeine on human health. Food Additives and Contaminants. 20: 1–30.

Patwardhan, R. V., Desmond, P. V., Johnson, R. F. and Schenker, S., 1980. Impaired elimination of caffeine by oral-contraceptive steroids. Journal of Laboratory and Clinical Medicine. 95: 603–608.

Rapoport, J. L., Berg, C. J., Ismond, D. R., Zahn, T. P. and Neims, A., 1984. Behavioral effects of caffeine in children: Relationship between dietary choice and effects of caffeine challenge. Archives of General Psychiatry. 41: 1073–1079.

Function and Effects

CHAPTER 14

Caffeine and the Brain: An Overview

ELIO ACQUAS*[1,3,4], MARIA ANTONIETTA DE LUCA[2,3], SANDRO FENU[2,3,4], ROSANNA LONGONI[1,3] AND LILIANA SPINA[1,3]

[1] Department of Life and Environment Sciences, Drug Sciences Section, University of Cagliari, Cagliari, - Via Ospedale, 72 - I-09124 Cagliari, Italy; [2] Department of Biomedicine; [3] INN - National Institute of Neuroscience, University of Cagliari, Via Ospedale, 72 - I-09124 Cagliari, Italy; [4] Center of Excellence for the Neurobiology of Dependence, University of Cagliari, - Via Ospedale, 72 – I-09124 Cagliari, Italy
*E-mail: acquas@unica.it

14.1 Introduction

Present in more than 60 known species of plants, caffeine (1,3,7-trimethyl-xanthine) is certainly the most consumed psychoactive substance in the world. Dietary sources of caffeine consist of different kinds of coffee and teas, chocolate, energy drinks and chewing gums. In addition, caffeine enters in the composition of many over-the-counter drugs used for treating asthma, nasal congestion, headache and sometimes for improving athletic endurance. Caffeine is mostly if not uniquely taken by the oral route; it is adsorbed very quickly and is converted enzymatically in active metabolites that share with caffeine some of its pharmacological properties.

Caffeine's central pharmacological effects are dose-dependently mediated by blockade of adenosine A_1 and A_{2A} receptors (Fredholm *et al* 1999).

Food and Nutritional Components in Focus No. 2
Caffeine: Chemistry, Analysis, Function and Effects
Edited by Victor R Preedy
© The Royal Society of Chemistry 2012
Published by the Royal Society of Chemistry, www.rsc.org

The most significant behavioral effects of caffeine occur after consumption of low to moderate doses and include increased alertness, energy feelings and ability to concentrate. In contrast, higher doses of caffeine may induce negative effects such as anxiety, restlessness, insomnia, and tachycardia.

Caffeine is considered an overall safe compound, it is not classified as a drug of dependence in the DSM-IV (APA, 1994) and the only restriction to its use refers to the IOC regulations. For these reasons, different kinds of dietary products containing caffeine are marketed to both adolescents and adults and moderate caffeine consumption rarely leads to health risks. Nevertheless, a health risk associated with caffeine consumption originates from the recently established habit of teenagers to mix energy drinks with alcoholic beverages (Malinauskas *et al* 2007). Furthermore, adverse effects have been reported, including cardiac arrest, after high doses of caffeine taken from energy drinks in combination with other substances (*i.e.* taurine) (Berger and Alford 2009). Over the years, the scientific literature on caffeine has generated an enormous amount of data which deal with caffeine from many points of view and, as far as the pharmacological properties of caffeine are concerned, besides the studies on the peripheral effects of caffeine, the research on its central properties covers a great number of fields which range from its psychomotor stimulant properties, to its role in cognition, in synaptic plasticity modulation and to its neuroprotective properties in neurodegenerative diseases. At the basis of the central effects of caffeine is the ability to modulate adenosine transmission and, indirectly, to affect the release of neurotransmitters such as DA, ACh and serotonin. Based on these premises, the present chapter aims to provide an overview of the literature on the effects of caffeine in the brain with particular emphasis on its central effects related to its psychomotor stimulant, arousing and neuroprotective properties.

14.2 Pharmacokinetic and Pharmacodynamic of Caffeine

Following its oral administration caffeine is adsorbed rapidly ($\sim$45 minutes) and almost completely ($\sim$99%) by the stomach and the small intestine. Concentrations of caffeine rise in plasma within 30 to 120 minutes as a function of dose and other factors related to the nature of the beverage or the food. Caffeine is a lipophilic molecule that easily penetrates biologic membranes and distributes to body tissues. The elimination half-life of caffeine in humans is in the range of 2.5 to 4.5 hours but it can be lowered by 30–50% in smokers as a consequence of smoke-elicited enzymatic induction. The metabolism of caffeine takes place mostly by the action of the isozyme CYP1A2 which converts it into paraxanthine (1,7-dimethylxanthine), although also theophylline, theobromine and the urinary metabolites, 1-methylxantine and 1-methyluric acid, must be taken into account. The four main molecular mechanisms responsible for central effects of caffeine are blockade of adenosine receptors, inhibition of phosphodiesterases, blockade of $GABA_A$ receptors and translocation of intracellular calcium. These effects are highly dose-dependent and the last three mechanisms cannot be detected at caffeine plasma concentrations obtained after regular–moderate consumption of coffee. Under this condition, blockade of adenosine receptors is responsible for the main central pharmacological effects of caffeine: these are exerted by opposing endogenous adenosine receptors activation and by removing the inhibitory adenosinergic tone. Indeed, adenosine actions are mediated through four distinct G protein-coupled receptor subtypes, namely A_1, A_{2A}, A_{2B} and A_3, whose stimulation leads to modifications involving the activity of adenylyl cyclase, phospholipase C, phospholipase D as well as to changes in the gating state of Ca^{++} and K^+ channels (Fredholm *et al* 1999). However, due to a significantly lower affinity of adenosine for A_{2B} and A_3 receptors, it is commonly described that, under physiological conditions, the actions of adenosine are mediated by A_1 and A_{2A} receptors and, for this reason, caffeine is referred to as a non-selective antagonist of these receptor subtypes (Fredholm *et al* 1999). In particular, adenosine A_1 receptors are coupled to inhibitory G proteins and are distributed throughout the brain with high density in the hippocampus, cerebral cortex, cerebellum and hypothalamus. By antagonizing adenosine actions onto these receptors, caffeine affects the release of neurotransmitters such as DA (this might be the mechanism of the psychomotor stimulant actions of caffeine) and ACh (this might be the mechanism of caffeine-elicited arousal and attention) (Acquas *et al* 2002). Interestingly, a positive interaction between A_1 receptors blockade and DA D_1 receptors stimulation is involved in the motor activating effects of caffeine and other adenosine antagonists and, in fact, blockade of A_1 receptors enhances motor effects of DA D_1 receptor agonists whereas activation of A_1 receptors influences the binding of D_1 agonists. In addition, some of the behavioral effects of caffeine might also take place by antagonizing the negative modulatory effects of adenosine on DA receptors-mediated actions in DA-rich regions such as the striatum, the Acb and the olfactory tubercles where A_{2A} receptors are highly expressed. These receptors are co-expressed with DA D_2 receptors in the striatum and Acb and a critical interaction, to form functional heterodimers, between A_{2A} and D_2 receptors, has been reported. Furthermore, such interaction between adenosine and DA receptors is not merely restricted to that between A_{2A} and D_2 receptors given that, throughout the basal ganglia circuitry, A_{2A} receptors blockade may also potentiate the responses induced by DA D_1 receptor agonists. In summary, the main central effects of caffeine such as increased motor activity, increased arousal, attention and alertness and psychomotor stimulation appear mediated by an action onto A_1 and A_{2A} adenosine receptors (Fredholm *et al* 1999).

14.3 Effects of Caffeine on *In Vivo* Neurotransmitters Release

The effects of caffeine on brain neurotransmission originate from its modulation of the actions of adenosine. Adenosine regulates the release of different neurotransmitters, such as DA, ACh, noradrenaline and serotonin, involved in motor activation and reward, in arousal and attention and in the sleep–wake cycle control. As previously reported, caffeine, at the plasma concentrations reached with dietary intake, acts mainly through the blockade of adenosine A_1 and A_{2A} receptors. A large number of studies have been performed to investigate the effects of caffeine as such, and with respect to the involvement of either A_1 and A_{2A} receptor subtypes, on the release of neurotransmitters. The majority of these studies have been performed by *in vivo* brain microdialysis which permits to analyze changes of concentrations of neurotransmitters in the synaptic cleft before and after the administration of drugs. Over the years, this approach has permitted to study, *in vivo*, the release of neurotransmitters in a wide range of animal models of neuropsychiatric disorders such as depression, schizophrenia and drug addiction. Here, we shortly review the literature on the effects of caffeine administration on *in vivo* DA, ACh, noradrenaline and serotonin neurotransmission in different brain regions as determined by microdialysis.

The understanding of the interactions between caffeine and DA transmission appears critical to characterize how caffeine may affect motor activation, psychomotor stimulation and motivated behaviors. In this regard, microdialysis studies, referring to three DA-rich brain areas, *i.e.* the striatum, the AcbSh and AcbC and the PFCx, have revealed that these interactions are quite complex as a consequence of adenosine and DA receptors co-localization and as a consequence of the opposite control on signal transduction mechanisms exerted by adenosine and DA. Thus, caffeine was reported to increase extracellular DA transmission (Okada *et al* 1996) but it was also reported to decrease it in the presence of an A_1 receptor antagonist (Okada *et al* 1996; 1997) suggesting that, under these conditions, caffeine requires both A_1 and A_{2A} receptors in order to activate DA transmission. Unfortunately, to our knowledge the effects of systemic administration of caffeine on *in vivo* DA transmission in the striatum have not been described and, since in the above studies caffeine was administered locally through the microdialysis probe, this evidence cannot be functional to characterize its effects after systemic administration. Contrasting results have also been reported by studies addressing the effects of systemic caffeine administration on DA transmission in the AcbSh and AcbC. The relevance of these results highly depends on the significance attributed to DA transmission in these forebrain DA-rich structures. In fact, preferential increases of DA transmission in the shell of the Acb, compared to the core, represent a neurochemical index of addictive properties of drugs (Di Chiara 2002), while drug-induced increases of DA transmission in the PFCx have been suggested to represent an index of the psychomotor stimulant properties but not of the addictive properties of drugs (Di Chiara 2002). Thus, while some authors reported failure to detect increases of DA in both shell and core compartments of the Acb (Acquas *et al* 2002; De Luca *et al* 2007), others reported the opposite (Solinas *et al* 2002; Borycz *et al* 2007).

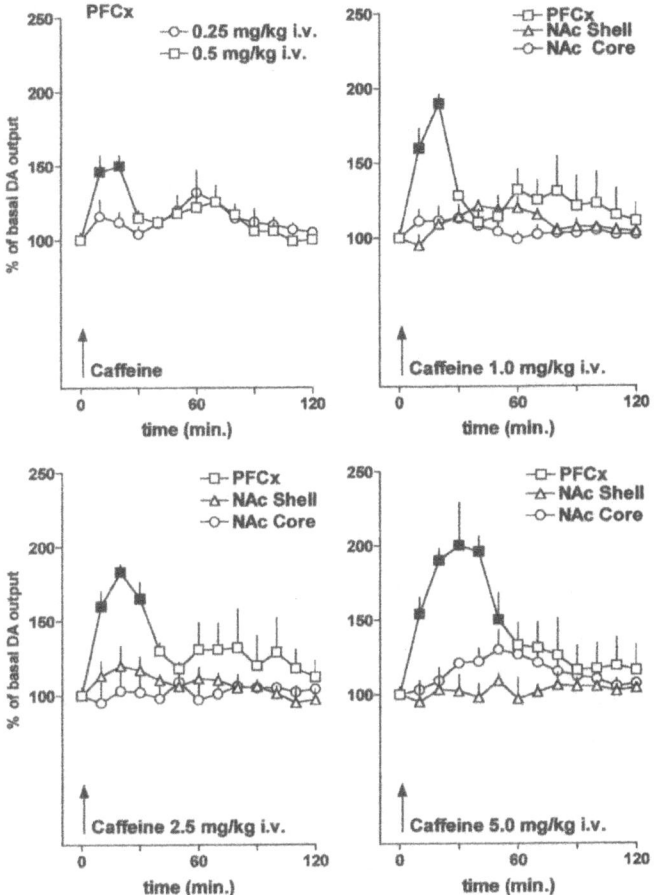

Figure 14.1 Effect of caffeine on *in vivo* DA transmission in the Acb and PFCx. Effect of intravenous administration of caffeine 0.25, 0.5, 1.0, 2.5 and 5.0 mg kg^{-1} (PFCx) and of caffeine 1.0, 2.5 and 5.0 mg kg^{-1} on AcbSh, AcbC and PFCx DA transmission. Values are expressed as percentage of baseline. Vertical bars represent S.E.M. Filled symbols indicate the sample points significantly different from baseline (p<0.05). Reproduced with permission from Acquas *et al* (2002), under Elsevier and RightsLink licence no. 2720761155768.

To settle these discrepancies, in the AcbSh, obtained after systemic administration of caffeine, it was concluded that these could be due to fine, but quite critical, sub-regional differences in the localization of microdialysis probes in the studies by Acquas *et al* (2002) and De Luca *et al* (2007) on one hand and in those by Solinas *et al* (2002) and Borycz *et al* (2007) on the other. It is now accepted that caffeine, at low doses, comparable to those reached by dietary sources in humans, fails to stimulate DA transmission in the ventral striatum (AcbSh and AcbC) and at least two lines of evidence might be considered to support this interpretation.

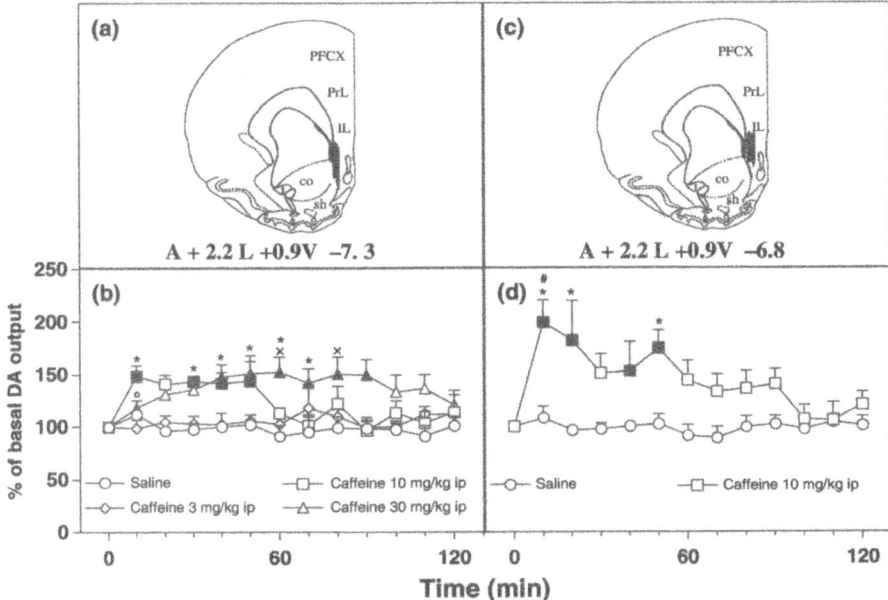

Figure 14.2   Effect of caffeine on *in vivo* DA transmission at the border between Acb shell and PFCx. Coronal brain sections showing the location of a 2 mm-long dialysing portion of a microdialysis probe in rats implanted at the border between the shell of the Acb and the PFCx and effects of caffeine (3–30 mg kg^{-1} i.p.) on dialysate DA. Filled symbols: $p<0.05$ with respect to basal; *$p<0.05$ with respect to saline group; $^{\times}p<0.05$ with respect to caffeine (3 mg kg^{-1}) group; $^{\#}p<0.05$ with respect to Acb shell–PFCx border group. Reproduced with permission from De Luca *et al* (2007), under John Wiley and Sons and RightsLink licence n° 2665971060808.

The first arises from the observation that caffeine at the doses at which fails to stimulate DA transmission in the Acb/ventral striatum, is able to stimulate it in the PFCx (see Figures 14.1, 14.2 and 14.3). A second line of evidence arises from a recent study showing that at the post-synaptic level, and at doses identical to those used in the microdialysis experiments, caffeine activates ERK in the PFCx but not in the AcbSh and AcbC (Acquas *et al* 2010). In addition, these data are in agreement with the exclusion of caffeine from the list of substances with abuse potential (APA, 1994; Di Chiara 2002) and clinical evidence obtained from a SPECT study further supports this conclusion (Nehlig *et al* 2010). The psychomotor stimulant effects of caffeine, on the other hand, can be ascribed to a functional interaction between adenosine and DA at the striatal–accumbal level, mediated by an opposite control exerted by A$_{2A}$ *vs* D$_2$ receptors on adenylate cyclase. This interpretation is supported by the finding that, in medium spiny neurons which represent the majority of the striatal-accumbal neuronal type, DA inhibits adenylate cyclase by stimulating

D_2 receptors. This action is mediated also by caffeine while blocking the stimulatory action exerted by adenosine via A_{2A} receptors (Fredholm *et al* 1999).

Adenosine is an endogenous sleep factor and it may affect sleep and EEG arousal through modulation of basal forebrain cholinergic projections to the cortex. However, an alternative hypothesis has recently suggested the involvement in these effects of an A_1 and A_{2A} receptors-mediated mechanism of modulation of ACh transmission in the PFCx as responsible of the activation of a descending pathway to the pontine reticular formation (Van Dort *et al* 2009). The effects of systemic administration of caffeine on ACh transmission have been studied in freely moving rats implanted with concentric microdialysis probes in the PFCx (Acquas *et al* 2002). Administration of caffeine at doses that elicit behavioral stimulation (0.25–5.0 mg kg^{-1} i.v.) dose-dependently increases ACh dialysate concentrations (Figure 14.3), an effect also reproduced by the selective A_1 and A_{2A} receptor antagonists, DPCPX and SCH 58261 respectively (Acquas *et al* 2002). Interestingly, the ability of the acute administration of caffeine to stimulate ACh transmission in the PFCx is also present after a regimen of chronic (25 mg kg^{-1} twice a day for 7 days) caffeine administration whereas the stimulation of DA transmission and of motor activity by caffeine is prevented by such chronic treatment.

These findings suggest that caffeine-induced stimulation of DA transmission in the PFCx might be related with a role of motor activation in the effects of caffeine on PFCx DA (Acquas *et al* 2002). In animals tolerant to the locomotor stimulant effects of caffeine, on the other hand, the unmodified ability of caffeine to stimulate PFCx ACh transmission might be related to the EEG desynchronizing effects of caffeine (Acquas *et al* 2002). *In vivo* brain microdialysis studies have also shown that caffeine (3–30 mg kg^{-1} p.o.) dose-dependently increases ACh transmission from rats' hippocampus suggesting that also hippocampal ACh transmission is under tonic inhibitory control of endogenous adenosine.

The effects of caffeine on *in vivo* neurotransmission of noradrenaline have been studied as well. Thus, in a microdialysis study in the hippocampus it was shown that neither systemic administration of caffeine (30 mg kg^{-1} p.o.) nor its local perfusion (1 mmol L^{-1}) could alter the extracellular concentrations of noradrenaline of awake, freely moving rats (Carter 1997). In this study, it was also shown that baseline and K$^+$-stimulated extracellular concentrations of noradrenaline are regulated by α_2 adrenoceptors and not by adenosine receptors (Carter 1997).

The ascending serotoninergic system, which originates in the brainstem, is considered a portion of the ascending reticular activating system and contributes to behavioral inhibition. Interestingly, although the serotoninergic system is not directly involved in arousal, its activity is maximal during the waking state, decreases during slow-wave sleep and stops during fast-wave sleep. In particular, the activity of pontine and medullary serotoninergic neurons facilitates motor output and coordinates autonomic and neuroendo-

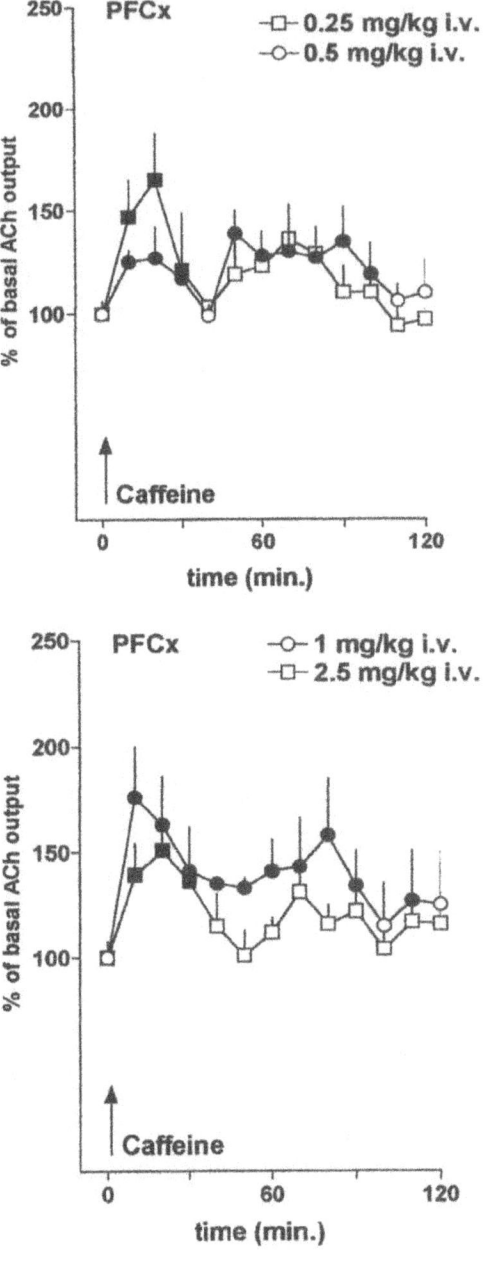

Figure 14.3   Effect of caffeine on *in vivo* ACh transmission in the PFCx. Effect of intravenous administration of caffeine 0.25, 0.5, 1.0 and 2.5 mg kg^{-1} on PFCx ACh transmission. Vertical bars represent S.E.M. Filled symbols indicate the sample points significantly different from baseline (p<0.05). Reproduced with permission from Acquas *et al* (2002), under Elsevier and RightsLink licence n° 2720761155768.

crine function. Moreover, serotonin exerts its effects on functions such as pain, feeding, memory and mood and for all these reasons it was considered of interest to study whether caffeine affects serotonin neurotransmission *in vivo*. Among the studies that addressed this issue, it was reported that caffeine, administered locally, significantly increases hippocampal serotonin transmission (Okada *et al* 1997). However, in another study it was shown that caffeine (1.7 mg kg^{-1} i.p.) could reduce the increases of serotonin in the hippocampus after restraint stress (Yamato *et al* 2002), suggesting that caffeine may exert a modulatory effect on this neurochemical response to stress.

14.4 Effects of Caffeine on Mechanisms of Synaptic Plasticity

Caffeine affects synaptic activities, either pre- and post-synaptically, by regulating excitatory neurotransmitters release and by affecting NMDA and non-NMDA receptors activities (He *et al* 2004). By acting onto adenosine A_1 receptors, *via* cAMP-dependent long-term potentiation (Lu *et al* 1999), caffeine also affects repetitive synaptic activities at the basis of long-term changes in synaptic efficiency in CA1 hippocampal neurons. Recently, and in addition to these mechanisms that involve caffeine in synaptic plasticity, caffeine has also been demonstrated to activate ERK in rodents' brain *in vivo* (Valjent *et al* 2004; Acquas *et al* 2010). ERK kinases are enzymes that act by sequential phosphorylation (MAPK–ERK-kinase-kinase activates MAPK–ERK-kinase which, in turn, activates MAPK, a kinase family of which phosphorylated ERK are the best known and characterized) (Sweatt 2004). The interest on these kinases ubiquitously present in the mature brain originated from the observation of their involvement in many central functions such as memory formation and consolidation in associative learning-based tasks, memory formation-related synaptic plasticity (Huang *et al* 2010), early- and late-phase of long-term potentiation, as well as in synapse formation and function and after physiological stimulation. Caffeine has been shown to significantly increase pERK expression in the superficial and deep layers of the PFCx and in the Cg_1-Cg_2 cortices (Acquas *et al* 2010) while failing to elicit pERK in M_2 cortex (Acquas *et al* 2010; Valjent *et al* 2004) and in other brain regions such as piriform cortex, basolateral amygdala, hypothalamus, lateral septum, ventral tegmental area, AcbSh and AcbC (Acquas *et al* 2010; Valjent *et al* 2004). This differential activation of ERK by caffeine in distinct brain regions suggests that there may be a differential impact of caffeine on these structures in terms of ERK-mediated plasticity. In particular, failure of caffeine to activate ERK in M_2 cortex indicates that the psychomotor properties of caffeine do not require ERK activation in this cortical region. On the other hand, the observation that caffeine activates ERK in the PFCx and Cg_1-Cg_2 cortex, in a DA D_1 receptor-dependent and -independent manner, respectively, supports to the interpretation that D_1 receptor-dependent phosphorylation of ERK in the PFCx by caffeine is specific for that cortical region. Interestingly, despite the fact that D_1 receptors are mostly present in the PFCx and Cg_1-Cg_2 cortices, whereas D_4 receptor subtypes are mostly present in the M_2 cortex, these observations are in agreement with the finding that blockade of D_1 receptors prevents caffeine-elicited ERK activation in the PFCx on one hand, whereas, on the other, the lack of D_1 receptors-mediated control of caffeine-elicited pERK in the Cg_1-Cg_2 cortices indicates that ERK activation might be the result of multisynaptic mechanisms (Acquas *et al* 2010). Thus, although the Cg_1-Cg_2 cortex is a D_1 receptor-rich cortical region, blockade of D_1 receptors fails to critically regulate caffeine-mediated ERK phosphorylation (Acquas *et al* 2010). These observations indicate that multisynaptic mechanisms must be taken into account in order to interpret the mechanism(s) of caffeine activated ERK phosphorylation in different cortical regions and overall suggest to investigate further the molecular mechanism(s) by which caffeine-mediated synaptic plasticity may take place in distinct cortical regions.

Recent theories on the neurobiological basis of addiction indicate that this behavioral disturbance may have at his basis a failure of the molecular mechanisms responsible of impulse control and associative learning and, interestingly, pERK, has been attributed a critical role in these mechanisms (Lu *et al* 2006). Thus, as a further example of the functional significance that can be highlighted by detection of activated ERK in different brain regions, it is worth mentioning, in this regard, the lack of ERK activation in the AcbSh and AcbC following the administration of caffeine, at doses that significantly enhance pERK expression in the PFCx and Cg_1-Cg_2 cortices (Acquas *et al* 2010). The Acb and the extended amygdala have been indicated as anatomical transition structures between motor and motivational responses and DA transmission in these regions has been pointed out as the main neurochemical mechanism that mediates both motor and motivational properties of addictive drugs (Di Chiara 2002). Thus, based on the evidence that DA transmission in the Acb and extended amygdala is increased by drugs with addictive potential, it was postulated that the preferential increases of DA transmission in the AcbSh compared to the AcbC represent the common neurochemical trait of addictive drugs belonging to different pharmacological classes (Di Chiara 2002). Interestingly, on the post-synaptic side, the activation of ERK was suggested to be differentially activated by addictive and non-addictive drugs, in distinct regions of the mouse brain (Valjent *et al* 2004). Thus, based on these premises, and in order to contribute to the ongoing debate on the possibility to exclude (or include) caffeine in the list of the addictive substances, we found that while failing to stimulate DA transmission in the AcbSh and AcbC, at doses able to stimulate DA and ACh transmissions in the PFCx (Acquas *et al* 2002), caffeine also fails to elicit pERK in the AcbSh and AcbC (Acquas *et al* 2010). These results brought us to conclude that failure to stimulate DA transmission and elicit pERK in the Acb justifies, from the preclinical point of view, the exclusion of caffeine from the list of the restricted substances (APA

1994), *i.e.* from the list of substances with addictive potential (Acquas *et al* 2010; Di Chiara 2002; Valjent *et al* 2004).

In summary, the understanding of the mechanisms of ERK activation in the brain might offer the possibility to cast fresh light on the molecular mechanisms at the basis of the behavioral outcomes of the pharmacological effects of caffeine in terms of memory and cognition as well as in terms of the relationship between cortical ERK activation and the attentional, arousing and mnemonic properties of caffeine. In addition, despite the demonstration that caffeine differentially activates ERK in distinct brain regions and despite the established role of caffeine in synaptic plasticity, very few data are available in the literature on the role of ERK in the ability of caffeine to elicit synaptic plasticity. In conclusion, while direct evidence involves the activation of ERK in the effects of caffeine and correlative evidence points to activated ERK as a possible critical player in the mechanisms of caffeine-mediated synaptic plasticity, these observations open to an extremely interesting and promising avenue for future research.

14.5   Caffeine in Neurodegenerative Diseases: Parkinson's and Alzheimer's

Caffeine, by virtue of its nutraceutical properties, together with other substances has, since 2003, been included in the list of compounds with potential neuroprotective actions as one of the more promising compounds in clinical trials by CINAPS (Ravina *et al* 2003). Indeed, epidemiological studies in humans and experimental data from animals have provided evidence that the consumption of coffee and of other sources of caffeine, such as black, Japanese and Chinese teas, can have beneficial effects in neurodegenerative disorders such as PD, AD, Huntington's disease (Tanaka *et al* 2011) and in other brain injuries such as ischemia.

In PD, characterized by motor symptoms like tremors, bradykinesia and muscular rigidity, a loss of dopaminergic neurons is present in the SNc leading to a reduced dopaminergic input to the striatum, contrasted therapeutically with DA agonists and levodopa. In addition to these more frequently recognized symptoms, memory impairments are also observed in a substantial number of PD patients, which also present other non-motor symptoms, such as cognitive and psychological disturbances and psychiatric symptoms. These latter are also present in AD patients in which the progressive loss of memory and the decline of cognitive functions parallels the neuronal degeneration characterized by Aβ peptide accumulation and by formation of intraneuronal neurofibrillary tangles. Among the factors involved in these neurodegenerative diseases are environmental toxins, genetic factors, blood brain barrier disruption, mitochondrial dysfunctions, inflammatory processes and oxidative stress.

The first significant set of data related to a neuroprotective role of caffeine in PD refers to a study, done in Hawaii (USA), that involved 8004 men examined

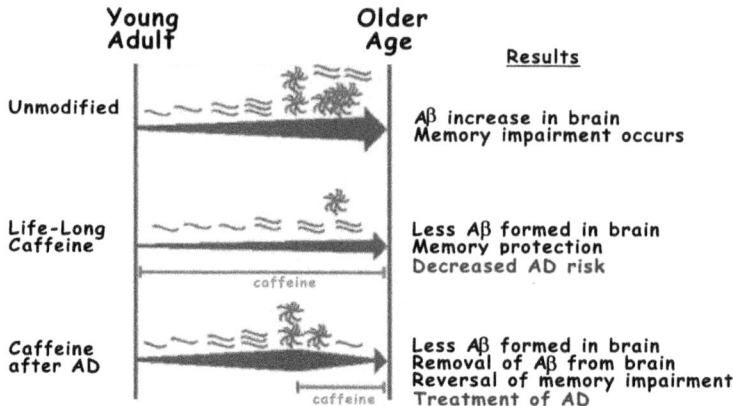

Figure 14.4   Schematic representation of the neuroprotective effects of caffeine. Schematic representation of the protective–preventive effects of caffeine intake on Aβ production–aggregation and memory deficits in AD transgenic mice when caffeine consumption is started during young adulthood (life-long caffeine) or after development of AD pathology–memory impairment. Reprinted from Journal of Alzheimer's Disease, Vol. 20, Suppl. 1, S116–S126, by Arendash G.W. and Cao C., Copyright (2010), with permission from IOS Press.

in 30 years. The authors observed an inverse association between coffee (and dietary caffeine intake) and the risk of PD concluding that caffeine protects against the development of the disease by increasing the central dopaminergic tone (Ross *et al* 2000). A similar lowered risk of PD in 47 000 men and 88 000 women in USA was later reported (Ascherio *et al* 2001). More recently, in a Finnish study with 101 incident cases of PD during 22 years, it was confirmed that daily consumption of coffee (10 or more cups) reduced the risk of PD but it was also observed that this effect was influenced by other factors like serum cholesterol, smoking and obesity (Saaksjarvi *et al* 2008). In this regard it is worth mentioning that a decreased incidence of the risk to develop PD was described in both fast and slow caffeine metabolizers, suggesting that neuroprotection can also be attributed to its major metabolite paraxanthine (Tan *et al* 2007). However also opposite results have been shown in a study reporting failure to detect any association with reduced risk of PD in consumers of coffee or of total caffeine (Checkoway *et al* 2002), in a study involving women under hormone-replacement therapy *versus* men (Ascherio *et al* 2003), and in a study reporting a non-consistent relationship between caffeine intake and rate of progression of PD in 413 early PD subjects (Simon *et al* 2008). Despite these contrasting results a recent meta-analysis study by Costa and colleagues (2010) has confirmed a strong relationship between caffeine intake and PD prevention with an overall 25% reduction of risk of developing the disease.

Epidemiological evidence on the ability of coffee and caffeine intake to prevent or delay AD has been documented as well. Thus, in the Canadian Study of Health and Aging prospective analysis of risk factors for AD in elderly people (>65 years old) it was reported that daily coffee drinking (among other factors) significantly reduced the risk of AD by 31% during a 5 years follow up (Lindsay *et al* 2002). Similarly, in a case–control study (54 patients with probable AD and 54 cognitively normal controls with an average daily caffeine intake of 74 and 199 mg, respectively, during the previous 20 years) a decreased risk for AD was reported (Maia and de Mendonça 2002). This finding is particularly relevant in the light of the slow onset of AD symptoms which supports the possibility that habitual caffeine consumption early in life may have an overall preventive effect on development of the disease. This is also suggested by a recent study reporting a 65% decreased risk of developing AD, as assessed 21 years later (late-life), in a study involving over 1400 people aged 65–79 who drank daily 3 to 5 cups of coffee at their midlife (Eskelinen *et al* 2009). However, failure to establish such correlation is also reported in the literature (Ritchie *et al* 2007).

Effects comparable to those found in humans are also observed in many animal models of neurodegenerative diseases. Chen and colleagues (2001), in the MPTP mouse model of PD, demonstrated that caffeine dose-dependently reduces the depletion of the nigrostriatal afferences by an A_{2A} receptor-mediated mechanism. Interestingly, and in contrast to the motor stimulant effects of caffeine, these neuroprotective effects do not show tolerance to chronic treatment, suggesting that distinct mechanisms of action might be involved in neuroprotective and motor stimulant effects (Xu *et al* 2002). In addition, in unilateral intrastriatal 6-OHDA-lesioned rats, pretreated with caffeine, rotational behavior induced by apomorphine was attenuated and histochemical evidence indicated a neuroprotective effect (Joghataie *et al* 2004). Furthermore, and in agreement with the clinical evidence (Ascherio *et al* 2003), it was shown, in the MPTP model, that estrogens could prevent the neuroprotective effect of caffeine, thus providing the biological basis for the interaction between estrogens and caffeine in modifying the risk of PD. In addition, the neuroprotection exerted by caffeine, in the MPTP model, may also be extended to its major metabolites, theophylline and paraxanthine, which both may act as A_{2A} receptor antagonists (Xu *et al* 2010). Interestingly, a recent report demonstrated that coffee, but not caffeine, exerts neuroprotective effects in a fly PD model and that caffeine-containing and decaffeinated coffees have neuroprotective effects also evident in a Drosophila model of AD (Trinh *et al* 2010).

Adenosine antagonists are also protective against cognitive disorders and a relationship between adenosine and memory has been widely documented (Takahashi *et al* 2008). Accordingly, animal models of AD revealed a strict relationship between chronic administration of caffeine and prevention of behavioral impairments and of memory deficits. One of the first *in vivo* studies that demonstrated these preventive effects of caffeine was done in mice with

Aβ-induced cognitive impairments in the inhibitory avoidance and spontaneous alternation tasks showing that prolonged (12 days) caffeine administration could provide protection against memory deficits in these tasks (Dall'Igna *et al* 2007). On the same vein, in a transgenic mouse model of AD, Arendash and colleagues (2009) reported memory restoration and reversal of histopathological signs (hippocampal Aβ) after 4 to 5 weeks of caffeine administration in their drinking water.

In summary, caffeine attenuates either motor and non-motor symptoms (PD and animal models) and cognitive impairments (AD and animal models) (see Figure 14.4). Further investigations are necessary for an insight and complete understanding of the molecular mechanism(s) implicated in these effects. Furthermore, beneficial effects might originate, besides from caffeine, also from other beverages (tea, chocolate) rich of coffee-components (phenolic acids, Mg^{++}, chlorogenic acid, cafestol, trigonelline) with antioxidant actions. This is supported by *in vitro* (Chu *et al* 2009) and *in vivo* studies (Trinh *et al* 2010) in fly models of PD and AD. In conclusion, caffeine-mediated prevention of the appearance of pathophysiological signs characteristics of PD and AD might represent in the future the most useful experimental approach to explain the epidemiologically documented efficacy of caffeine, caffeinated, and perhaps decaffeinated, beverages on these neurodegenerative diseases.

Key Facts

Alzheimer's disease: This neurodegenerative and terminal disease was described for the first time by Alois Alzheimer in 1906. Alzheimer's disease is commonly diagnosed in people over 65 years of age, but early-onset Alzheimer's disease can also occur. Although the development of Alzheimer's disease is peculiar to every person affected, there are some common symptoms. In an early phase, the most common symptom is inability to acquire new memories, with difficulty in recalling recently observed events. As the disease progresses, symptoms include confusion, irritability and aggression, mood swings, language breakdown, memory loss. Bodily functions are progressively lost and this ultimately leads to death.

In vivo Brain Microdialysis: Sampling technique used for continuous measurement of the concentration of free, unbound molecules in the extracellular fluid. These might virtually be any endogenous compound (*e.g.* neurotransmitter, ions, *etc.*) that satisfy the criteria to be unbound in the extracellular fluid and to have a molecular weight below the cut-off of the dialytic membrane. Brain microdialysis can also be used to assess the concentration in the extracellular fluid of exogenous compounds (*e.g.* medications), and of metabolites of either endogenous and exogenous compounds. The microdialysis technique requires the insertion of a small microdialysis probe into the tissue in order to reach the brain nuclei of interest (usually under stereotaxic coordinates). The microdialysis probe is designed to mimic a blood capillary and consists of a shaft with a semi-permeable hollow membrane at its tip, which is connected to inlet and outlet tubing. This probe is continuously perfused with an aqueous solution that strictly mimics the ionic composition of the extracellular fluid at a very slow $(0.1–5~\mu L~min^{-1})$ and constant flow rate. Small molecules can cross the membrane by passive diffusion under a concentration gradient and this principle allows the application of microdialysis probes either as sampling or as delivery tools. The solution leaving the probe (dialysate, outlet) is collected at certain time intervals for analysis. In the experimental setup, *in vivo* brain microdialysis can be performed in awake, freely moving animals.

IOC–International Olympic Committee: The International Olympic Committee is the supreme authority of the Olympic Movement (http://www.olympic.org/about-ioc-institution).

Parkinson's disease: Neurodegenerative disorder that results from the loss of DA-containing cells in the midbrain region substantia nigra pars compacta. The disease is named after James Parkinson, who first described it in 1817. The cause of death of these neurons is unknown. Early in the course of the disease, the symptoms are mostly movement related (shaking, rigidity, slowness of movement and difficulty with walking). Late stages of the disease present cognitive and behavioral problems (including dementia). Other symptoms may also include sensory, sleep and emotional problems.

Pharmacokinetic: The branch of pharmacology devoted to the determination of the fate of administered substances and includes the study of absorption, distribution, metabolism and excretion. Pharmacokinetic also studies the rate at which a drug action begins, the profiles of its plasma concentration and the duration of its effect.

Pharmacodynamic: The branch of pharmacology devoted to the study of the effects of drugs on the body, of the mechanisms of drug actions and of the relationship between drug concentration and effects. Pharmacodynamic is sometimes defined as the study of what a drug does to the body, whereas pharmacokinetic is the study of what the body does to a drug.

Summary Points

- Caffeine is the most famous and widely consumed psychoactive substance worldwide. Its large diffusion and consumption reflects its safety although recent evidence suggests caution in particular when caffeine is taken in large quantities in combination with other substances present in energy drinks and in over-the-counter medications.
- Caffeine exerts its central actions (increased motor activity, increased arousal, attention, alertness and psychomotor stimulation) mostly by acting as an adenosine receptor antagonist at A_1 and A_{2A} receptors.
- Caffeine, indirectly, affects the release of central neurotransmitters such as DA, ACh, and serotonin and its ability to affect the release of these

neurotransmitters might represent the neurochemical basis of its psycho-pharmacological properties.

- Studying caffeine-elicited ERK activation in different brain regions will offer the possibility to cast fresh light on the molecular mechanisms at the basis of its effects in terms of memory and cognition as well as in terms of the relationship between cortical ERK activation and the attentional, arousing and mnemonic properties of caffeine.
- Experimental evidence indicates that caffeine (but also caffeinated and decaffeinated beverages) may affect neurodegenerative disorders by reducing motor and cognitive impairments. These data, corroborated by epidemiological studies, represent an important and promising avenue for future research and overall improvement of quality of life.

Definitions of Words and Terms

Adenosine: Central inhibitory neurotransmitter which plays important roles in sleep promotion and arousal suppression.

Adverse effects: Unwanted effects. Other terms used are adverse drug events or side-effects. The term adverse drug reaction is technically more appropriate for adverse drug effects that are unwanted, unpleasant, noxious, and potentially harmful.

Dose-dependent pharmacological effects: Beneficial, desired effects which depend upon the dose administered. For instance, a given compound can have either stimulatory or inhibitory properties, depending on the dose.

Drug dependence: Indicates the condition of dependence on a drug to maintain a normal body functioning. Usually it is accompanied by the appearance of withdrawal symptoms upon abrupt interruption of drug-intake. A person may have a physical dependence on a drug without having an addiction (see below). For example, certain blood pressure medications do not cause addiction but they may cause physical dependence.

Drug addiction: Refers to the compulsive use of a substance, despite its short- and long-term negative consequences.

Long-term potentiation: Long-term strengthening of the synapses between two neurons that are activated simultaneously.

Neuroprotection: Neuroprotection can be defined as an action that may result in salvage, recovery and–or regeneration of the nervous system at the cellular, structural and functional level.

Neurotransmitter release: The process of freeing the chemicals "neurotransmitters" in the synaptic cleft (*i.e.* the space between a pre-synaptic and a post-synaptic element) from the vesicles in which they are stored in the pre-synaptic element.

Over-the-counter drugs (OTC): A drug (with one or more pharmacological principles) that is sold without any prescription. In contrast, a prescription medicine refers to a drug that is available only with written instructions from a physician, veterinarian or dentist to a pharmacist.

Synaptic plasticity: Synapses are the sites of connections between neurons. These connections are changeable and can mutate, in the sense that connections between neurons can change their shape or their function over periods of time (from a few seconds to a lifetime). Synaptic plasticity refers to changes in the structure and function of synaptic connections.

List of Abbreviations

6-OHDA	6-hydroxy-dopamine
Aβ	Amyloid-β
AcbSh	Nucleus accumbens shell
AcbC	Nucleus accumbens core
ACh	Acetylcholine
AD	Alzheimer's Disease
Cg_1-Cg_2	Cingulate cortex
CINAPS	Committee to Identify Neuroprotective Agents in Parkinson's Disease
DSM	Diagnostic and Statistical Manual of Mental Disorders
ERK	Extracellular signal Regulated Kinase
DA	Dopamine
IOC	International Olympic Committee
M_2	Secondary motor cortex
PFCx	Medial prefrontal cortex
MPTP	1-methyl-4-phenyl-1,2,3,6-tetrahydropyridine
PD	Parkinson's Disease
pERK	phosphorylated ERK
SNc	Substantia Nigra, pars compacta
SPECT	Single-photon emission computed tomography

References

Acquas, E., Tanda, G., and Di Chiara, G., 2002. Differential effects of caffeine on dopamine and acetylcholine transmission in brain areas of drug-naive and caffeine-pretreated rats. Neuropsychopharmacol. 27: 182–193.

Acquas, E., Vinci, S., Ibba, F., Spiga, S., De Luca, M. A. and Di Chiara, G., 2010. Role of dopamine D(1) receptors in caffeine-mediated ERK phosphorylation in the rat brain. Synapse. 64: 341–349.

American Psychiatric Association (APA) 1994. Diagnostic and statistical manual of mental disorders. 4th Edition. American Psychiatric Association, Washington, DC.

Arendash, G. W., Mori, T., Cao, C., Mamcarz, M., Runfeldt, M., Dickson, A., Rezai-Zadeh, K., Tane, J., Citron, B. A., Lin, X., Echeverria, V. and Potter, H., 2009. Caffeine reverses cognitive impairment and decreases brain amyloid-beta levels in aged Alzheimer's disease mice. J. Alzheimer's Dis. 17(3): 661–680.

Arendash, G. W. and Cao, C., 2010. Caffeine and coffee as therapeutics against Alzheimer's Disease. Journal of Alzheimer's Dis. 20: S117–S126.

Ascherio, A., Zhang, S. M., Hernán, M. A., Kawachi, I., Colditz, G. A., Speizer, F. E. and Willett, W. C., 2001. Prospective study of caffeine consumption and risk of Parkinson's disease in men and women. Ann Neurol. 50(1): 56–63.

Ascherio, A., Chen, H., Schwarzschild, M. A., Zhang, S. M., Colditz, G. A. and Speizer, F. E., 2003. Caffeine, postmenopausal estrogen, and risk of Parkinson's disease. Neurology. 60(5): 790–795.

Berger, A. J. and Alford, K., 2009. Cardiac arrest in a young man following excess consumption of caffeinated "energy drinks". Med. J. Aust. 190(1): 41–43.

Borycz, J., Pereira, M. F., Melani, A., Rodrigues, R. J., Köfalvi, A., Panlilio L., Pedata F., Goldberg S. R., Cunha R. A. and Ferré, S. 2007. Differential glutamate-dependent and glutamate-independent adenosine A1 receptor-mediated modulation of dopamine release in different striatal compartments. J. Neurochem. 101(2): 355–363.

Carter, A. J. 1997. Hippocampal noradrenaline release in awake, freely moving rats is regulated by alpha-2 adrenoceptors but not by adenosine receptors. J. Pharmacol. Exp. Ther. 281(2): 648–654.

Checkoway, H., Powers, K., Smith-Weller, T., Franklin, G. M., Longstreth, W. T. Jr, and Swanson, P. D., 2002. Parkinson's disease risks associated with cigarette smoking, alcohol consumption, and caffeine intake. Am. J. Epidemiol. 155(8): 732–738.

Chen, J. F., Xu, K., Petzer, J. P., Staal, R., Xu, Y. H., Beilstein, M., Sonsalla, P. K., Castagnoli, K., Castagnoli, N. Jr, and Schwarzschild, M. A., 2001. Neuroprotection by caffeine and A(2A) adenosine receptor inactivation in a model of Parkinson's disease. J. Neurosci. 21(10): RC143.

Chu, Y. F., Brown, P. H., Lyle, B. J., Chen, Y., Black, R. M., Williams, C. E., Lin, Y. C., Hsu, C. W., and Cheng, I. H., 2009. Roasted coffees high in lipophilic antioxidants and chlorogenic acid lactones are more neuroprotective than green coffees. J. Agric. Food Chem. 57(20): 9801–9808.

Costa, J., Lunet, N., Santos, C., Santos, J. and Vaz-Carneiro, A., 2010. Caffeine exposure and the risk of Parkinson's disease: a systematic review and meta-analysis of observational studies. J. Alzheimer's Dis. (Suppl 1): S221–S238.

Dall'Igna, O. P., Fett, P., Gomes, M. W., Souza, D. O., Cunha, R. A., and Lara, D. R., 2007. Caffeine and adenosine A(2a) receptor antagonists prevent beta-amyloid (25–35)-induced cognitive deficits in mice. Exp. Neurol. 203(1): 241–245.

De Luca, M. A., Bassareo, V., Bauer, A. and Di Chiara, G., 2007. Caffeine and accumbens shell DA. J. Neurochem. 103(1): 157–163.

Di Chiara, G. 2002. Nucleus accumbens shell and core dopamine: differential role in behavior and addiction. Behav. Brain. Res. 137(1–2): 75–114.

Eskelinen, M. H., Ngandu, T., Tuomilehto, J., Soininen, H. and Kivipelto, M., 2009. Midlife coffee and tea drinking and the risk of late-life dementia: a population-based CAIDE study. J. Alzheimer's Dis. 16(1): 85–91.

Fredholm, B. B., Bättig, K., Holmén, J., Nehlig, A. and Zvartau, E. E., 1999. Actions of caffeine in the brain with special reference to factors that contribute to its widespread use. Pharmacol. Rev. 51(1): 83–133.

He, S. J., Xiao, C., Wu, Z. Y. and Ruan, D. Y., 2004. Caffeine-dependent stimulus-triggered oscillations in the CA3 region of hippocampal slices from rats chronically exposed to lead. Exp. Neurol. 190: 525–534.

Huang, C-H., Chiang, Y-W., Liang, K-C., Thompson, R. F. and Liu, I. Y., 2010. Extra-cellular signal-regulated kinase 1/2 (ERK1/2) activated in the hippocampal CA1 neurons is critical for retrieval of auditory trace fear memory. Brain. Res. 1326: 143–151.

Joghataie, M. T., Roghani, M., Negahdar, F. and Hashemi, L. 2004. Protective effect of caffeine against neurodegeneration in a model of Parkinson's disease in rat: behavioral and histochemical evidence. Parkinsonism Relat. Disord. 10(8): 465–468.

Lindsay, J., Laurin, D., Verreault, R., Hébert, R., Helliwell, B., Hill, G.B. and McDowell, I., 2002. Risk factors for Alzheimer's disease: a prospective analysis from the Canadian Study of Health and Aging. Am. J. Epidemiol. 156(5): 445–453.

Lu, K. T., Wu, S. P., and Gean, P. W., 1999. Promotion of forskolin induced long-term potentiation of synaptic transmission by caffeine in area CA1 of the rat hippocampus. Chin. J. Physiol. 42: 249–253.

Lu, L., Koya, E., Zhai, H., Hope, B.T. and Shaham, Y., 2006. Role of ERK in cocaine addiction. Trends Neurosci. 29(12): 695–703.

Maia, L. and de Mendonça, A., 2002. Does caffeine intake protect from Alzheimer's disease? Eur. J. Neurol. 9(4): 377–382.

Malinauskas, B. M., Aeby, V. G., Overton, R. F. and Barber-Heidal, K., 2007. A survey of energy drink consumption patterns among college students. Nutr. J. 6(35): 1–7.

Nehlig, A., Armspach, J. P. and Namer, I. J., 2010. SPECT assessment of brain activation induced by caffeine: no effect on areas involved in dependence. Dialogues Clin. Neurosci. 12(2): 255–263.

Okada, M., Mizuno, K. and Kaneko, S., 1996. Adenosine A1 and A2 receptors modulate extracellular dopamine levels in rat striatum. Neurosci. Lett. 212(1): 53–56.

Okada, M., Kawata, Y., Kiryu, K., Mizuno, K., Wada, K., Tasaki, H. and Kaneko, S. 1997. Effects of adenosine receptor subtypes on hippocampal extracellular serotonin level and serotonin reuptake activity. J Neurochem. 69(6): 2581–2588.

Ravina, B. M., Fagan, S. C., Hart, R. G., Hovinga, C. A., Murphy, D. D., Dawson, T. M. and Marler, J. R., 2003. Neuroprotective agents for clinical trials in Parkinson's disease: a systematic assessment. Neurology. 60(8): 1234–1240.

Ritchie, K., Carrière, I., de Mendonca, A., Portet, F., Dartigues, J. F., Rouaud, O., Barberger-Gateau, P. and Ancelin, M. L., 2007. The neuroprotective effects of caffeine: a prospective population study (the Three City Study). Neurology. 69(6): 536–545.

Ross, G. W., Abbott, R. D., Petrovitch, H., Morens, D. M., Grandinetti, A., Tung, K. H., Tanner, C. M., Masaki, K. H., Blanchette, P. L., Curb, J. D., Popper, J. S. and White, L. R., 2000. Association of coffee and caffeine intake with the risk of Parkinson disease. JAMA. 283(20): 2674–2679.

Sääksjärvi, K., Knekt, P., Rissanen, H., Laaksonen, M. A., Reunanen, A. and Männistö, S., 2008. Prospective study of coffee consumption and risk of Parkinson's disease. Eur. J. Clin. Nutr. 62(7): 908–915.

Simon, D. K., Swearingen, C. J., Hauser, R. A., Trugman, J. M., Aminoff, M. J., Singer, C., Truong, D., Tilley, B. C. and NET-D Investigators, 2008. Caffeine and progression of Parkinson disease. Clin Neuropharmacol. 31(4): 189–196.

Solinas, M., Ferré, S., You, Z. B., Karcz-Kubicha, M., Popoli, P. and Goldberg, S. R., 2002. Caffeine induces dopamine and glutamate release in the shell of the nucleus accumbens. J. Neurosci. 22(15): 6321–6324.

Takahashi, R. N., Pamplona, F. A., Prediger, R. D., 2008. Adenosine receptor antagonists for cognitive dysfunction: a review of animal studies. Front. Biosci. 13: 2614–2632.

Tan, E. K., Chua, E., Fook-Chong, S. M., Teo, Y. Y., Yuen, Y., Tan, L. and Zhao, Y., 2007. Association between caffeine intake and risk of Parkinson's disease among fast and slow metabolizers. Pharmacogenet. Genomics. 17(11): 1001–1005.

Tanaka, K., Miyake, Y., Fukushima, W., Sasaki, S., Kiyohara, C., Tsuboi, Y., Yamada, T., Oeda, T., Miki, T., Kawamura, N., Sakae, N., Fukuyama, H., Hirota, Y., Nagai, M. and the Fukuoka Kinki Parkinson's Disease Study Group, 2011. Intake of Japanese and Chinese teas reduces risk of Parkinson's disease. Parkinsonism Relat. Disord. 17(6): 446–450.

Trinh, K., Andrews, L., Krause, J., Hanak, T., Lee, D., Gelb, M. and Pallanck, L., 2010. Decaffeinated coffee and nicotine-free tobacco provide neuroprotection in Drosophila models of Parkinson's disease through an NRF2-dependent mechanism. J. Neurosci. 30(16): 5525–5532.

Van Dort, C. J., Baghdoyan, H. A. and Lydic, R., 2009. Adenosine A1 and A2A receptors in mouse prefrontal cortex modulate acetylcholine release and behavioral arousal. J. Neurosci. 29(3): 871–881.

Valjent, E., Pages, C., Herve, D., Girault, J. A. and Caboche, J., 2004. Addictive and non-addictive drugs induce distinct and specific patterns of ERK activation in mouse brain. Eur. J. Neurosci. 19: 1826–1836.

Xu, K., Xu, Y. H., Chen, J. F. and Schwarzschild, M. A., 2002. Caffeine's neuroprotection against 1-methyl-4-phenyl-1,2,3,6-tetrahydropyridine toxicity shows no tolerance to chronic caffeine administration in mice. Neurosci. Lett. 322(1): 13–16.

Xu, K., Xu, Y. H., Chen, J. F. and Schwarzschild, M. A., 2010. Neuroprotection by caffeine: time course and role of its metabolites in the MPTP model of Parkinson's disease. Neuroscience. 167(2): 475–481.

Yamato, T., Yamasaki, S., Misumi, Y., Kino, M., Obata, T. and Aomine, M., 2002. Modulation of the stress response by coffee: an *in vivo* microdialysis study of hippocampal serotonin and dopamine levels in rat. Neurosci. Lett. 332(2): 87–90.

CHAPTER 15

Caffeine and Cognitive Performance

ANA ADAN*[1] AND JOSEP MARIA SERRA-GRABULOSA[2]

[1] Department of Psychiatry and Clinical Psychobiology, University of Barcelona and Institute for Brain, Cognition and Behaviour (IR3C), Pg. Vall Hebron 171, 08035 Barcelona, Spain; [2] Department of Psychiatry and Clinical Psychobiology, University of Barcelona and Institut d'Investigacions Biomèdiques August Pi i Sunyer, IDIBAPS, Pg. Vall Hebron 171, 08035 Barcelona, Spain
*E-mail: aadan@ub.edu

15.1 Introduction

Caffeine is part of our everyday lives and probably the most widely-used psychoactive substance around the world. Thousands of people use caffeine as a way of getting ready for the day. It is present in coffee as well as in a wide variety of other drinks (*i.e.* tea, soft drinks). Because caffeine use is so widespread, several studies have been carried out on the effects of caffeine on human behavior and cognition. Many of these works have found improvements in cognitive performance thanks to caffeine, although in the last decades multiple factors have been suggested which may affect the results and account for some inconsistent findings. Among them, the dose and form of administration, the subjects' characteristics regarding caffeine (usual level of caffeine consumption, caffeine withdrawal) and the specific circumstances in which the measures are taken (low arousal or activation, sleep deprivation, *etc.*).

Food and Nutritional Components in Focus No. 2
Caffeine: Chemistry, Analysis, Function and Effects
Edited by Victor R Preedy
© The Royal Society of Chemistry 2012
Published by the Royal Society of Chemistry, www.rsc.org

The amount and form of administration of caffeine is a key element. Low doses are below 200 mg, moderate range between 200–400 mg, and high doses are above 400 mg. Although there is no consensus with clear cut-off points in the categorization of dosage, most studies select a dose between 100–400 mg. The most usual form of administering caffeine is in a single dose (acute), although if the dose is moderate or high this form of administration is not in accordance with natural conditions of caffeine consumption. Moreover, in research the preferred preparations are in the form of beverage, capsule, tablet, bar, and caffeine gum instead of a cup of coffee (espresso, filter, instant, *etc.*) or tea, which are the two most common forms of consumption. These methodological aspects set some limitations on generalizing the results to real-life conditions.

Since caffeine consumption is very common in daily life, another important variable is the consumption habits of the participants. This fact has raised the controversy over whether the beneficial effects of caffeine really exist. Thus, the withdrawal reversal hypothesis has been developed, stating that regular caffeine consumers experience acute negative cognitive and behavioral effects after overnight caffeine abstinence, these negative effects being relieved by caffeine intake. Therefore, the apparent effects of caffeine might in fact reflect withdrawal reversal or the simple restitution of cognitive ability to its normal level. Although some studies give support to this hypothesis (James and Rogers 2005; Rogers 2007; Rogers *et al* 2003), the majority of studies support the idea that caffeine has psychoactive effects even in the absence of withdrawal (Childs and de Wit 2006; Christopher *et al* 2005; Haskell *et al* 2005; Hewlett and Smith 2006, 2007, to mention a few works which highlight this aspect in their discussion). There are three pieces of evidence supporting the idea that the beneficial effects of caffeine are not explicable solely in terms of reversal of withdrawal effects and that there are real benefits on cognitive performance after caffeine ingestion. First, there is much literature on animal research showing beneficial effects, and the animals clearly were not withdrawn from caffeine. Second, there are also positive effects on performance in non-consumers of caffeine. Third, the administration of caffeine in non-withdrawn individuals also presents benefits. Smith's (2002) revision may be consulted to go further on this topic.

15.2   The Difficulty of Assessing Cognitive Performance

A problem faced by the various studies conducted in this research area is the complexity to select sensitive cognitive tasks which at the same time conveniently assess performance. Cognitive performance encompasses many specific functions or skills (*e.g.* attention, motor control, learning, memory and executive functions) and the execution of a task, no matter how simple it is, relies on many of them. Moreover, although there are hundreds of standard neuropsychological tasks, even the simplest ones often require the use of more than one function, and there is not always a consensus on their sensitivity. The performance of a task can be assessed using estimates of speed (time of response) and–or accuracy (correct answers, errors and lapses), which may not all be affected equally by caffeine administration.

Furthermore, many of the standard neuropsychological tasks for attention, memory and executive functions are very useful for detecting pathological conditions in clinical examination. Nevertheless, this is not the case when they involve healthy individuals, as their performance is often in the normal range of the current scales. The selection of specific tasks carried out in psychopharmacological research might be more useful and sensitive to detect the effects associated with caffeine intake. However, one problem is that these tasks are usually too simple, making it difficult to extrapolate the results to the cognitive social and occupational demands compared with those that individuals are subjected to in reality. In the future it would be interesting to make an effort in this field of study to introduce more real-life tasks, such as videogames of skill or driving, which at the same time allow for detailed assessment of performance (processing and response time, accuracy, orienting and executive control networks, *etc.*).

In addition, individual variability is important in the assessment of performance, which can result in a high dispersion rate that does not provide significant results, even when controlling the numerous socio-demographic and individual factors known to influence cognitive performance. Individual variability is minimized when designing repeated measurements, rendering this an excellent choice for studies on biological parameters. Nevertheless, the use of these designs in the assessment of cognitive performance may lead to fatigue and–or learning effects that are impossible to determine *a posteriori*, which is a problem in obtaining reliable and valid results (Adan *et al* 2008b; Schmidt *et al* 2007). This aspect is minimized if we work with wide samples of participants and control the highest number of variables which are known to affect the results.

15.3   Effects of Caffeine on Attention and Psychomotor Tasks

We regard attention tasks as repetitive and monotonous tasks that involve perceptual and motor skills, and whose common characteristic is to prioritize the reaction time. We further distinguish between those of vigilance or sustained attention, defined as the ability to maintain an adequate status monitoring certain events or stimuli prolonged in time (*i.e.*, visual search, digit substitution, rapid visual information tests) and reaction time (simple and choice). Most of these works which include attention tasks obtain benefits after the administration of caffeine when compared to placebo. In fact, these tasks hold the largest evidence of their sensitivity to caffeine benefits.

In sustained attention tasks we find more speed for encoding new information and responding, as well as more accuracy (more true positives, fewer omissions or errors) after caffeine consumption. This has been observed both at low or moderate doses as well as high doses (Brice and Smith 2001; Childs and de Wit 2006; Hewlett and Smith 2007; Hogervorts *et al* 2008; Smith 2009), in habitual consumers (Attwood *et al* 2007; Christopher *et al* 2005; Harrell and Juliano 2009) and non-habitual consumers (Christopher *et al* 2005; Haskell *et al* 2005). Caffeine has also shown beneficial effects on reaction time tasks, both simple and of choice, reaction time being the most sensitive execution parameter (Adan and Serra-Grabulosa 2010; Attwood *et al* 2007; Smith, 2009). This is seen independently from the participants being habitual or non-habitual caffeine consumers (Haskell *et al* 2005).

Caffeine administration has no effect on the orienting network of attention, understood as the region of space to which we selectively attend (Brunyé *et al* 2010a; 2010b) nor on visuospatial tasks (Adan and Serra-Grabulosa 2010). This result is independent from the dose administered and the habitual caffeine intake of the participants.

In tasks which require lower levels of attention, the administration of caffeine in a dose inferior to 200 mg does not usually produce perceptible benefits in non-habitual or low caffeine consumers (Adan and Serra Grabulosa 2010; Brunyé *et al* 2010a; Hogervorts *et al* 2008). In habitual caffeine consumers the effective dose to obtain benefits is of 400 mg (Brunyé *et al* 2010b). At these doses, the benefits in the execution of attention tasks are higher for the high consumers than for the moderate consumers of caffeine (Attwood *et al* 2007), although we still have to ascertain why this is so.

The effects of caffeine on attention have been assessed by neurophysiologic and neuroimaging techniques, which give us information on the neural mechanisms underlying these effects. In the first study to use functional magnetic resonance imaging (fMRI) to evaluate caffeine cognitive effects on sustained attention, it was found that a low dose of caffeine modulates neuronal activity during performance without impact on behavioral recordings (Koppelstaetter *et al* 2008).

Psychomotor tasks are those which rely mainly on manual dexterity, such as finger tapping or the Purdue pegboard. There are not many experimental studies examining psychomotor functions in healthy participants, and most of the motor assessments are inferred from the motor reaction time during tests of attention. Some works do not obtain significant differences in motor executions with the administration of caffeine in different dosages when compared to placebo in healthy young people (Adan and Serra-Grabulosa 2010; Childs and de Wit 2006), although Harrell and Juliano (2009) observe a better execution in a finger tapping task with the ingestion of 280 mg of caffeine when compared to placebo. Moreover, caffeine improves motor-skill performance in real-life motor tasks such as handwriting in skilled subjects (Tucha *et al* 2006), as well as in a simulated driving task in healthy participants (Brice and Smith 2001). Caffeine benefits in the tasks of simulated driving translated into fewer steering wheel movements, making for a safer, accurate and more stable drive after caffeine intake compared with placebo. Moreover, the effect was maintained continually throughout the time in task (1 hour).

In general, more beneficial effects on attentional and–or psychomotor tasks were observed with the administration of low doses of caffeine in situations with a deficit in activation, such as in the case of fatigued subjects (Hogervorst *et al* 2008; Smith *et al* 2005), working at night or during sleep deprivation (Killgore *et al* 2006; Kohler *et al* 2006; Sun *et al* 2007; Wyatt *et al* 2004; Wesensten *et al* 2005). As suggested by Wyatt *et al* (2004), more studies should be carried out in target populations (*e.g.*, hospital residents, truck drivers) assessing how useful and secure it is in the long run to use caffeine as a countermeasure against sleep and circadian related performance impairment.

15.4   Effects of Caffeine on Memory and Executive Function tasks

Memory tasks involve the presentation of a material that the subject must recall after a certain amount of time, or use it to reply to an exercise that he is presented with, or freely narrate its content. These tasks are differentiated according to the capacity and duration of the information storage. In the immediate or short-term memory tasks, the interval between presentation and retrieval is short, and the capacity of storage is limited (*e.g.*, retrieving lists of digits or words). Working memory tasks are those which require a longer duration of the information storage and a limited but larger capacity than in the short-term tasks (arithmetic, verbal reasoning and comprehension). Finally, there are the long-term memory tasks, where the interval between presentation and retrieval ranges from several hours to days or even months, and where the duration of the information storage is long and of unlimited capacity. These latter tasks are not usually included in the cognitive tasks batteries selected by the works which study the effects of caffeine consumption.

The administration of caffeine does not improve either speed or precision in learning and memory tasks of healthy subjects (immediate and delayed recall) in most of the works published (Adan and Serra-Grabulosa 2010; Childs and de Witt, 2006; Haskell *et al* 2005; Hogervorst *et al* 2008; Oei and Hartley, 2005). This is found independently of the dose used and of the participants being habitual or non-habitual consumers of caffeine.

Similarly, the results do not show benefits of caffeine consumption in the execution of verbal and spatial working memory tasks in healthy subjects (Haskell *et al* 2005; Valladares *et al* 2009). Haskell *et al* (2005) did obtain benefits from caffeine administration in the speed of execution in a task of numeric working memory, both in habitual and in non-habitual caffeine consumers. Perhaps the modality of the material in these tasks is relevant to showing beneficial results. The studies using fMRI show some changes in neural activity related with caffeine intake (Koppelstaetter *et al* 2010). Thus, in absence of effects on execution estimations, caffeine modulates neuronal activity during performance in a working memory task in regions related to working memory and executive functions (prefrontal dorsolateral and cingulated cortex in the right hemisphere (Koppelstaetter *et al* 2008).

Higher-order processes are those considered to be involved in the active monitoring, guidance, and coordination of behavior (Brunyé *et al* 2010a). An important component of these is the ability to inhibit inappropriate actions, and reduce the influence of interfering information. The Stroop color-word tasks and the Wisconsin card sorting test are two neuropsychological tasks which assess inhibitory control (executive function). In executive control or higher-order processes tasks, the administration of low doses of caffeine does not show significant effects independently of the profile of caffeine consumption of the participants (Adan and Serra-Grabulosa 2010; Brunyé *et al* 2010a; Hogervorst *et al* 2008; Sun *et al* 2007; Tieges *et al* 2009). Some benefits are found only with the administration of doses from 200 mg in non-habitual consumers (Brunyé *et al* 2010a) and of 400 mg in habitual consumers of caffeine (Brunyé *et al* 2010b).

15.5 Effects of Caffeine in Subjective Activation and Affect

The measurement of the subjective state of the individuals may be carried out using several different self-assessment questionnaires. Among the instruments most widely used to study the effects of caffeine there are the Stanford Sleepiness Scale (SSS), the Profile of Mood States (POMS) and the Visual Analogue Scales (VAS). All of them have been shown to be sensitive to the effects of caffeine intake.

The VAS are faster to answer and facilitate the collection of information in cases of repeated measures. Using a set of eight VAS (see Figure 15.1) designed by our team, four scales refer to the dimension of subjective activation (alertness, vigor, weariness and sleepiness) and four refer to affect (happiness, calmness, sadness and tension). These scales allow us to measure adequately and completely the participants' subjective state (Adan *et al* 2008b). Each scale is 100 mm long and asks the subject to put a cross at points between "very little" (left) and "very much" (right) based on perception at the moment of response (scores range from 0 to 100). Each group of four scales can then be added algebraically to give a simple global value of activation [(alert) + (vigorous) + 200 − (weariness) − (sleepiness) / 4] for activation, and [(happiness) + (calm) + 200 − (sadness) − (tense) / 4] for affect. Each scale can also be analyzed independently to determine whether they differ in sensitivity to the variable studied.

The effects of caffeine on subjective state have been widely studied in human subjects. Research has shown that caffeine consumption is reliably associated with positive subjective effects even in the absence of acute withdrawal effects. The subjects report that they feel more alert, efficient, jittery, vigorous, able to concentrate and motivated to work. In contrast, the subjective perception of sleepiness, fatigue, drowsiness, tiredness, and negative affect decreases. This has been observed independently of the instrument used to assess the subjective state, and has also been observed with a wide variety of doses (Adan *et al*

How are you feeling right now?		
	Very little	Very much
Alert	———————————————	
Sad	———————————————	
Tense	———————————————	
Vigorous	———————————————	
Happy	———————————————	
Weary	———————————————	
Calm	———————————————	
Sleepy	———————————————	

Figure 15.1 A set of eight Visual Analogue-Scales (VAS) to measure subjective activation and subjective affect. Size of the horizontal lines must be exactly 100 mm long. Subjective activation was measured using the adjectives alert, vigorous, weary, and sleepy. Subjective affect was measured using the adjectives sad, tense, happy and calm. A score of zero corresponds to the lowest rating in reference to positive scales (*e.g.*, alertness) or the optimal response in reference to negative scales (*e.g.*, sleepiness).

2008b; Brice and Smith 2001; Brunyé *et al* 2010a; 2010b; Childs and de Witt 2006; Christopher *et al* 2005; Haskell *et al* 2005; Hewlett and Smith 2007; Smith 2002; 2009). In general, there seems to be a dose-dependent effect, where the most effective dose range goes from 100 to 400 mg.

More beneficial subjective effects were observed with administration of low doses of caffeine in situations with a deficit in activation, such as in the case of fatigued subjects (Hogervorst *et al* 2008; Smith *et al* 2005), and working at night or during sleep deprivation (Killgore *et al* 2006; Kohler *et al* 2006; Wesensten *et al* 2005; Wyatt *et al* 2004). Moreover, although the subjective activation and affect-enhancing effects of caffeine occur both in habitual and in non-habitual consumers, high-caffeine consumers are more likely to perceive broadly positive effects (Attwood *et al* 2007; Hewlett and Smith 2006; Rogers *et al* 2003).

On the negative side, there are some adverse effects of caffeine, particularly its tendency to increase anxiety, tension, nervousness and dysphoria, as well as producing tremor, which can negatively affect performance in tasks with a motor component. This has been associated with the administration of high doses of caffeine in subjects vulnerable to the adverse behavioral effects of this substance (Attwood *et al* 2007; Childs and de Wit 2006; Smith 2002).

The research by Adan *et al* (2008) is of interest because it highlights the existence of an early effect of caffeinated coffee (100 mg; 10–30 min post-consumption) and decaffeinated coffee consumption in the subjective state of healthy participants. The administration of a caffeinated espresso coffee induced arousing effects (lesser somnolence and greater activation) in all post-

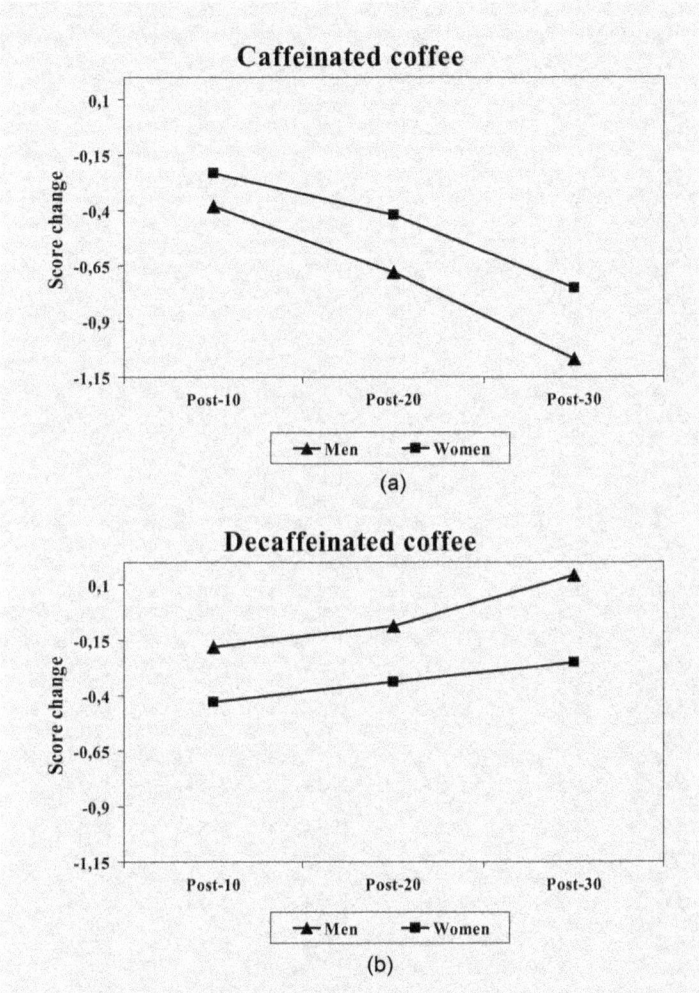

Figure 15.2 Change from basal line scores of sleepiness at 10, 20 and 30 min post-consumption of an espresso caffeinate coffee (A) or an espresso decaffeinate coffee (B) according to the sex of healthy young participants. Sleepiness was measured by the Stanford Sleepiness Scale (SSS), sensitive to the effects of caffeine. The scale consists of seven statements describing the state of alertness, ranging from "1-feeling active and vital; alert; wide-awake", to "7- almost in reverie; sleep onset soon; losing struggle to remain awake". Caffeine contents in caffeinate coffee was 100 mg. Data revised from Adan *et al* (2008).

consumption records, while the effects of a decaffeinated drink (also espresso) were only apparent at 10 min. Because caffeine reaches maximum plasma concentration (T_{max}) around 30–45 min after oral administration in humans, most studies investigating its psychopharmacological effects assessed measures

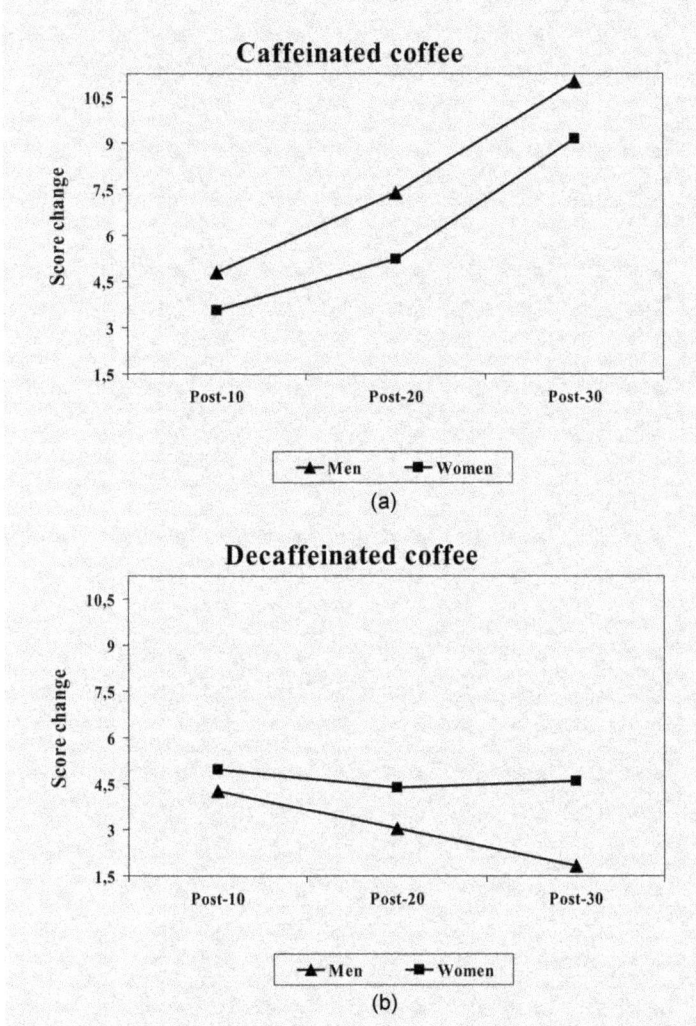

Figure 15.3   Change from basal line scores of subjective activation at 10, 20 and 30 min post-consumption of an espresso caffeinate coffee (A) or an espresso decaffeinate coffee (B) according to the sex of healthy young participants. Subjective activation was measured by four Visual Analogue Scales sensitive to the effects of caffeine. In order to obtain the self-assessment scores of activation we used the formula: (alert) + (vigorous) + 200 − (weariness) − (sleepiness) / 4. Caffeine contents in caffeinate coffee was 100 mg. Data revised from Adan *et al* (2008).

around 30–120 min post-consumption. However, levels of caffeine may already be 50% of T_{max} within 10 min post-consumption and this may be sufficient to trigger the rapid responses observed. Moreover, this study is the first to find differences in gender with respect to arousing effects after consumption of caffeinated and decaffeinated beverages. Caffeinated coffee effects were greater in men, whereas the decaffeinated beverage produced greater effects in women. Figure 15.2 shows the change from baseline in sleepiness and Figure 15.3 in subjective activation for the three post-consumption records (10, 20 and 30 min) in men and women. The results show the relevance of taking the measures before 30 minutes have passed, in order to increase the knowledge about effects of beverages containing caffeine and without an active component.

15.6 Caffeine Combined With Other Substances

A very common form of caffeine ingestion is in cola drinks which mostly contain high levels of glucose and low doses of caffeine (42 grams sugar + 30 mg caffeine per can). We will not deal here with the energy drinks that contain a number of other psychostimulant ingredients other than caffeine, such as taurine, ginseng, and ginkgo biloba. There are very few published works studying the impact of

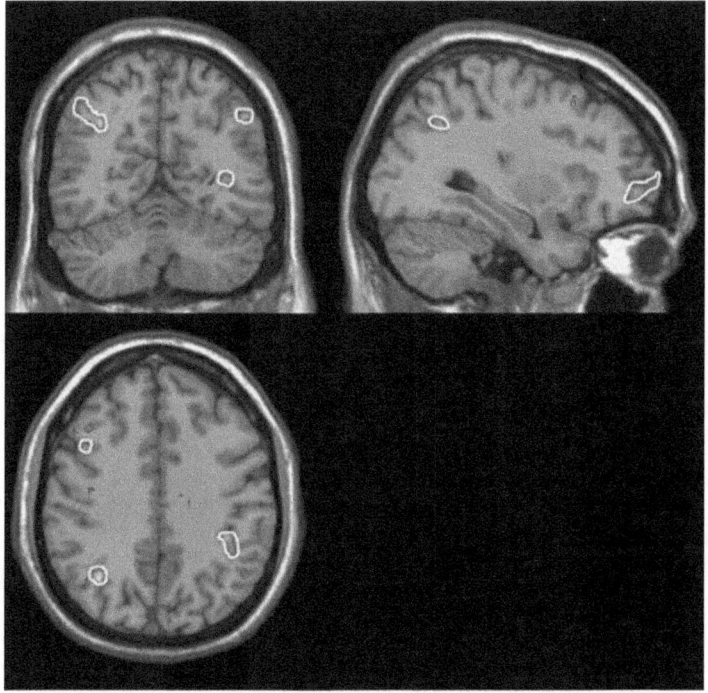

Figure 15.4 Cortical areas that showed a decrease of activation after the consumption of combined caffeine and glucose during the performance of a high-demand sustained attention task. There were changes in the bilateral parietal cortex and left prefrontal cortex. These areas sustain the processes of attention and working memory. This change in the activity pattern was observed without differences in task performance before consumption, which suggests an increase in brain effectiveness. Unpublished data from our research team.

this combination of caffeine and glucose on cognitive performance. Results indicate that the combination of caffeine and glucose produces synergistic effects when compared to administering each substance alone in healthy adults. The benefits are observed especially in learning (immediate memory) and memory consolidation (Adan and Serra-Grabulosa 2010).

In addition, using the technique of evoked potentials, it has been shown that combined caffeine and glucose produce an improvement in performance in attention tasks, modulating the components N1, N2 and P3 (Rao *et al* 2005). On the other hand, using fMRI, there have been beneficial effects from the consumption of both substances in high-demand sustained attention processes (Serra-Grabulosa and Adan 2010), with an increase in effectiveness in the brain areas that sustain attention and working memory (lesser activation in the bilateral parietal and left prefrontal cortex) for a performance similar to that of the subjects who took placebo, or caffeine or glucose alone (see Figure 15.4).

However, further research is needed to evaluate the potential pharmacokinetic interactions of the co-administration of caffeine and glucose and their impact on behavior. Maybe not all the effects of taking drinks containing high sugar and low caffeine are beneficial. Thus, the work by Anderson and Horne (2006) has shown that these drinks increase sleepiness after an extended period of time rather than alleviate it. Moreover, caffeine combined with high levels of glucose could also act synergistically to dopamine reward pathways and increase the reinforcing properties of both substances when taken alone (Temple *et al* 2009).

15.7   Caffeine Use in Children and Aged Subjects

Caffeine consumption has been increasing in children and adolescents in the last decades, although its safety during long-term periods is understudied. Thus, pharmacological treatments with caffeine mention that they are not advised for children under 12, except in apneal episodes in premature infants. The main motivations in children and adolescents who consume caffeine are different from those of adults. In young people, enhancement of sports performance and peer pressure are more common.

Temple (2009), in his recent revision, explains that children may be more likely to experience the positive effects of caffeine but may also be more vulnerable to the negative effects of caffeine than adults. In the first case, Temple's argument relies on the consumption pattern of this group, which is of smaller amounts and less frequent. In the second case, the possible effects on brain development, sleep deprivation or its contribution to both a poor diet and weight gain are relevant aspects. Finally, Temple also suggests an increased risk in that habitual caffeine use may lead to cross-sensitization of the dopaminergic neural reward to other substances and drugs.

There is evidence to suggest that caffeine consumption may lower slightly the risk of cognitive decline in older subjects, and there is a plausible mechanism for this, which somehow involves the neuroprotective functions of the adenosine system (Rogers 2007). The observational field study carried out by Johnson-Kozlow *et al* (2002) showed that lifetime coffee consumption and current coffee consumption may be associated with better cognitive performance only among women, and especially among those aged 80 or older. The lifetime coffee consumption was related to better scores in several recall tests, the Mini Mental State examination and category fluency. However, a limitation of this study is that self-reported lifetime coffee intake may be inaccurate. The recent work by Ritchie *et al* (2010) using structural magnetic resonance imaging, confirms the protective effect of caffeine consumption limited to women, white matter lesion–cranial volume ratios found to be significantly lower in women consuming more than three units of caffeine per day (one unit = 100 mg).

There have also been published works where there are very few or no benefits of habitual caffeine intake in cognitive functions of aged participants. Boxtel *et al* (2003), with a longitudinal 6 years study, found only a small association in motor speed in a reaction time task. Corley *et al* (2010) have obtained a positive effect in general cognitive ability and memory in both sexes. However, when the intelligence quotient of the participants was considered—assessed during their school period—most of the protective effects of caffeine in cognition were absent. An interesting piece of data from this study is that only drinkers of ground coffee (filter and espresso) presented better performance in tests of verbal intelligence when compared to tea or instant coffee drinkers. Both works assessed only current caffeine consumption, and thus the valuation of the lifetime consumption may have been more sensitive in the prediction of beneficial effects in performance decline related to age. The research about the possible implication of the caffeine consumption in successful and–or pathological aging is very interesting, with a great potential in the future if robust results were confirmed that could be translated into areas of prevention or treatment.

15.8   Methodological Difficulties to Solve in the Future

It is well established that there are diurnal variations in both physical and cognitive performance associated with circadian rhythm expression, which also depend on the skill(s) required for a specific task to be performed (Adan *et al* 2008a; Schmidt *et al* 2007). It is estimated that intra-individual variation in cognitive test performance, depending on the time of day, may be 20% when comparing the best and worst moments of performance. Similarly, the time of day has an influence both on the pharmacokinetics and on the pharmacodynamics of psychoactive substances. However, studies that take into account the time of day when the caffeine is administered are limited in that they only evaluate the potential diurnal antifatigue effects or the sleep deprivation effect of caffeine.

Gender has been found to be a determining factor in relation to providing differences in cognitive performance. However, most studies include either only men in their samples or participants of both genders, without studying them separately. Little interest has been given to gender in researching the behavioral effects of caffeine, even though females show lesser dopamine release and subjective response following stimulant drug administration (Munro *et al* 2006), and lesser sensitivity than men to the subjective effects produced by caffeine (Adan *et al* 2008b). Women may also be less prone to noticing the effects of caffeine on cognitive performance (Brunyé *et al* 2010). In the future, more research should be carried out on gender differences, and suitable control of menstrual cycle should also be included, as it may constitute a modulating factor in the results. There is evidence that the menstrual cycle affects cognitive task performance, which is better in the luteal phase with regard to attention tasks and in the ovulatory phase for visuospatial memory tasks.

In recent decades, circadian typology has proven to be the most important individual difference in the functioning of individuals, which includes three groups: morning, intermediate (or neither-type) and evening. This typology determines the differences in the best moments of biological and behavioral parameters, which can reach 12-hour time gaps in the performance of attention tasks (Adan *et al* 2008a; 2008b). It is necessary to monitor the circadian typology when analyzing the effects of caffeine on cognitive performance, since this could be a decisive factor in explaining some of the existing contradictory results. Likewise, the normal sleep pattern—duration and quality—of the participants is another variable related to the quality of the waking period, although the studies that consider it are limited to analyzing the night(s) immediately prior to the experimental measurements.

Although most studies check for the consumption of psychoactive substances (alcohol, nicotine, *etc.*) for a few hours before and during the experiments, the participants' usual consumption is rarely taken into account. Thus, the results of performance and subjective condition can be influenced. The use of medicines for various pathological reasons can also have both positive and negative effects on cognitive performance. Therefore, it should be properly assessed and controlled when selecting participants. Moreover, it is very useful to include objective analytical measurements (urine, blood or saliva) to confirm the proper implementation of the dietary and–or pharmacological precautions requested in the studies, with emphasis on the possible consumption of caffeine.

There are other participant-related variables that are known to influence performance in cognitive tasks, such as the level of education, personality traits that are related to stress levels and to the response to stress (*i.e.* anxiety), as well as the motivation to take part in the study in order to obtain financial compensation. All of these should be specified in the relevant methodology sections of the research because they could be factors that explain the differences observed in the effects of caffeine on the performance of very similar tasks.

Finally, in order to better understand the effects of caffeine on cognition it is necessary to use neuroimaging techniques that allow us to record the brain changes associated to the execution of the different cognitive skills. The study of the effects of caffeine consumption that add information on the brain areas involved and their degree of activation, together with behavioral recordings, will undoubtedly be beneficial both to increase our theoretical knowledge and to be more specific about the practical applications that may arise.

Summary Points

- This chapter focuses on the findings, mainly during the last decade, on the effects of caffeine on the cognitive performance of healthy subjects.
- Caffeine enhances the performance in attention tasks (speed and accuracy) even with the administration of low doses, both in habitual and non-habitual consumers.
- Caffeine improves motor-skill performance in real-life motor tasks such as handwriting and simulated driving.
- More beneficial effects of caffeine in attentional and–or psychomotor tasks were observed with the administration of low doses of caffeine in situations with a deficit in activation, and working at night or during sleep deprivation.
- The administration of caffeine does not improve either speed or precision in learning and memory tasks of healthy subjects (immediate and delayed recall).
- Research has shown that caffeine consumption is reliably associated with positive subjective effects even in the absence of acute withdrawal effects, this being dose-dependent between 100–400 mg without practically any adverse effects.
- Although there are few studies on the possible sex differences in caffeine effects, women seem to be less prone than men to noticing the effects of caffeine on cognitive performance.
- Caffeine consumption is not recommended in children and adolescents since these may be more vulnerable than adults to the negative effects of caffeine and to develop cross-sensitization of the dopaminergic neural reward to other substances and drugs.
- The lifetime coffee consumption may have a protective effect on cognitive performance, especially among women. The data suggest that the consumption has to be higher than 300 mg of caffeine per day and that it is preferable to drink ground coffee (filter and espresso).
- We need more systematic research on the effects of caffeine, incorporating the study of factors known to affect cognitive performance, such as circadian rhythmicity, level of education or personality traits that are related to stress, to mention a few.

Key Facts

- The study of cognitive performance undoubtedly has a practical interest on work and school situations.
- Assessing cognitive performance is more difficult than assessing biological aspects, because even in simple cognitive tasks there are several skills involved in their resolution (attention, motor control, *etc.*).
- It is difficult to adequately control the effects of fatigue if we measure the execution for a long period of time or repeatedly.
- Many performance tasks, especially the more complex ones, involve learning or a better resolution the more they have been practiced.
- The performance of a task can be assessed using estimates of speed or accuracy (correct responses, errors and lapses) and not always responding faster implies a better resolution.
- Individual variation (level of education, motivation, *etc.*) has a notorious influence on cognitive performance, and it has to be controlled in order not to reach wrong conclusions.

Definition of Words and Terms

Arousal or activation: Generalized physical and mental state which prepares an organism to respond.

Attention: Complex concept that represents a cohesive set of processes, which include sensory, motor and cognitive processing.

Inhibition: Mechanisms preventing the entrance of non pertinent information into working memory and suppressing information that has become irrelevant for the current tasks.

Immediate or short-term memory: Memory in which the lapse between presentation and retrieval is short, and the capacity of storage is limited (*e.g.*, retrieving lists of digits or words). The information in this memory is easily erased.

Long-term memory: Memory established in the associative cortex or in subcortical regions. It may be retrieved and managed as needed. It is time-resistant, so once formed it is difficult to erase.

Maximum plasma concentration (T_{max}): Pharmacokinetic parameter which refers to the time that a substance takes to reach its peak plasma concentration. For caffeine it is reached between 30 and 45 minutes after oral consumption.

Placebo: Substance or procedure that has no inherent power to produce a rought or expected effects.

Vigilance or sustained attention: Ability to attend over long and generally continuous periods of time for the purpose of detecting and responding to relevant stimuli.

Withdrawal: A set of signs and symptoms that appear when the consumption of a drug is considerably reduced or suddenly discontinued. It may also appear if a specific antagonist is administered which blocks the effects of the drug.

Working memory: Ability to hold and manipulate information in the mind over short periods of time. Mental workspace that is used to store important information in the course of our everyday lives.

List of Abbreviations

fMRI	functional magnetic resonance imaging
mg	milligrams
POMS	Profile of Mood States
SSS	Stanford Sleepiness Scale
T_{max}	maximum plasma concentration
VAS	Visual Analogue Scales

References

Adan, A., Natale, V. and Caci, H., 2008a. Cognitive strategies and circadian typology. In: Léglise, A. L. (ed.). Progress in circadian rhythms research. Nova Science Publishers, Inc., New York, pp. 141–161.

Adan, A., Prat, G., Fabbri, M. and Sánchez-Turet, M., 2008b. Early effects of caffeinated and decaffeinated coffee on subjective state, and gender differences. Progress in Neuropsychopharmacology & Biological Psychiatry. 32: 1698–1703.

Adan, A. and Serra-Grabulosa, J. M., 2010. Effects of caffeine and glucose, alone and combined, on cognitive performance. Human Psychopharmacology. Clinical and Experimental. 25: 310–317.

Anderson, C. and Horne, J., 2006. A high sugar content, low caffeine drink does not alleviate sleepiness but may worsen it. Human Psychopharmacology. Clinical and Experimental. 21: 299–303.

Attwood, A. S., Higgs, S. and Terry, P., 2007. Differential responsiveness to caffeine and perceived effects of caffeine in moderate and high regular caffeine consumers. Psychopharmacology. 190: 469–77.

Boxtel, M. P. J., Schmitt, J. A. J., Bosma, H. and Jolles, J., 2003. The effects of habitual caffeine use on cognitive change: a longitudinal perspective. Pharmacology, Biochemistry and Behavior. 75: 921–927.

Brice, C. and Smith, A., 2001. The effects of caffeine on simulated driving, subjective alertness and sustained attention. Human Psychopharmacology. Clinical and Experimental. 16: 523–531.

Brunyé, T. T., Mahoney, C. R., Lieberman, H. R. and Taylor, H. A., 2010a. Caffeine modulates attention network functions. Brain and Cognition. 72: 181–188.

Brunyé, T. T., Mahoney, C. R., Lieberman, H. R., Giles, G. E. and Taylor, H.A., 2010b. Acute caffeine consumption enhances the executive control of visual attention. Brain and Cognition. 72: 186–192.

Childs, E. and de Wit, H., 2006. Subjective, behavioral, and physiological effects of acute caffeine in light, nondependent caffeine users. Psychopharmacology. 185: 514–523.

Christopher, G., Sutherland, D. and Smith, A., 2005. Effects of caffeine in non-withdrawn volunteers. Human Psychopharmacology. Clinical and Experimental. 20: 47–53.

Corlei, J., Jia, X., Kyle, J. A. M., Gow, A. L., Brett, C. E., Starr, J. M., McNeill, G. and Deary, I. J., 2010. Caffeine consumption and cognitive function at age 70: the Lothian birth cohort study. Psychosomatic Medicine. 72: 206–214.

Harrel, P. T. and Juliano, L. M., 2009. Caffeine expectancies influence the subjective and behavioural effects of caffeine. Psychopharmacology. 207: 335–342.

Haskell, C. F., Kennedy, D. O., Wesnes, K. A. and Scholey, A. B., 2005. Cognitive and mood improvements of caffeine in habitual consumers and habitual non-consumers of caffeine. Psychopharmacology. 179: 813–825.

Hewlett, P. and Smith, A., 2006. Acute effects of caffeine in volunteers with different patterns of regular consumption. Human Psychopharmacology. Clinical and Experimental. 21: 167–180.

Hewlett, P. and Smith, A., 2007. Effects of repeated doses of caffeine on performance and alertness: new data and secondary analyses. Human Psychopharmacology. Clinical and Experimental. 22: 339–350.

Hogervorst, E., Bandelow, S., Schmitt, J., Jentjens, R., Oliveira, M., Allgrove, J., Carter, T. and Gleeson, M., 2008. Caffeine improves physical and cognitive performance during exhaustive exercise. Medicine and Science in Sports and Exercise. 40: 1841–1851.

James, J. E. and Rogers, P. J., 2005. Effects of caffeine on performance and mood: withdrawal reversal is the most plausible explanation. Psychopharmacology. 182: 1–8.

Johnson-Kozlow, M., Kritz-Silverstein, D., Barrett-Connor, E. and Morton, D., 2002. Coffee consumption and cognitive function among older adults. American Journal of Epidemiology. 156: 842–850.

Killgore, W. D. S., McBride, S. A., Killgore, D. B. and Balkin, T. J., 2006. The effects of caffeine, dextroamphetamine, and modafinil on humor appreciation during sleep deprivation. Sleep. 29: 841–847.

Kohler, M., Pavy, A. and van den Heuvel, C., 2006. The effects of chewing *versus* caffeine on alertness, cognitive performance and cardiac autonomic activity during sleep. Journal of Sleep Research. 15: 358–368.

Koppelstaetter, F., Poeppel, T. D., Siedentopf, C. M., Ischebeck, A., Verius, M., Haala, I., Mottaghy, F. M., Rhomberg, P., Golaszewski, S., Gotwald, T., Lorenz, I. H., Kolbitsch, C., Felber, S. and Krause, B. J., 2008. Does caffeine modulate verbal working memory processes? An fMRI study. Neuroimage. 39: 492–499.

Koppelstaetter, F., Poeppelb, T. D., Siedentopf, C. M., Ischebeck, A., Kolbitsch, K., Mottaghy, F. M., Felber, S. R., Jaschke, W. R. and Krause, B. J., 2010. Caffeine and cognition in Functional Magnetic Resonance Imaging. Journal of Alzheimer's Disease. 20: S71–S84.

Munro, C. A., McCaul, M. E., Wong, D. F., Oswald, L. M., Zhou, Y., Brasic, J., Kubawara, H., Kumar, A., Alexander, M., Ye, W. and Wand, G. S., 2006. Sex differences in striatal dopamine release in healthy adults. Biological Psychiatry. 59: 966–974.

Oei, A. and Hartley, L. R., 2005. The effects of caffeine and expectancy on attention and memory. Human Psychopharmacology. Clinical and Experimental. 20: 193–202.

Rao, A., Hu, H. and Nobre, A. C., 2005. The effects of combined caffeine and glucose drinks on attention in the human brain. Nutritional Neuroscience. 8: 141–153.

Ritchie, K., Artero, S., Portet, F., Brickman, A., Muraskin, J., Beaino, E., Ancelin, M-L and Carrière, I., 2010. Caffeine, cognitive functioning, and white matter lesions in the elderly: establishing causality from epidemiological evidence. Journal of Alzheimer's Disease. 20 (Suppl 1): S161–S166.

Rogers, P. J., Martin, J., Smith, C., Heatherley, S. V. and Smit, H. J., 2003. Absence of reinforcing, mood and psychomotor performance effects of caffeine in habitual non-consumers of caffeine. Psychopharmacology. 167: 54–62.

Rogers, P. J., 2007. Caffeine, mood and mental performance in everyday life. Nutrition Bulletin. 32: 84–89.

Schmidt, C., Collette, F., Cajochen, C. and Peigneux, P., 2007. A time to think: circadian rhythms in human cognition. Cognitive Neuropsychology. 24: 755–789.

Serra-Grabulosa, J. M., Adan, A., Falcón, C. and Bargalló, N., 2010. Glucose and caffeine effects on sustained attention: an exploratory fMRI study. Human Psychopharmacology. Clinical and Experimental. 25: 543–552.

Smith, A. 2002. Effects of caffeine on human behavior. Food and Chemical Toxicology. 40: 1243–1255.

Smith, A. 2009. Effects of caffeine in chewing gum on mood and attention. Human Psychopharmacology. Clinical and Experimental. 24: 239–247.

Smith, A., Sutherland, D. and Christopher, G., 2005. Effects of repeated doses of caffeine on mood and performance of alert and fatigued volunteers. Journal of Psychopharmacology. 19: 620–626.

Sun, Y., Zhang, Y., He, N., Liu, X. and Miao, D., 2007. Caffeine and placebo expectation. Effects on vigilance, cognitive performance, heart rate, and blood pressure during 28 hours of sleep deprivation. Journal of Psychophysiology. 21: 91–99.

Temple, J. L., 2009. Caffeine use in children: what we know, what we have left to learn, and why we should worry. Neuroscience and Biobehavioral Reviews. 33: 793–806.

Tieges, Z., Snel, J., Kok, A. and Ridderinkhof, J. R., 2009. Caffeine does not modulate inhibitory control. Brain and Cognition. 69: 316–327.

Tucha, O., Walitza, S., Mecklinger, L., Stasik, D., Sontag, T. A. and Lange, K. W., 2006. The effect of caffeine on handwriting movements in skilled writers. Human Movement Sciences. 25: 523–535.

Valladares, L. 2009. Effects of caffeine on cognitive tasks. Masters Thesis RMIT University. http://researchbank.rmit.edu.au/eserv/rmit:6833/Valladares.pdf

Wesensten, N. J., Killgore, W. D. and Balkin, T. J., 2005. Performance and alertness effects of caffeine, dextroamphetamine, and modafinil during sleep deprivation. Journal of Sleep Research. 14: 255–66.

Wyatt, J. K., Cajochen, C., Ritz-De Cecco, A., Czeisler, C. A. and Dijk, D-J., 2004. Low-dose repeated caffeine administration for circadian-phase-dependent performance degradation during extended wakefulness. Sleep. 27: 374–381.

CHAPTER 16

Neuroprotective Effects of Caffeine in Sleep Deprivation

KARIM A. ALKADHI*[1], MUNDER A. ZAGAAR[1], IBRAHIM A. ALHAIDER[2] AND KAREM H. ALZOUBI[3]

[1] Department of Pharmacological and Pharmaceutical Sciences, College of Pharmacy, University of Houston, Houston, TX, USA; [2] College of Clinical Pharmacy, King Faisal University, Saudi Arabia; [3] Department of Clinical Pharmacy, Faculty of Pharmacy, Jordan University of Science and Technology, Irbid, Jordan
*E-mail: kalkadhi@uh.edu

16.1 Introduction

Sleep is a state of temporary lack of consciousness that can be interrupted by various stimuli. Although the function of sleep is not well understood, it seems to be essential for the wellbeing and survival of the organism. Prolonged sleep loss, and insufficient sleep can results in dire health consequences in humans. For example, insufficient sleep is linked to increased risk of obesity, type-2 diabetes, heart problems and dementia (Wilson 2005). Animal studies suggest that prolonged sleep deprivation (SD) causes severe behavioral, synaptic, and membrane excitability alterations in the hippocampus (McDermott et al 2003), a brain structure involved in learning and memory. However, the immediate effect of SD is usually seen as deterioration of mental functions. The alertness associated with sleep loss leads to diminished attention and motivation, deterioration of mental concentration and decline of intellectual capacity leading to poor performance. This problem is clearly seen in overworked

Food and Nutritional Components in Focus No. 2
Caffeine: Chemistry, Analysis, Function and Effects
Edited by Victor R Preedy
© The Royal Society of Chemistry 2012
Published by the Royal Society of Chemistry, www.rsc.org

healthcare providers (Gaba 2002). Medical errors brought about by sleepiness of healthcare providers are known to cause thousands of deaths and cost billions of dollars every year.

In general, the performance of sleep-deprived individuals is significantly lower than that of those with adequate sleep (Pilcher 1996). Even partial SD, which leads to moderate levels of sleepiness *(e.g.* 5 hours of sleep in a 24 hour period) can downgrade performance to a level similar to or greater than that caused by alcohol intoxication or treatment with sedatives (Lieberman *et al* 2005). Sleepiness consistently increased the risk of driving accidents and occupational injury (Chau 2004; Robb 2008). In fact, disastrous human accidents including the Three Mile Island nuclear plant incident, Chernobyl reactor explosion and the space shuttle Challenger explosion are believed to be due to human errors caused by sleepiness (Wilson 2005).

The safest and most effective measure for preventing the harmful effects of SD is adequate and quality sleep. However, in modern industrial societies there is always a need for individuals to be alert at times when they are supposed to be sleeping. Consequently there is a need for interventions that can keep individuals such as airline pilots, physicians, nightshift workers and the military personnel in the battlefield, vigilant in doing their tasks. It is a general knowledge that stimulants such as caffeine and nicotine stabilize the deleterious cognitive effects of SD, owing to their neuroprotective properties. Hence, these stimulants are used for self-medication (Poston *et al* 2008). Caffeine-containing beverages are the most commonly used interventions for promoting alertness and alleviating the symptoms of sleepiness. Caffeine is the most widely used psychoactive drug that is found naturally in coffee and tea, and as added ingredient in many soft drinks. However, excessive consumption of caffeine is associated with negative side effects, including anxiety, stress and consequent sleep disruption.

Caffeine can influence a variety of signaling molecules with diverse affinities. High (millimolar) concentrations of caffeine may act as inhibitor of cyclic nucleotide phosphodiesterase enzymes in the central nervous system (CNS) and other tissues (Cardinali 1980). Additionally, millimolar concentrations of caffeine are required for GABAa receptor blockade or to release calcium from intracellular stores, through activation of ryanodine-sensitive channels (Fredholm *et al* 1999; Cunha and Agostinho 2010). The psychostimulant properties of caffeine are believed to be through inhibition of adenosine receptors.

Interestingly, adenosine, a byproduct of metabolism of adenosine triphosphates (ATP), does not fit the criteria normally used to define a neurotransmitter; it is not accumulated in vesicles, and is not released from nerve terminals in a calcium-dependent fashion (Fisone *et al* 2004). Adenosine concentration rises wherever there is activity that involves metabolism of ATP. Inhibition of adenosine A1 receptors in the CNS increases neuronal firing whereas inhibition of A2a receptors suppresses firing. Caffeine binds with high affinity to adenosine A1 and A2a receptors, which are activated by nanomolar concentrations of adenosine, approximating those normally present in the brain (Lloyd and Fredholm 1995). In fact, adenosine receptors are blocked after drinking even a single cup of coffee, which can produce plasma concentrations of up to 10 mM (Fredholm *et al* 1999). In this review, we summarize the effects of caffeine on the harmful consequences of SD on the brain function.

16.2 Sleep and Sleep Cycle

Sleep in the broadest sense, is a phylogenetically preserved process characterized by diminished responses to external stimuli and relative reversibility. It was not until the early twentieth century that the newly discovered electroencephalography (EEG), which measured the sum of local electric field potentials as a correlate of cortical activity, was utilized to characterize differences between the waking and sleeping brain. Sleep is represented by the cyclical occurrence of 2 main sleep phases, each characterized by different EEG profiles and cellular processes: 1) rapid eye movement (REM) sleep also known as paradoxical sleep and 2) non-rapid eye movement (NREM) sleep, also known as delta sleep or slow wave sleep, which includes four sub-stages (stages 1 to 4).

A definitive function for sleep remains elusive although many hypotheses have been offered ranging from temperature regulation to memory consolidation (Kalia 2006). Sleep is a homeostatically regulated process whose disruption produces an anxiety response in the organism. This is particularly relevant in modern societies, as social and occupational demands require sacrificing sleep in order to increase productivity. Indeed, the percentage of Americans that claim less than 7 h of sleep has increased in the past 4 decades to an estimated 37.1% (Anonymous 2011). With sleep, the brain plays the role of an unforgiving accountant that keeps track of lost sleep as a debt to be repaid. If one sacrifices too much sleep, the interest on this sleep debt eventually manifests as fatigue or illness, which forces the debtor to bed. The emerging repercussions of irregular sleep quality and duration are based on increasing epidemiological evidence that suggests a corresponding increased risk of metabolic and cardiovascular diseases such as obesity and type II diabetes mellitus (Penev 2007). In order to understand the dynamics of caffeine effect on sleep, we must understand the subtleties of the phenomenon of sleep and the consequences of sleep loss on brain function.

16.2.1 Neurochemistry During Sleep

The brain functions differently during wakefulness and sleep; however, the complex neurochemical interplay between acetylcholine (ACh) and monoamine producing neurons allows for the same brain regions to mediate both sleep and wakefulness. For instance, there are brain stem neurons that release ACh from the basal telencephalon, serotonin (5-HT) from the dorsal raphe nucleus (DRN) and norepinephrine (NE) from the locus coeruleus (LC) to generate the activity necessary to maintain wakefulness (Porkka-Heiskanen *et al* 2007). The same brain stem regions involved in wakefulness show diminished activity during NREM sleep with the exception of serotonergic neurons of the DRN, which paradoxically play a role in triggering sleep. At NREM sleep onset, cholinergic and noradrenergic activity in the brain stem subsides so as to allow the DRN neurons to stimulate the anterior hypothalamus and block waking brain activity from reaching higher brain regions such as the thalamus and cortex (McGinty and Harper 1976). The brain's neurochemical environment changes once more during the transition from NREM to REM sleep. Specifically, REM sleep manifests when both of the monoamine releasing neurons of the LC and DRN are simultaneously quiescent, which in turn permits the large cholinergic neurons in the brain stem to produce the excessive activity similar to that of wakefulness. This intricate neurochemical modulation across different brain regions is what produces the neuronal activity characteristic of sleep, which can be visualized with polysomnography recordings.

16.2.2 Sleep Architecture

Polysomnography studies show that throughout the course of the night, different sleep stages manifest and alternate regularly in a composed manner classified as sleep cycles. A complete night of sleep consists of five to six major cycles, and although all sleep cycles in a night last about 90 minutes, the duration of each of the two major sleep phases changes as the night progresses with REM sleep increasing and NREM sleep decreasing. The deepest, most restorative sleep occurs during NREM stage 4 and is characterized by a low overall neuronal activity. In contrast, REM sleep, which begins after NREM sleep, is a period of increased brain activity similar to that observed during wakefulness. In addition, REM sleep is hallmarked by a generalized loss of skeletal muscle tone that occurs in temporal association with rapid eye movements (Porkka-Heiskanen *et al* 2007).

16.3   Nutrition, Metabolism and Sleep

Humoral factors such as cytokines, endocrine hormones in addition to macronutrients and metabolic byproducts such as adenosine have been linked to sleep homeostasis (Penev 2007). These findings seem to support the notion of a reciprocal relationship between energy metabolism and sleep, specifically in the context of SD and dietary intake (Nicolaidis 2006). Studies suggest that increased consumption of energy-rich foods increases sleep directly *via* vagal nerve stimulation of higher order CNS neurons or indirectly through gastrointestinal hormones such as cholecystokinin. Furthermore, other studies have suggested that sleep quality and quantity depend on the macronutrient content of food as these components (*e.g.* amino acids) serve as synthetic precursors of crucial neurotransmitters involved in sleep regulation (*e.g.* 5-HT). For instance, it has been shown that REM sleep depends on dietary amino acid and subsequent protein synthesis and metabolism whereas, restricted food intake increases sleep latency and diminishes NREM sleep (Karklin *et al* 1994). These findings suggest that the CNS can sense nutritional status and alter sleeping patterns accordingly. Increased energy status favors sleep whereas diminished energy reserves, initially, increases slow wave sleep (stages 3 and 4 NREM sleep) in order to conserve energy but later leads to sleep loss and increased vigilance as energy stores approach depletion (Nicolaidis 2006). The increase in vigilance reflects a shift in priorities for the brain from conserving energy to actively searching for metabolic fuel to replenish energy stores. Tryptophan, a 5-HT precursor, seems to plays crucial role in modulating sleep latency and brain development in infants. In line with this finding, α-lactalbumin, a type of milk with 80% less casein than whole milk improves sleep even after 3–4 days of total food restriction by increasing tryptophan transport and subsequent 5-HT synthesis in the brain. Moreover, rats deprived of dietary tryptophan show a reorganization of sleep architecture and diminished sleep circadian rhythm (Lanoir *et al* 1981). In addition, it has been shown that SD increases brain serotonin levels which may be responsible for subsequent sleep rebound (Grossman *et al* 2000).

16.4   Sleep Deprivation

Sleep and its function have attracted the attention of scientists and philosophers throughout history and its influence is pervasive in modern societies as certain occupations or social constraints demand the prospect of inadequate sleep. The ability to voluntarily sacrifice sleep for more time awake is unique to humans and the health implications are far reaching.

Although sleep function is not completely understood at the moment, there is no doubt that sleep is homeostatically regulated. Thus, the physiological need for sleep is undoubtedly one of the strongest drives found in animals and humans. Indeed, such a process must be vital if it is worth the risk of decreased vigilance and vulnerability to predatory attack. Accordingly, the detrimental effects of SD on different physiological aspects such as: immunological factors, body temperature regulation, endocrine secretion patterns and learning and memory processes reflect the homeostatic imbalance of the organism. As the organism repeatedly attempts to restore homeostasis throughout the period of "sleepy" wakefulness the stress of being continuously awake becomes greater and an allostatic load develops. The allostatic load represents a change in the set point for homeostasis and was first characterized in blood pressure dynamics. Rechstaffen's group were the first to characterize the SD syndrome in rats that leads to death after 3–4 weeks of SD using the disk-over-water method (Rechtschaffen *et al* 1983). After continuous total SD for 2–3 weeks, a constellation of symptoms appears including an initial increase in body temperature without a concomitant increase in energy expenditure followed by a drastic drop in body temperature right before death. Furthermore, skin ulcerations developed on the footpads of the animals in addition to an overall unkempt appearance. During the SD period, there were no stress symptoms such as stress ulcers, increased corticosterone levels or decreased metabolic rate indicating that SD produces distinct detrimental effects on homeostasis compared to those of stress related hypothalamic-pituitary-adrenal activation (Rechtschaffen *et al* 1983).

However, during these SD experiments a rather intriguing finding emerged that seemed to tie in with the reciprocal relationship between mammalian sleep and energy homeostasis. There was a paradoxical decrease in weight with the increased food intake indicating that food derived energy was being utilized in a non-thermogenic capacity. This negative energy balance could be the result of decreases in key metabolic hormones (*e.g.* insulin) (Everson and Crowley 2004) or increased activity in hypothalamic feeding centers (Koban *et al* 2006) or increases in brown adipose tissue thermogenesis in animals. Yet, this catabolic effect of SD seems to only appear in small mammals thus far while human findings remain inconclusive. With that said, several epidemiological studies have suggested a link between short sleep duration and increased obesity or adiposity (Penev 2007). This supports other studies showing that short sleep periods can influence the neuroendocrine control of energy balance by altering levels of leptin and ghrelin, two hormones involved in energy metabolism, in a way that promotes feeding in the absence of increased caloric need, which could eventually lead to weight gain. Shorter sleep has been suggested to alter glucose metabolism and result in diabetic complications such as insulin resistance in healthy young adults (Gottlieb *et al* 2005). This metabolic derangement could be due to increased circulation of counter-regulatory catabolic and anabolic hormones (*e.g.* cortisol and growth hormone) (Yaggi *et al* 2006) or even elevated secretion of both pro- and anti-inflammatory cytokines (*e.g.* tumor necrosis factor and interleukin-6) (Vgontzas *et al* 2006). Thus, the finding that SD leads to increased energy expenditure and weight loss is consistent with the role of sleep in energy conservation possibly explaining its role as a risk factor for metabolic disorders such as diabetes mellitus and obesity.

16.4.1 Sleep Disruption and Neural Function

Both partial and total SD experiments play key roles in elucidating many of the effects that sleep exerts on memory consolidation (Smith and Rose 1996). In other words, as synaptic plasticity is thought to underlie memory formation, blocking the consolidation that occurs during these sleep stages using SD models reveals certain impairments. Sleep deprivation studies are generally divided in to partial SD, which is defined as one night of reduced or fragmented sleep and total SD, which involves no sleep during a normal sleep–wake cycle. For example, the disc-on-water method can be considered forms of chronic total SD (Rechtschaffen *et al* 1983) whereas other, less stressful techniques prevent sleep for shorter durations (*e.g.* gentle handling, treadmill) (Guzman-Marin *et al* 2006; Vecsey *et al* 2009) are considered to be partial SD. Many partial SD studies have shown that suppression of certain sleep stages negatively impacts different types of memory. For instance, declarative memory is largely hippocampus dependent, representing episodic spatial and temporal characteristics of a memory trace, and seems to be strengthened in the presence of NREM sleep (Diekelmann and Born 2010). The opposite seems true for procedural memory, which depends more on striatal circuits and reflects activity-response habits. For example, blocking REM sleep is detrimental to procedural memory consolidation (Diekelmann and Born 2010). Yet, in animal studies of REM SD using the columns in water method which relies on REM muscle atonia and subsequent water exposure to motivate the animals to stay awake, spatial memory is impaired as tested by either the Morris water maze (MWM) or radial arm water maze (RAWM). These two tasks require the animals to find and remember the location of a hidden platform by seemingly utilizing spatial cues on the walls and primarily test spatial working and reference memory in the case of the RAWM. Sleep deprivation for 24 or 48 hours before a learning task in the RAWM impairs acquisition, short-term (30 min) memory in addition to long-term (24 h) memory (Alhaider *et al* 2010a; 2011; Aleisa *et al* 2010; 2011). Additionally, SD for 72 hours using the column-in-water paradigm can decrease spatial learning and memory in the MWM (Wang *et al* 2009).

In rats, *in vivo* long term potentiation (LTP), a cellular correlate of learning and memory, is impaired after 24 and 48 hours of SD indicating synaptic plasticity does not occur due to weakened associations between synapses in the hippocampus, a key area for temporary memory storage (Aleisa *et al* 2011). The negative impact that both partial SD and total SD have on synaptic plasticity is thought to be a product of the underlying deleterious changes in intracellular signaling. For instance, N-methyl-D-aspartate receptors (NMDARs), which are critical to LTP induction, show negative alterations in receptor subunit composition and turnover after 24 hours of REM SD (Chen *et al* 2006). In addition, 12 h of partial SD impairs phosphorylation and membrane trafficking of hippocampal glutamate alpha-amino-3-hydroxy-5-methyl-4-isoxazolepropionic acid (AMPA) receptors, which are crucial in initiating synaptic plasticity by mediating postsynaptic ion flux (Hagewoud *et al* 2010). Molecular studies looking further in to the synapse at the protein level, have shown that the expression of key signaling molecules and growth factors (*e.g.* cAMP response element binding protein, CREB, and Brain-derived Neurotrophic Factor, BDNF) involved in LTP and memory are impaired in the hippocampus after 8, 24 and 48 hours of SD (Guzman-Marin *et al* 2006; Alhaider et al 2010a; 2011). In particular, the expression of phosphorylated CaMKII (P-CaMKII), the gatekeeper for LTP induction, is markedly decreased after 24 hours of SD whereas the level of calcineurin, which dephosphorylates CaMKII and suppresses LTP, is increased (Alhaider *et al* 2010a; 2010b). This finding suggests that increased phosphatase levels, in particular, could mediate the decreased P-CaMKII expression before and after LTP induction and the detrimental effects of SD on memory (Alhaider *et al* 2010a). Additionally, SD negatively impacts intracellular signaling pathways namely the cyclic adenosine mono-phosphate (cAMP)–PKA pathway, which plays an important role in LTP and memory. Sleep deprivation as short as 5 hours increases levels of phosphodiesterase IV (PDEIV), which decreases the levels of cAMP and impairs its subsequent role in the LTP signaling cascade (Vecsey *et al* 2009).

On the genetic level, expression of genes during sleep is consistent with its proposed role in protein synthesis and synaptic plasticity. However, during prolonged wakefulness there is an initial up-regulation of gene expression, which gradually drops off as SD is prolonged. Noradrenalin (norepinephrine, NE) is not only a key neurotransmitter in sleep-wake regulation; it also stimulates gene expression and activity-dependent plasticity during wakefulness (Cirelli 2002). Increased levels of monoamines (*e.g.* NE and 5-HT) during and after prolonged wakefulness could partially explain the impairment of memory as a result of the continuous activity of these molecules, which have diffuse projections across the brain. This is of interest as the enzyme arylsulfotranferase, which sulfonates and inactivates these monoamines is, in fact, up regulated in proportion to the length of SD. The up regulation of this enzyme may indicate a last effort by the brain to cope with high catecholamine activation (Cirelli 2002).

16.5   Caffeine and Adenosine

Although caffeine exerts a significant stimulant effect that enhances attention and psychomotor abilities, the focus of this section will be primarily on the pharmacology of caffeine and a closely related molecule known as adenosine. In the previous sections we have discussed sleep physiology and its role in neural function, but in order to understand the mechanisms by which caffeine exerts its neuroprotective effects we must first discuss caffeine pharmacology. Caffeine influences cellular communication and neural activity by antagonizing the effect of adenosine, a ubiquitous byproduct of energy metabolism (Stenberg 2007). Notably, adenosine levels are higher during active waking periods in the brain than during resting or sleeping periods. Accordingly, during extended periods of wakefulness the demand on cellular metabolism is increased resulting in elevated adenosine levels. This coincides with the fact that adenosine is actually the degradation product of ATP, whose high-energy phosphate bonds provide the energy necessary for various cellular functions. Thus, the amount of adenosine produced in the brain during wakefulness reflects the activity level of all cells that utilize ATP as an energy source. By virtue of the fact that the intense activity required to maintain wakefulness consumes large amounts of ATP, areas of the brain directly involved in maintaining wakefulness such as the basal forebrain begin to develop an energy deficit, which causes adenosine to accumulate. Increasing extracellular adenosine levels act as a triggering mechanism for

NREM sleep, during which the brain is less active and is thereby better able to enter into a recovery phase (Porkka-Heiskanen *et al* 2003). During this phase, glycogen stores are replenished and adenosine is continuously metabolized by the enzyme adenosine deaminase leading to conditions that favor wakefulness (Huang *et al* 2011).

16.5.1   Adenosine Pharmacology

The pharmacological behavior of adenosine is complex as the activation of its various receptor subtypes can produce different downstream effects. More specifically, adenosine binds and activates a group of purinergic G-protein coupled adenosine receptor subtypes A_1, A_{2A}, A_{2B} and A_3. Adenosine binding to A_1 receptors present in wake promoting areas such as the basal forebrain promotes sleep. Activation of A_1 receptors inhibits the enzyme adenylate cyclase, which then suppresses the influx of calcium ions into the presynaptic terminals. Under normal physiological conditions, this calcium influx promotes the release of neurotransmitters, thus with decreased calcium there is smaller amounts of neurotransmitters secretion by many neurons resulting in an overall inhibitory effect.

In the areas of the brain containing neurons that promote sleep such as the midbrain reticular formation and basal forebrain, adenosine injection produces a sleep-promoting effect. At first glance, this finding would seem paradoxical, however, this effect can be explained by the fact that there are different adenosine receptor subtypes, which can enhance or decrease adenylate cyclase levels along with subsequent neuronal activity. For example, activation of A_2 adenosine receptors in the preoptic nucleus leads to increased activity of sleep promoting neurons (Dunwiddie and Masino 2001). This allows for one particular molecule to diversify its effects on the already complex system of sleep-wake regulation.

16.5.2   Caffeine Pharmacology

In general, caffeine produces long-term cerebral hypoperfusion while at the same time producing its alerting effect *via* nonselective blockade of adenosine A_1 and A_{2a} receptors in the basal forebrain and midbrain reticular formation (Fredholm 1995). However, the pharmacological effects of caffeine reach beyond that of sleep-wake regulation. Specifically, caffeine, independent of its stimulant effects, modulates adenosine pharmacology to induce beneficial changes in molecular signaling cascades that mediate synaptic plasticity. In fact, studies have shown that caffeine has a neuroprotective action against cognitive decline in neurodegenerative disorders such as Alzheimer's disease (Dall'Igna *et al* 2007).

Normally, adenosine disrupts the underlying processes of learning and memory at the synaptic level whereas caffeine seems to reverse many of adenosine effects on sleep propensity and even learning and memory by blocking adenosine receptor signaling (Alhaider *et al* 2010a; 2011). One of the multiple mechanisms by which caffeine affects synaptic plasticity is thought to stem from its different affinities for distinct types of receptors present on the synaptic membranes and cytoplasmic calcium stores. For instance, on the synaptic membrane level, caffeine blocks mainly adenosine A_1 and adenosine A_{2a} in the midbrain reticular formation and basal forebrain (Fredholm *et al* 1999). On the other hand, caffeine at higher concentration also activates ryanodine receptors, which cause an increase in calcium release from cytoplasmic calcium stores (McPherson *et al* 1991), leading to an increase in calcium-dependent signaling pathways. Additionally, caffeine inhibits the phosphodiesterase IV enzyme (Smellie *et al* 1979), which increases the level of intracellular cAMP and enables caffeine to enhance the cAMP signaling cascades involved in LTP and memory. It further enhances striatal glutamate-dependent and glutamate-independent release of dopamine. This is achieved by targeting adenosine receptors, namely A_1 and A_{2a} receptors on the presynaptic striatal glutamatergic, and A_1 receptors on the presynaptic striatal dopaminergic terminals (Ferre 2010). In particular, a growing body of evidence suggests that endogenous adenosine interferes with synaptic plasticity through activation of the highly expressed adenosine A_1 receptors in the hippocampus. For example, adenosine produces an inhibitory effect on LTP in rat hippocampal slices and disrupts the process of learning and memory at the synaptic level (de Mendonca and Ribeiro 1994). Additionally, studies show that adenosine inhibits the release of glutamate from nerve terminals, and prevents NMDA receptor mediated activity at the post-synaptic membrane (Dunwiddie and Masino 2001).

Moreover, it is thought that the increase in the levels of adenosine during prolonged wakefulness adversely affects hippocampus-dependent learning and memory through its action on A_1 receptors. In contrast, the diverse effects of caffeine have been shown to be neuroprotective against SD induced insults to synaptic plasticity and memory function. Although, the precise mechanism by which caffeine prevents the effects of SD on memory and LTP is not clearly understood, it is postulated that caffeine protects against SD-induced LTP impairment by preventing the decrease in the levels of key signaling molecules such as CaMKII and CREB (Alhaider *et al* 2010a; 2010b; 2011). In view of this, it is quite possible that caffeine can prevent learning and memory impairment associated with SD through several mechanisms including antagonism of adenosine receptors, phosphodiesterase enzyme inhibition (Smellie *et al* 1979), and increasing calcium induced-calcium release (McPherson *et al* 1991). However, studies indicate that the concentration of caffeine typically consumed by humans (*i.e.* 40–180mg per cup) may act to improve learning and memory primarily by inhibiting adenosine receptors (Fredholm 1995). Thus, by antagonizing the abundant A_1 receptors, caffeine disrupts the deleterious signaling cascades mediated by adenosine on both the pre-synaptic neurons and post-synaptic neurons leading to overall enhancement in synaptic plasticity.

16.5.3 Caffeine and Cognitive Function

Based on the contribution of a variety of brain structures and neural circuits during valuations of free recall and recognition, memory can be divided into two main classes (Diekelmann and Born 2010). Implicit, or non-declarative memory is remembrance of information for tasks and skills that result from frequent practice, not required for conscious recall, *e.g.,* driving a bicycle. Implicit memory is said to depend on the striatum and cerebellum, although current findings also suggest a function for hippocampus in implicit memory (Diekelmann and Born 2010). The second type of memory, the explicit, or declarative memory includes intentional recall of faces, spatial assemblies, objects, and events. This type of memory depends on the hippocampus and associated medial temporal lobe structures, along with neocortical areas for long-term storage (Diekelmann and Born 2010).

Caffeine is implicated in the modulation of the learning and memory function (Dall'Igna *et al* 2004). Several lines of evidence have shown that caffeine improves the cognitive function in clinical studies as well as in experimental animal models (Angelucci *et al* 2002). However, there is some disagreement as to whether caffeine produces a positive or negative effect, if any, in cognitive function. While some investigators report enhanced cognition as a consequence of caffeine use (Angelucci *et al* 2002; Smith 2009), others report no effect (Warburton *et al* 2001; Alhaider *et al* 2010a). Still others report impaired memory after caffeine treatment (Fisher and Guillet 1997). These inconsistent findings may be due to differences in the duration of the treatment, dosage form and dose level of caffeine, type of memory tested, experimental procedure used, or time of use. The caffeine dose-response curve is an inverted U shape, which shows that at low doses (0.3–3 mg kg^{-1}), caffeine improves memory consolidation, whereas at high doses (30–100 mg kg^{-1}), it disrupts the acquisition phase of the water maze reference task, which measures hippocampus-dependent learning and memory (Angelucci *et al* 2002).

Our work reveals that chronic caffeine treatment has no effect on learning and memory in normal animals. However, in SD, which interferes with learning and memory, caffeine antagonizes the deleterious effect of SD on learning, short-term memory, and long-term memory (Alhaider *et al* 2010a; 2011; Figure 16.1). These findings support our view that caffeine acts only in a need-dependent manner (*i.e.,* when there is memory impairment). Therefore, caffeine appears to be a protector rather than a promoter of memory function.

16.5.4 Caffeine, Sleep Deprivation and LTP Impairment

Neurons communicate with each other to form networks, which are arranged in functionally related circuits. The performance at the behavioral level is a result of the integrated function of these neuronal systems rather than simply the outcome of individual neurons. Although electrophysiological studies are critical for understanding details of these processes, they do not integrate such

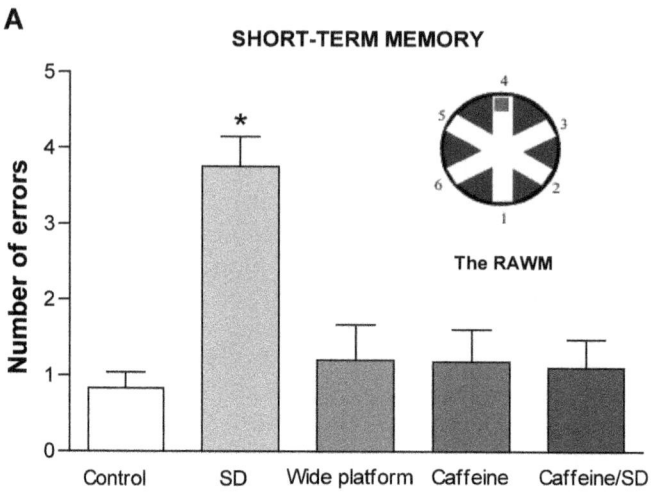

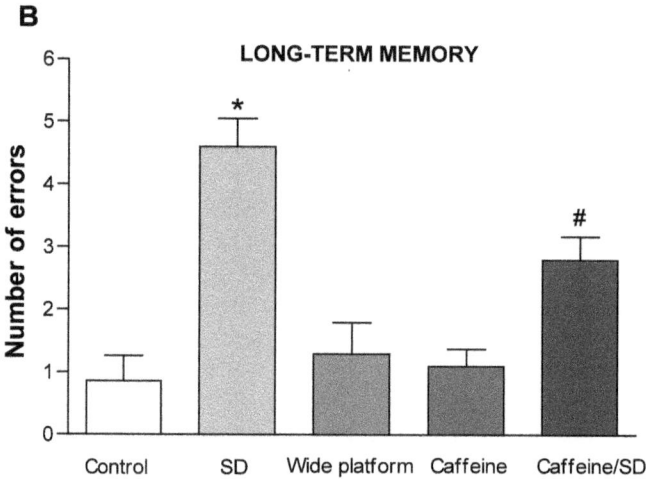

Figure 16.1   Impairment of memory by sleep deprivation and prevention by chronic caffeine treatment. Summary of the average number of errors made by each of the 5 groups in (A) short-term memory test (30 min after last learning test) and (B) long-term memory test (24 h after last learning test). Chronic caffeine treatment before sleep deprivation prevented short-term impairment and reduced long-term memory suppression. Each point is the mean ± SEM of 10 to 12 rats. *Indicates significant difference from all other groups (P<0.01) and # indicates significant difference from all other groups (P<0.05). (Modified from Alhaider *et al* 2010b; 2011 with permissions).

findings to ultimately and completely describe cognitive performance. Therefore, an understanding of neuronal electrophysiology is essential to test the functional integrity of neural pathways on the cellular level.

The activity of the neural networks within the hippocampus can undergo alterations in the strength of synaptic transmission as an outcome of repetitive activity (Burgess *et al* 2002). These changes, known as synaptic plasticity, represented by LTP, are believed to be correlates of learning and memory. Two major phases of LTP have been established, the early-phase (E-LTP), which utilizes calcium–calmodulin-dependent protein kinase II (CaMKII) phosphorylation, and the late-phase (L-LTP), which requires protein synthesis through activation of transcription factors such as CREB. The two phases are widely considered as hypothetical models for short-term and long-term memory, respectively (Malenka and Bear 2004).

Electrophysiological experiments in the CA1 and DG of anesthetized rats indicate that chronic caffeine treatment does not alter normal basal synaptic transmission (Alhaider *et al* 2010a; 2011). Additionally, chronic administration of caffeine in normal rats has no effect on either E-LTP or L-LTP in both CA1 and DG regions of the hippocampus (Alhaider *et al* 2010a; 2011; Figure 16.2). However, a positive effect of caffeine on plastic changes in synaptic transmission has been reported. It has been shown that caffeine induces an LTP-like response in the CA1 pyramidal neuron synapses in rat hippocampal slices (Martin and Buno 2003). Consistent with this observation, caffeine can mediate structural changes in cultured-hippocampal neurons by enhancing the size of dendritic spines (Korkotian and Segal 1999). It is believed that caffeine enhances the excitability of rat hippocampal neurons by antagonizing the effects of adenosine on A_1 receptors. Adenosine produces an inhibitory effect on LTP in area CA1 in rat hippocampal slices (de Mendonca and Ribeiro 1994).

16.6 Caffeine and Neuroprotection

Although a low dose of chronic caffeine treatment in rats had no significant effect on normal memory, it prevented memory loss of recent information associated with acute SD (Alhaider *et al* 2010b; 2011, Figure 16.1). Thus, it seems that at low chronic dose, caffeine worked as a protector rather than a promoter of memory function. Interestingly, previous reports have demonstrated the neuroprotective effects of chronic intake of caffeine in animal models of ischemia (Sutherland *et al* 1991) as well as in human subjects (Maia and de Mendonca 2002). Furthermore, the positive impact of caffeine on learning and memory is supported by epidemiological studies, which demonstrate an inverse correlation between coffee intake and the incidence of Parkinson's disease and Alzheimer's disease later in life (Ascherio *et al* 2001; Maia and de Mendonca 2002; Ritchie *et al* 2007). Consistent with this view, a large body of evidence has shown that caffeine administration may be a protective factor against memory impairment resulting from several animal models of brain disorders including Alzheimer's disease (Dall'Igna *et al* 2007; Arendash *et al* 2009; Cao *et al* 2009), Parkinson's disease (Gevaerd *et al* 2001), attention deficit hyperactivity disorder (Prediger *et al* 2005), age-related

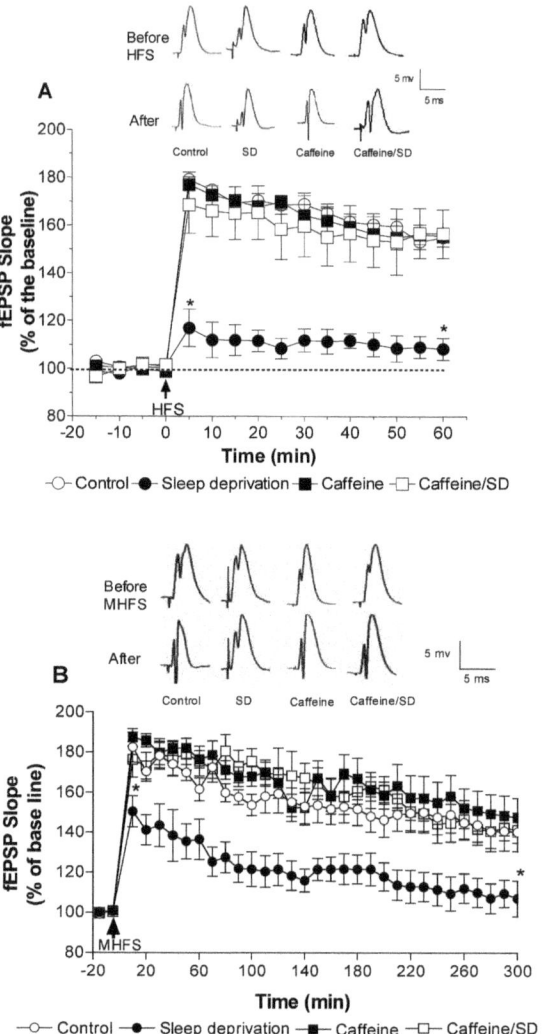

Figure 16.2 Chronic caffeine treatment prevents sleep deprivation-induced impairment of synaptic plasticity of the dentate gyrus. Hippocampal long-term potentiation (LTP) evoked by high-frequency stimulation applied to the Schaffer collateral–commissural pathway of anesthetized rats. (A) Early-phase LTP (E-LTP; evoked by HFS at zero time: arrow) in the DG area. (B) Late-phase LTP (L-LTP; evoked by MHFS applied at time zero: arrow) in area CA1. LTP was measured as increases in the slope of fEPSP expressed as percentage of the baseline values (before stimulation). The fEPSP slope values in the sleep-deprived rats are significantly lower than those of control, caffeine, and caffeine–sleep deprivation groups at all time points after stimulation, but chronic caffeine treatment prevents fEPSP slope suppression. All the points between the two (*) are significantly different from other groups (p<0.05). Insets are representative experiments; calibrations, 5 mV 5 ms^{-1}, apply to all traces. (Modified from Alhaider *et al* 2010a; 2011, with permissions).

cognitive decline (Riedel and Jolles 1996; Costa *et al* 2008a), scopolamine-induced amnesia (Riedel *et al* 1995), and epilepsy (Cognato *et al* 2010).

16.7 Effects of Caffeine on Sleep-Related Neural Signaling Pathways

Numerous studies have shown that caffeine and SD produce different effects on neural signaling pathways that are important for learning and memory. Recently, the combined effects of caffeine and SD on these signaling molecules have been investigated (Alhaider *et al* 2010a; 2010b; 2011; Figures 16.3, 16.4). Western blot analysis of the levels of signaling molecules reveals that SD reduces the levels of total-CaMKII and P-CaMKII in CA1 (unpublished data), and DG areas (Alhaider *et al* 2010a; Figure 16.3), and that chronic caffeine treatment prevents the decrease in the basal protein levels of P-CaMKII, and total-CaMKII in sleep-deprived rats (Alhaider *et al* 2010a; Figure 16.3). The significance of CaMKII in the induction of LTP is reflected in the findings that active CaMKII produces LTP-like synaptic potentiation in hippocampal slices (Pettit *et al* 1994). Furthermore, mutant mice lacking CaMKII show impairment in spatial memory and LTP (Silva *et al* 1992b). The transient increase in intracellular Ca^{2+}, on activation of NMDA receptors, frees calmodulin from the calmodulin-neurogranin complex to form a Ca^{2+}–calmodulin complex. The Ca^{2+}–calmodulin complex rapidly binds to and triggers CaMKII autophosphorylation generating a constitutively active CaMKII (P-CaMKII) that phosphorylates AMPA receptors. The phosphorylated form of AMPA receptors is important for the expression of LTP (Nayak *et al* 1996). In our recent molecular studies, we found that the beneficial effect of caffeine in sleep-deprived rats may be attributed to its ability to prevent the decrease in the basal levels of P-CaMKII. Analysis in the sleep-deprived rats suggests an overall decrease in the total protein levels of CaMKII. Thus, maintaining the basal levels of t-CaMKII by caffeine during SD is probably responsible for maintaining the production of adequate P-CaMKII levels, which positively impacts short-term memory as well as E-LTP expression in hippocampal CA1 and DG areas (Aleisa *et al* 2011, Alhaider *et al* 2010a; 2010b). Additionally, chronic caffeine administration averts the failure of high frequency stimulation (HFS) to increase the levels of P-CaMKII during SD (Alhaider *et al* 2010a; 2010b). These molecular findings stress the essential function of P-CaMKII in the expression of LTP, which is in agreement with previous findings (Pettit *et al* 1994, Lledo *et al* 1995).

Calcineurin, a phosphatase enzyme, is an essential signaling molecule for the regulation of memory and synaptic plasticity through dephosphorylation of kinases. Evidence indicates that calcineurin reduces post-synaptic activity and impairs LTP in the hippocampus (Wang and Kelly 1997). It has been reported that over-expression of calcineurin in the hippocampus impairs LTP (Winder *et al* 1998) and that pharmacological inhibitors of calcineurin facilitate LTP in hippocampal slices (Wang and Kelly 1997). Calcineurin exerts its action by

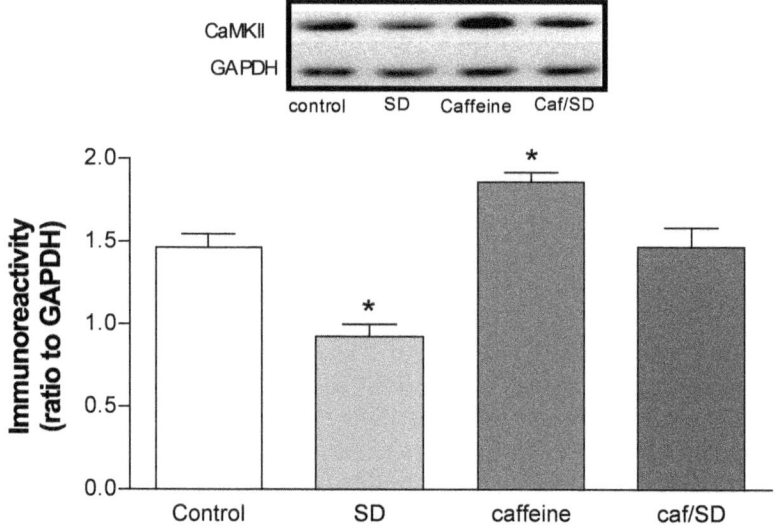

Figure 16.3   Caffeine treatment prevents the sleep deprivation-induced decrease in the levels of calcium calmodulin kinase-II. Effect of chronic caffeine treatment and–or sleep deprivation (SD) on the basal expression of total CaMKII, measured in the DG area and expressed as a ratio to the GAPDH. Note that chronic caffeine increased CaMKII level in normal rats (caffeine group) and prevented SD-induced decrease in CaMKII (Caf–SD group). Each point is the mean ± SEM of 4–6 rats. *Indicates significant difference from all other groups (p<0.05). (Modified from Alhaider *et al* 2010a, with permission).

suppressing natural inhibitor-1, which leads to stimulation of protein phosphatase 1 (PP1) (Mulkey *et al* 1994). The activated PP1 dephosphorylates P-CaMKII, which results in reduction of memory retention (Wang and Kelly 1997). It has been reported that the activity, but not the basal protein levels, of calcineurin is increased after 72 h of SD (Wang *et al* 2009). We have recently showed that the protein levels of calcineurin in the hippocampal CA1 and DG areas are not affected by 24 h of SD or by caffeine administration (Alhaider *et al* 2010a). Furthermore, it has been proposed that SD does not influence the level of PP1, which is activated by calcineurin (Guan *et al* 2004).

The protein family of neurotrophins, including BDNF, is well known to regulate the survival of neurons and encourage growth and differentiation of new neurons and synapses (Huang and Reichardt 2001). In the last decades, accumulated evidence suggests that BDNF plays a pivotal role in hippocampal synaptic plasticity and spatial learning (Kesslak *et al* 1998). BDNF influences synaptic plasticity by raising the efficiency of synaptic transmission through activation of CaMKII and CREB in the hippocampus (Boulanger and Poo 1999). Acute SD for 8 or 48 h periods decrease the gene expression and protein levels of BDNF in the hippocampus (Guzman-Marin *et al* 2006). In parallel, it has been shown that 24 h SD reduces the basal levels of BDNF in the CA1 and

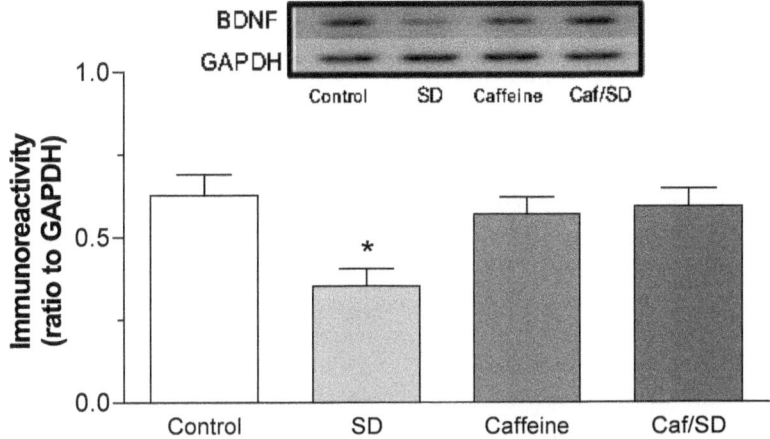

Figure 16.4   Decline of levels of brain derived neurotrophic factor (BDNF) in dentate gyrus of sleep-deprived rats and prevention by chronic caffeine treatment. Effect of chronic caffeine treatment and–or sleep deprivation (SD) on the protein levels of BDNF measured in the DG area and expressed as a ratio to the GAPDH. Note that SD caused a marked decrease in BDNF level (SD group) whereas chronic caffeine treatment (Caf–SD group) prevented SD-induced effect. Each point is the mean ± SEM of 4–6 rats. *Indicates significant difference from all other groups ($p<0.05$) (Modified from Alhaider *et al* 2010a, with permission).

DG areas (Alhaider *et al* 2010a; Figure 16.4). Caffeine treatment prevents reduction of BDNF protein levels in sleep-deprived rats (Alhaider *et al* 2010a). This decrease in the basal levels of BDNF in sleep-deprived rats may be considered a factor involved in the reduction of both P-CaMKII and P-CREB basal levels. It appears that by preventing the reduction in the basal levels of BDNF, caffeine prevents the SD-induced reduction of P-CaMKII and P-CREB protein levels and the subsequent impairment of E-LTP and L-LTP in brain hippocampal regions (Alhaider *et al* 2010a).

The activity dependent BDNF can be increased by HFS of the hippocampus (Gartner and Staiger 2002). Five hours after stimulation, the protein levels of BDNF are not increased in area CA1 of untreated sleep-deprived rats, but are markedly enhanced in sleep-deprived caffeine treated rats. Thus, chronic caffeine treatment prevents the SD-induced decrease in BDNF protein levels (Alhaider *et al* 2011). It has been revealed that BDNF activates CREB-dependent protein synthesis, and that the BDNF gene is considered to be one of the CREB targets (Tao *et al* 1998). As a consequence, BDNF synthesis can be stimulated upon CREB activation. We have recently shown a normal activation of CREB in caffeine and caffeine-treated sleep-deprived rats (Alhaider *et al* 2011). It appeared that the levels of BDNF during the expression of L-LTP follow those of P-CREB (Alhaider *et al* 2011). Applying BDNF after stimulation seems adequate to express L-LTP in slices treated with a protein synthesis inhibitor (Pang *et al* 2004). Therefore, the inability of multiple HFS to enhance the BDNF levels in sleep-deprived rats and the significantly higher levels of BDNF in stimulated caffeine treated sleep-deprived rats could be responsible for the positive effect of caffeine against L-LTP deficit induced by SD (Alhaider *et al* 2011).

CREB mediates the synthesis of proteins that are important for synaptic plasticity and long-term memory. The gene expression of CREB decreases after 8 and 48 h periods of sleep loss in the hippocampus (Guzman-Marin *et al* 2006). This reduction could be attributed to the significant decrease in the upstream modulators of CREB including extracellular signal-regulated kinase (ERK) and calcium–calmodulin-dependent protein kinase IV (CaMKIV) (unpublished data). In fact, it has been reported that sleep loss decreases the levels of ERK phosphorylation (Guan *et al* 2004). Additionally, acute, 24 h SD decreases the basal protein levels of total-CREB and P-CREB in the hippocampal CA1 as well as in DG areas. On the other hand, chronic low dose of caffeine treatment prevents SD-induced decrease in the basal levels of total-CREB and P-CREB in the hippocampus without affecting their levels in normal rats (unpublished data). In parallel, caffeine has been shown to induce widespread elevation in cAMP levels in the brain (Wu *et al* 2009). After expression of L-LTP, the levels of P-CREB are increased in the stimulated caffeine, and caffeine treated sleep deprived animals, but not in the untreated sleep deprived group. The inability of multiple HFS to increase the phosphorylation of CREB may be responsible for the impairment of hippocampal L-LTP in sleep-deprived rats (Alhaider *et al* 2011). Furthermore, the ability of caffeine to prevent the decrease in P-CREB levels after multiple HFS emphasizes the role of P-CREB in the L-LTP decline associated with 24 h SD (Alhaider *et al* 2011).

Mice with mutation in CaMKIV show impairment in L-LTP, long-term memory and CREB phosphorylation in the hippocampus (Kang *et al* 2001). However, these mutant mice have normal E-LTP and short-term memory. Recent work reveals reduced basal protein levels of CaMKIV in CA1 and DG areas of the hippocampus in the sleep-deprived rats. This decrease is prevented by chronic caffeine treatment of sleep-deprived rats (unpublished data). It seems that decreased CaMKIV level is a factor in the reduced levels of CREB phosphorylation in sleep-deprived rats leading to impairment of L-LTP. Conversely, the normal levels of CaMKIV in caffeine treated sleep deprived rats could contribute to normal CREB phosphorylation and intact L-LTP seen in these animals. After MHFS, the levels of CaMKIV are increased in all stimulated groups (Alhaider *et al* 2011). The increase in the levels of CaMKIV has also been observed after using a theta burst stimulation protocol in area CA1 (Tokuda *et al* 1997). This enhancement supports the role of CaMKIV in the expression of L-LTP.

Summary Points

- Sleep deprivation can result in increased risk of obesity, type-2 diabetes, heart problems and dementia.

- Caffeine has a neuroprotective action against cognitive decline in neurodegenerative disorders including Alzheimer's disease.
- The protective mechanism of chronic caffeine treatment against SD-induced cognitive impairment is not clearly understood.
- Although caffeine can act through several mechanisms, it has been suggested that the concentration of caffeine typically consumed by humans, which is comparable to the caffeine dose used in most of the relevant studies, may act mainly by inhibiting adenosine receptors.
- By antagonizing the abundant adenosine A_1 receptors, caffeine disrupts the signaling cascades mediated by adenosine on the pre-synaptic neurons as well as on the post-synaptic neurons, leading to caffeine-induced protective changes in important signaling molecules.
- Caffeine provides its neuroprotective actions through antagonizing the deleterious effects of SD on learning, short-term memory, and long-term memory. On the cellular level, chronic caffeine treatment in animals protects against SD-induced E-LTP and L-LTP impairment, which may be attributed to its ability to prevent SD-induced alterations in signaling molecules such as P-CaMKII, and P-CREB and BDNF, during the expression of E-LTP, and L-LTP, respectively.
- Caffeine could be acting by antagonizing adenosine receptors, thus, disrupting the signaling cascades mediated by adenosine on the presynaptic neurons as well as on the postsynaptic neurons.

Definitions of Words and Terms

AMPA receptor: is a non-NMDA receptor that is both a glutamate receptor and ion channel found mainly in the brain where it mediates fast synaptic transmission and the initiation of synaptic plasticity in the post-synaptic membrane.

Basal forebrain: is a structure located in the front bottom part of the frontal lobe of the brain. It is considered to be the major cholinergic neuronal (ACh producing neurons) site of output of the central nervous system.

Brain derived neurotrophic factor (BDNF): is a protein that mediates growth, proliferation and differentiation of neurons and synapses in the CNS and periphery. BDNF can also produce and enhance LTP expression.

Calcineurin: is a phosphatase responsible for the dephosphorylation and deactivation processes of previously phosphorylated molecules (*i.e.* CaMKII).

Calcium calmodulin kinase II (CaMKII): is a serine–threonine kinase regulated by the Ca^{2+}–calmodulin complex and involved in many signaling cascades especially those mediating learning and memory.

Columns in water (CIW): is an SD model adapted from the modified multiple platform model and is comprised of a large water-filled aquarium containing 20 columns (spaced 7 cm apart with a diameter of 5 cm each) upon which the animals are able to balance on to avoid falling in the water. Although this model disrupts both NREM and REM phases of sleep the loss of muscle tone during REM sleep causes the rats to fall into the water and awaken.

Cyclic AMP response element binding (CREB): is a nuclear transcription factor that when activated (*i.e.* (P-CREB) mediates the expression of genes involved in a number of physiological processes in organisms namely long term memory processes.

Disc-on-water (DOW): is an SD model comprised of a rotating metal disk in a Plexiglas container of water wherein an animal falling asleep is forced awake by the movement of the disk and the threat of falling in the water.

EEG: is a recording device used to record the results of all of the activity of the cortical pyramidal neurons by means of electrodes attached to the scalp at standard locations.

Long term potentiation (LTP): is an activity dependent on strengthening of the connection between two neurons which lasts for an extended period of time; LTP is widely considered the cellular basis of learning and memory.

Midbrain reticular formation: is a structure in the core of the upper part of the brainstem. It is a part of a larger system called the Reticular Activating System (RAS), which is responsible for regulating arousal and sleep-wakefulness transition.

NMDA receptor: is a glutamate receptor linked to an ion channel that is mainly found on the post-synaptic membrane and when activated is responsible for the calcium influx that mediates synaptic plasticity signaling cascades.

Polysomnography: is a diagnostic tool used in the study of sleep and sleep disorders. It involves a complete recording of the biophysiological modifications that occur in sleep.

List of Abbreviations

ACh	Acetylcholine
AMPA	Alpha-amino-3-hydroxy-5-methyl-4-isoxazole propionic acid
ATP	Adenosine triphosphate
BDNF	Brain-derived neurotrophic factor
cAMP	Cyclic adenine monophosphate
CaMKII	Calcium–calmodulin-dependent kinase II
CaMKIV	Calcium–calmodulin-dependent kinase IV
CNS	Central nervous system
CREB	Cyclic-AMP response element binding protein
EEG	Electroencephalogram
DG	Dentate Gyrus
DRN	Dorsal Raphe nucleus
E-LTP	Early-long term potentiation
ERK	Extracellular signal-regulated kinase
HFS	High frequency stimulation
5-HT	Serotonin
LC	Locus Coeruleus

L-LTP	Late-long term potentiation
LTP	Long term potentiation
MHFS	Multiple high frequency stimulation
NMDA	N-methyl-D-aspartate
MRF	Midbrain reticular formation
NE	Norepinephrine–Noradrenaline
NREM	Non-rapid eye movement
P-CaMKII	Phosphorylated calcium–calmodulin-dependent kinase II
P-CREB	Phosphorylated cyclic-AMP response element binding protein
PDEIV	Phosphodiestrase IV enzyme
PKA	cAMP-dependent protein kinase
PP1	Protein phosphatase-1
PP2B	Protein phosphatase 2B (calcineurin)
RAWM	Radial arm water maze
REM	Rapid eye movement
SD	Sleep deprivation

References

Aleisa, A. M., Helal, G., Alhaider, I. A., Alzoubi, K. H., Srivareerat, M., Tran, T. T., Al-Rejaie, S. S. and Alkadhi, K. A., 2011. Acute nicotine treatment prevents REM sleep deprivation-induced learning and memory impairment in rat. Hippocampus. 21: 899–909.

Aleisa, A. M., Alzoubi, K. H. and Alkadhi, K. A., 2011. Post-learning REM sleep deprivation impairs long-term memory: Reversal by acute nicotine treatment. Neuroscience Letters. 499: 28–31.

Alhaider, I. A., Aleisa, A. M., Tran, T. T. and Alkadhi, K. A., 2010a. Caffeine prevents sleep loss-induced deficits in long-term potentiation and related signaling molecules in the dentate gyrus. European Journal of Neuroscience. 31: 1368–1376.

Alhaider, I. A., Aleisa, A. M., Tran, T. T., Alzoubi, K. H. and Alkadhi, K. A., 2010b. Chronic caffeine treatment prevents sleep deprivation-induced impairment of cognitive function and synaptic plasticity. Sleep. 33: 437–444.

Alhaider, I. A., Aleisa, A. M., Tran, T. T. and Alkadhi, K. A., 2011. Chronic caffeine prevents impairment of spatial memory and late phase LTP in acutely sleep-deprived rats. Molecular and Cellular Neuroscience. 46: 742–751.

Angelucci, M. E., Cesario, C., Hiroi, R. H., Rosalen, P. L. and Da Cunha C., 2002. Effects of caffeine on learning and memory in rats tested in the Morris water maze. Brazilian Journal of Medical and Biological Research. 35: 1201–1208.

Arendash, G. W., Mori, T., Cao, C., Mamcarz, M., Runfeldt, M., Dickson, A., Rezai-Zadeh, K., Tane, J., Citron, B. A., Lin, X., Echeverria, V. and

Potter, H., 2009. Caffeine reverses cognitive impairment and decreases brain amyloid-beta levels in aged Alzheimer's disease mice. Journal of Alzheimer's Disease. 17: 661–680.

Ascherio, A., Zhang, S. M., Hernan, M. A., Kawachi, I., Colditz, G. A., Speizer, F. E. and Willett, W. C., 2001. Prospective study of caffeine consumption and risk of Parkinson's disease in men and women. Annals of Neurology. 50: 56–63.

Boulanger, L. and Poo, M. M., 1999. Gating of BDNF-induced synaptic potentiation by cAMP. Science. 284: 1982–1984.

Burgess, N., Maguire, E. A. and O'Keefe J., 2002. The human hippocampus and spatial and episodic memory. Neuron. 35: 625–641.

Cao, C., Cirrito, J. R., Lin, X., Wang, L., Verges, D. K., Dickson, A., Mamcarz, M., Zhang, C., Mori, T., Arendash, G. W., Holtzman, D. M. and Potter, H., 2009. Caffeine suppresses amyloid-beta levels in plasma and brain of Alzheimer's disease transgenic mice. Journal of Alzheimer's Disease. 17: 681–697.

Cardinali, D. P. 1980. Methylxanthines: possible mechanisms of action in the brain. Trends in Pharmacological Science. 1: 405–407.

Chau, N., Mur, J.-M., Touron, C., Benamghar, L. and Dehaene, D., 2004. Correlates of occupational injuries for various jobs in railway workers: a case-control study. Journal of Occupational Health. 46: 272–280.

Chen, C., Hardy, M., Zhang, J., LaHoste, G. J. and Bazan, N. G., 2006. Altered NMDA receptor trafficking contributes to sleep deprivation-induced hippocampal synaptic and cognitive impairments. Biochemical and Biophysical Research Communications. 340: 435–440.

Cirelli, C. 2002. How sleep deprivation affects gene expression in the brain: a review of recent findings. Applied Physiology. 92: 394–400.

Cognato, G. P., Agostinho, P. M., Hockemeyer, J., Muller, C. E., Souza, D. O. and Cunha, R. A., 2010. Caffeine and an adenosine A(2A) receptor antagonist prevent memory impairment and synaptotoxicity in adult rats triggered by a convulsive episode in early life. Journal of Neurochemistry. 112: 453–462.

Costa, M. S., Botton, P. H., Mioranzza, S., Souza, D. O. and Porciuncula, L. O., 2008. Caffeine prevents age-associated recognition memory decline and changes brain-derived neurotrophic factor and tirosine kinase receptor (TrkB) content in mice. Neuroscience. 153: 1071–1078.

Cunha, R. A. and Agostinho, P. M., 2010. Chronic caffeine consumption prevents memory disturbance in different animal models of memory decline. Journal of Alzheimers Disease. 20: S95–S116.

Dall'Igna, O. P., Fett, P., Gomes, M. W., Souza, D. O., Cunha, R. A. and Lara, D. R., 2007. Caffeine and adenosine A(2a) receptor antagonists prevent beta-amyloid (25–35)-induced cognitive deficits in mice. Experimental Neurology. 203: 241–245.

de Mendonca, A. and Ribeiro, J. A., 1994. Endogenous adenosine modulates long-term potentiation in the hippocampus. Neuroscience. 62: 385–390.

Diekelmann, S. and Born, J., 2010. The memory function of sleep. Nature Reviews Neuroscience. 11: 114–126.

Dunwiddie, T. V. and Masino, S. A., 2001. The role and regulation of adenosine in the central nervous system. Annual Review of Neuroscience. 24: 31–55.

Ferre, S. 2010. Role of the central ascending neurotransmitter systems in the psychostimulant effects of caffeine. Journal of Alzheimers Disease. 20: S35–S49.

Fisher, S. and Guillet, R. 1997. Neonatal caffeine alters passive avoidance retention in rats in an age-and gender-related manner. Developmental Brain Research. 98: 145–149.

Fisone, G., Brogkvist, A. and Usiello, A., 2004. Caffeine as a psychomotor stimulant: mechanism of action. Cellular and Molecular Life Sciences. 61: 857–872.

Fredholm, B. B. 1995. Astra Award Lecture. Adenosine, adenosine receptors and the actions of caffeine. Basic and Clinical Pharmacology and Toxicology. 76: 93–101.

Fredholm, B. B., Battig, K., Holmén, J., Nehlig, A. and Zvartau, E. E., 1999. Actions of caffeine in the brain with special reference to factors that contribute to its widespread use. Pharmacological Reviews. 51: 83–133.

Gaba, D. M. and Howard, S. K., 2002. Patient safety: fatigue among clinicians and the safety of patients. *New England Journal of Medicine.* 347: 1249–1255.

Gartner, A. and Staiger, V., 2002. Neurotrophin secretion from hippocampal neurons evoked by long-term-potentiation-inducing electrical stimulation patterns. Proceedings of the National Academy of Sciences of the United States of America. 99: 6386–6391.

Gevaerd, M. S., Takahashi, R. N., Silveira, R. and Da Cunha, C., 2001. Caffeine reverses the memory disruption induced by intra-nigral MPTP-injection in rats. Brain Research Bulletin. 55: 101–106.

Gottlieb, D. J., Punjabi, N. M., Newman, A. B., Resnick, H. E., Redline, S., Baldwin, C. M. and Nieto, F. J., 2005. Association of sleep time with diabetes mellitus and impaired glucose tolerance. Archives of Internal Medicine. 165: 863–867.

Grossman, G. H., Mistlberger, R. E., Antle, M. C., Ehlen, J. C. and Glass, J. D., 2000. Sleep deprivation stimulates serotonin release in the suprachiasmatic nucleus. Neuroreport. 11: 1929–1932.

Guan, Z., Peng, X. and Fang, J., 2004. Sleep deprivation impairs spatial memory and decreases extracellular signal-regulated kinase phosphorylation in the hippocampus. Brain Research. 1018: 38–47.

Guzman-Marin, R., Ying, Z., Suntsova, N., Methippara, M., Bashir, T., Szymusiak, R., Gomez-Pinilla, F. and McGinty, D., 2006. Suppression of hippocampal plasticity-related gene expression by sleep deprivation in rats. The Journal of Physiology. 575: 807–819.

Hagewoud, R., Havekes, R., Novati, A., Keijser, J. N., Van der Zee, E. A. and Meerlo, P., 2010. Sleep deprivation impairs spatial working memory and reduces hippocampal AMPA receptor phosphorylation. Journal of Sleep Research. 19: 280–288.

Huang, E. J. and Reichardt, L. F., 2001. Neurotrophins: roles in neuronal development and function. Annual Reviews of Neuroscience. 24: 677–736.

Huang, Z. L., Urade, Y. and Hayaishi, O., 2011. The role of adenosine in the regulation of sleep. Current Topics in Medicinal Chemistry. 11(8): 1047–1057.

Kalia, M. 2006. Neurobiology of sleep. Metabolism. 55: S2–S6.

Kang, H., Sun, L. D., Atkins, C. M., Soderling, T. R., Wilson, M. A. and Tonegawa, S., 2001. An important role of neural activity-dependent CaMKIV signaling in the consolidation of long-term memory. Cell. 106: 771–783.

Karklin, A., Driver, H. S. and Buffenstein, R., 1994. Restricted energy intake affects nocturnal body temperature and sleep patterns. American Journal of Clinical Nutrition. 59: 346–349.

Kesslak, J. P., So, V., Choi, J., Cotman, C. W. and Gomez-Pinilla, F. 1998. Learning upregulates brain-derived neurotrophic factor messenger ribonucleic acid: a mechanism to facilitate encoding and circuit maintenance? Behavioral Neuroscience. 112: 1012–1019.

Koban, M., Le, W. W. and Hoffman, G. E., 2006. Changes in hypothalamic corticotropin-releasing hormone, neuropeptide Y, and proopiomelanocortin gene expression during chronic rapid eye movement sleep deprivation of rats. Endocrinology. 147: 421–431.

Korkotian, E. and Segal, M., 1999. Release of calcium from stores alters the morphology of dendritic spines in cultured hippocampal neurons. Proceeding of the National Academy of Sciences of the United States of America. 96: 12068–12072.

Lanoir, J., Ternaux, J. P., Pons, C. and Lagarde, J. M., 1981. Long-term effects of a tryptophan-free diet on serotonin metabolism and sleep-waking balance in rats. Experimental Brain Research. 41: 346–357.

Lieberman, H. R., Bathalon, G. P., Falco, C. M., Morgan, C. A. 3rd, Niro, P. J. and Tharion, W. J., 2005. The fog of war: decrements in cognitive performance and mood associated with combat-like stress. Aviation and Space Environmental Medicine. 76: C7–C14.

Lledo, P. M., Hjelmstad, G. O., Mukherji, S., Soderling, T. R., Malenka, R. C. and Nicoll, R. A., 1995. Calcium–calmodulin-dependent kinase II and long-term potentiation enhance synaptic transmission by the same mechanism. Proceedings of the National Academy of Sciences of the United States of America. 92: 11175–11179.

Lloyd, H. G. and Fredholm, B. B., 1995. Involvement of adenosine deaminase and adenosine kinase in regulating extracellular adenosine concentration in hippocampal slices. Journal of Neurochemistry. 26: 387–395.

McPherson P. S., Kim Y. K., Valdivia H., Knudson C. M., Takekura H., Franzini-Armstrong, C., Coronado, R. and Campbell, K. P., 1991. The brain ryanodine receptor: a caffeine-sensitive calcium release channel. Neuron. 7: 17–25.

Maia, L. and de Mendonca, A., 2002. Does caffeine intake protect from Alzheimer's disease? European Journal of Neurology. 9: 377–382.

Malenka, R. C. and Bear, M. F., 2004. LTP and LTD: an embarrassment of riches. Neuron. 44: 5–21.

Martin, E. D. and Buno, W., 2003. Caffeine-mediated presynaptic long-term potentiation in hippocampal CA1 pyramidal neurons. Journal of Neurophysiology. 89: 3029–3038.

McDermott, C. M., LaHoste, G. J., Chen, C., Musto, A., Bazan, N. G. and Magee, J. C., 2003. Sleep deprivation causes behavioral, synaptic, and membrane excitability alterations in hippocampal neurons. Journal of Neuroscience. 23: 9687–9695.

McGinty, D. J. and Harper, R. M., 1976. Dorsal raphe neurons: depression of firing during sleep in cats. Brain Research. 101: 569–575.

Mulkey, R. M., Endo, S., Shenolikar, S. and Malenka, R. C., 1994. Involvement of a calcineurin–inhibitor-1 phosphatase cascade in hippo-campal long-term depression. Nature. 369: 486–488.

Nayak, A. S., Moore, C. I. and Browning, M. D., 1996. Ca^{2+}–calmodulin-dependent protein kinase II phosphorylation of the presynaptic protein synapsin I is persistently increased during long-term potentiation. Proceedings of the National Academy of Sciences of the United States of America. 93: 15451–15456.

Nicolaidis, S. 2006. Metabolic mechanism of wakefulness (and hunger) and sleep (and satiety): Role of adenosine triphosphate and hypocretin and other peptides. Metabolism. 55: S24–29.

Pang, P. T., Teng, H. K., Zaitsev, E., Woo, N. T., Sakata, K., Zhen, S., Teng, K. K., Yung, W. H., Hempstead, B. L. and Lu, B., 2004. Cleavage of proBDNF by tPA–plasmin is essential for long-term hippocampal plasticity. Science. 306: 487–491.

Penev, P. D. 2007. Sleep deprivation and energy metabolism: to sleep, perchance to eat? Current Opinion Endocrinology Diabetes and Obesity. 14: 374–381.

Pettit, D. L., Perlman, S. and Malinow, R., 1994. Potentiated transmission and prevention of further LTP by increased CaMKII activity in postsynaptic hippocampal slice neurons. Science. 266: 1881–1885.

Pilcher, J. J. and Huffcutt, A. I., 1996. Effects of sleep deprivation on performance: a meta-analysis. Sleep. 19: 318–326.

Porkka-Heiskanen, T., Alanko, L. and Stenberg, D., 2007. Neurochemistry of Sleep. Handbook of Neurochemistry and Molecular Neurobiology: Behavioral Neurochemistry, Neuroendocrinology and Molecular Neurobiology Berlin: Springer pp. 869–894.

Porkka-Heiskanen, T., Kalinchuk, A., Alanko, L., Urrila, A. and Stenberg, D., 2003. Adenosine, energy metabolism, and sleep. The Scientific World Journal. 3: 790–798.

Poston, W. S., Taylor, J. E., Hoffman, K. M., Peterson, A. L., Lando, H. A., Shelton, S. and Haddock, C. K., 2008. Smoking and deployment: perspectives of junior-enlisted US Air Force and US Army personnel and their supervisors. Military Medicine. 173: 441–447.

Prediger, R. D., Pamplona, F. A., Fernandes, D. and Takahashi, R. N., 2005. Caffeine improves spatial learning deficits in an animal model of attention deficit hyperactivity disorder (ADHD)–the spontaneously hypertensive rat (SHR). International Journal of Neuropsychopharmacolology. 8: 583–594.

Rechtschaffen, A., Gilliland, M. A., Bergmann, B. M. and Winter, J. B., 1983. Physiological correlates of prolonged sleep deprivation in rats. Science. 221(4606): 182–184.

Riedel, W., Hogervorst, E., Leboux, R., Verhey, F., van Praag, H. and Jolles, J., 1995. Caffeine attenuates scopolamine-induced memory impairment in humans. Psychopharmacology (Berl). 122: 158–168.

Ritchie, K., Carriere, I., de Mendonca, A., Portet, F., Dartigues, J. F., Rouaud, O., Barberger-Gateau, P. and Ancelin, M. L., 2007. The neuroprotective effects of caffeine: a prospective population study (the Three City Study). Neurology. 69: 536–545.

Silva, A. J., Stevens, C. F., Tonegawa, S. and Wang, Y., 1992. Deficient hippocampal long-term potentiation in alpha-calcium-calmodulin kinase II mutant mice. Science. 257: 201–206.

Smellie, F. W., Davis, C. W., Daly, J. W. and Wells, J. N., 1979. Alkylxanthines: inhibition of adenosine-elicited accumulation of cyclic AMP in brain slices and of brain phosphodiesterase activity. Life Sciences. 24: 2475–2482.

Smith, A. P. 2009. Caffeine, cognitive failures and health in a non-working community sample. Human Psychopharmacology. 24: 29–34.

Smith, C. and Rose, G. M., 1996. Evidence for a paradoxical sleep window for place learning in the Morris water maze. Physiology & Behavior. 59: 93–7.

Stenberg, D. 2007. Neuroanatomy and neurochemistry of sleep. Cellular Molecular Life Sciences. 64: 1187–204.

Sutherland, G. R., Peeling, J., Lesiuk, H. J., Brownstone, R. M., Rydzy, M., Saunders, J. K. and Geiger, J. D., 1991. The effects of caffeine on ischemic neuronal injury as determined by magnetic resonance imaging and histopathology. Neuroscience. 42: 171–182.

Tao, X., Finkbeiner, S., Arnold, D. B., Shaywitz, A. J. and Greenberg, M. E., 1998. Ca^{2+} influx regulates BDNF transcription by a CREB family transcription factor-dependent mechanism. Neuron. 20: 709–726.

Tokuda, M., Ahmed, B. Y., Lu, Y. F., Matsui, H., Miyamoto, O., Yamaguchi, F., Konishi, R. and Hatase, O., 1997. Involvement of calmodulin-dependent protein kinases-I and -IV in long-term potentiation. Brain Research. 755: 162–166.

Vecsey, C. G., Baillie, G. S., Jaganath, D., Havekes, R., Daniels, A., Wimmer, M., Huang, T., Brown, K. M., Li, X. Y., Descalzi, G., Kim, S. S., Chen, T., Shang, Y. Z., Zhuo, M., Houslay, M. D. and Abel, T., 2009. Sleep deprivation impairs cAMP signalling in the hippocampus. Nature. 461(7267): 1122–1125.

Vgontzas, A. N., Bixler, E. O. and Chrousos, G. P., 2006. Obesity-related sleepiness and fatigue: the role of the stress system and cytokines. The New York Academy of Sciences. 1083: 329–344.

Unknown: C. f. D. C. a. P. 2011"Effect of short sleep duration on daily activities — United States, 2005–2008." Morbidity and Mortality Weekly Report" 60: 239–242.

Wang, G. P., Huang, L. Q., Wu, H. J., Zhang, L., You, Z. D. and Zhao, Z. X., 2009. Calcineurin contributes to spatial memory impairment induced by rapid eye movement sleep deprivation. Neuroreport. 20: 1172–1176.

Warburton, D. M., Bersellini, E. and Sweeney, E., 2001. An evaluation of a caffeinated taurine drink on mood, memory and information processing in healthy volunteers without caffeine abstinence. Psychopharmacology (Berl). 158: 322–328.

Wilson, J. F. 2005. Is sleep the new vital sign? Annuals of Internal Medicine. 142: 877–880.

Winder, D. G., Mansuy, I. M., Osman, M., Moallem, T. M. and Kandel, E. R., 1998. Genetic and pharmacological evidence for a novel, intermediate phase of long-term potentiation suppressed by calcineurin. Cell. 92: 25–37.

Wu, M. N., Ho, K., Crocker, A., Yue, Z., Koh, K. and Sehgal, A., 2009. The effects of caffeine on sleep in Drosophila require PKA activity, but not the adenosine receptor. Journal of Neuroscience. 29: 11029–11037.

Yaggi, H. K., Araujo, A. B. and McKinlay, J. B., 2006. Sleep duration as a risk factor for the development of type 2 diabetes. Diabetes Care. 29: 657–661.

Zepelin, H. 1983. A life span perspective on sleep. In: Mayes, A. (ed). Sleep Mechanisms and Functions in Humans and Animals: An Evolutionary Perspective. Cambridge University Press, pp. 126–160.

CHAPTER 17

Caffeine and Exercise Performance

TODD A. ASTORINO*[1] AND AILISH C. WHITE[2]

[1] Associate Professor, Department of Kinesiology, CSU—San Marcos, 333 S. Twin Oaks Valley Road, MH 352, San Marcos, CA 92096-0001 USA; [2] Graduate Assistant, Exercise Physiology Program, University of New Mexico, JC B143, Albuquerque, NM 87131-1251
*E-mail: astorino@csusm.edu

17.1   Introduction

Endurance exercise and repeated, high-intensity efforts require continued delivery of substrates to maintain exercise performance. This is critical as the primary substrates degraded to supply adenosine triphosphate (ATP), carbohydrate from muscle and liver glycogen and phosphocreatine (PCr), are limited, and cannot be resynthesized during exercise. In the absence of proper nutrition, fatigue is imminent. This emphasizes the importance of proper nutrition before and during exercise to optimize exercise performance. This is of utmost importance for competitive athletes in whom small changes in performance separate winners from losers.

Besides dietary manipulation is the use of ergogenic aids to enhance performance. One of the most widely-used ergogenic aids is caffeine, which has been used since the Stone Age (Escohotado and Symington 1999) and whose capacity to enhance muscular work was identified over 100 years ago (Rivers and Webber 1907). Caffeine is found naturally in the leaves, fruits, or seeds of plants, and contained in various foods, drinks, and medicinal products. Lovett (2005) reported that 90% of US adults consume caffeine daily in coffee, tea,

Food and Nutritional Components in Focus No. 2
Caffeine: Chemistry, Analysis, Function and Effects
Edited by Victor R Preedy
© The Royal Society of Chemistry 2012
Published by the Royal Society of Chemistry, www.rsc.org

soft drinks, and energy drinks. It is also widely-used by athletes (Forman *et al* 1995), as up to 27% of athletes ingest caffeine to enhance performance.

Due to the widespread availability and acceptability of caffeine intake worldwide, it is important to thoroughly examine how caffeine enhances performance in various athletic activities. Its multiple actions on the body, time course, dosing regimens, and effects on exercise performance will be discussed.

17.2 Mechanisms Explaining Ergogenic Effect of Caffeine

Early studies (Costill *et al* 1978; Ivy *et al* 1979) showed that caffeine increased lipolysis and spared glycogen during exercise. This finding has been corroborated (Wiles *et al* 1992; Engels *et al* 1999), yet Jackman *et al* (1996) and Graham *et al* (2000) revealed that caffeine does not spare glycogen during exercise. Alternatively, increased intracellular calcium concentration (Doherty *et al* 2004) and–or altered excitation-contraction coupling (Clausen 2003) have been postulated. However, a study by Rosser *et al* (2009) concluded that when ingested in physiological doses, caffeine's effects lie outside the muscle fiber. Davis *et al* (2003) reported that caffeine delays fatigue by acting as an adenosine antagonist. Adenosine is a normal component of the cell whose concentration increases with muscular contraction. *Via* binding to its receptors, it inhibits neuron excitability and synaptic transmission, thus decreasing arousal and increasing fatigue (Porkka-Heiskanen 1999). In male rats, run performance was 60% greater with caffeine *versus* an adenosine agonist (Davis *et al* 2003).

Caffeine ingestion also modifies perceptual responses that may alter performance. In a meta-analysis of 21 studies containing 202 subjects, Doherty and Smith (2005) revealed that caffeine decreased rating of perceived exertion (RPE) by 5.6% during prolonged exercise, which explained 33% of improved performance. They speculated that caffeine augments performance by masking the perception of fatigue and recruiting additional motor units during exercise. A pain-reducing effect of caffeine has also been documented. In young men (Motl *et al* 2003) and women (Gliottoni and Motl 2008) completing submaximal cycling, 5–10 mg kg^{-1} caffeine ingested 1 h pre-exercise significantly reduces leg pain *versus* placebo. Experimentally, pain influences motor unit recruitment, so enhanced motor unit activation was identified as a potential mechanism explaining the attenuated pain sensation with caffeine. These data oppose findings from Astorino *et al* (2010) showing improved force production, yet no change in pain perception or RPE, when caffeine was ingested prior to maximal knee extension–flexion exercise. However, high doses may not be practical as a pre-exercise intervention, and these experimental protocols do not simulate the demands of sport, so further study is merited.

Proposed mechanisms for caffeine's ergogenic effects during short-term, high-intensity exercise

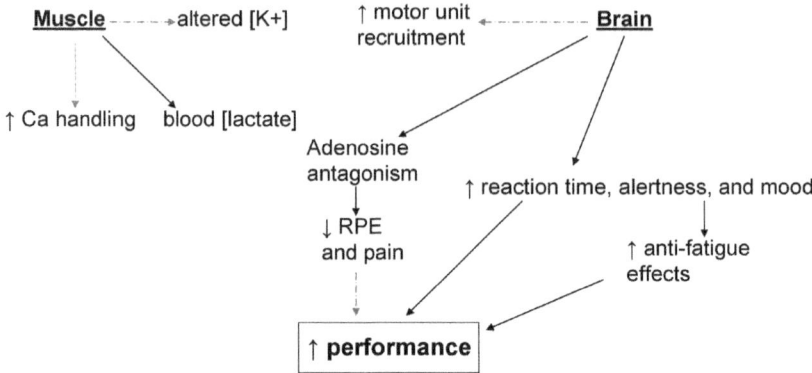

Figure 17.1   Proposed mechanisms for ergogenic effects of caffeine. This table identifies all mechanisms proposed to explain performance-enhancing effects of acute ingestion of caffeine.

Caffeine also affects CNS function by promoting serotonin release, increasing sympathetic activity, and decreasing the activity of inhibitory neurons through adenosine antagonism (Porkka-Heiskanen 1999). Consequently, reduced tiredness, improved mood, and enhanced alertness and reaction time occur with caffeine ingestion. In one study (Smith *et al* 2005), caffeine improved mood and performance on various tasks *versus* placebo. It also enhanced vigilance and running performance during sustained operations in soldiers (McLellan *et al* 2005).

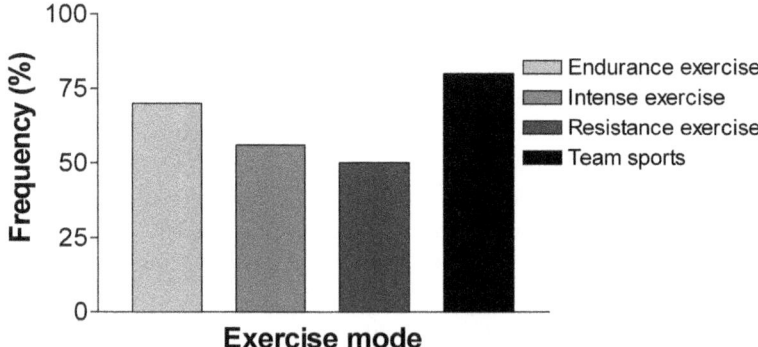

Figure 17.2   Frequency of studies reporting ergogenic effects of caffeine for endurance exercise, intense exercise, resistance training, and team sports exercise. This table reports the frequency of studies demonstrating ergogenic effects of caffeine for various exercise modes.

Overall, the ergogenic effects of caffeine are multifactorial and include central and peripheral mechanisms (see Figure 17.1). Currently, adenosine antagonism seems to be the most accepted mechanism, yet enhanced performance is also related to altered RPE, pain, cognition, and mood, especially in activities that are sustained or require repetitive execution of simple tasks.

17.3 Recommended Dosing, Timing, and Form of Caffeine Ingestion

Caffeine is consumed as an anhydrous powder, in chewing gum, or *via* soft drinks, coffee, or energy drinks. Doses ranging from 3–9 mg kg^{-1} body weight improved performance in many studies (MacIntosh and Wright 1995; Graham *et al* 1998; Bruce *et al* 2000), with a 5–6 mg kg^{-1} dose typically used. For a 70 kg male, this equals approximately four brewed cups of coffee. However, lower doses are also ergogenic. In combination with a carbohydrate-electrolyte beverage, 2.1 mg kg^{-1} of caffeine enhanced time-trial performance *versus* placebo, although higher doses had a greater effect upon performance (Kovacs *et al* 1998). Cox *et al* (2002) revealed that 1.5 mg kg^{-1} of caffeine as Coca-Cola increased cycling performance similar to a 6 mg kg^{-1} dose. Yet, doses greater than 9 mg kg^{-1} do not improve performance more than lower doses (3–6 mg kg^{-1}). Because of the wide range of ergogenic doses, it is important for the individual athlete to determine a minimal ergogenic dose of caffeine to use prior to competition.

A typical dosing regimen includes drug ingestion followed by a 60 min rest period before exercise. This allows caffeine concentration to peak in plasma (Graham 2001). Caffeine has a half-life of 4–6 h which may vary depending upon the size of the ingested dose, as doses above 300 mg elicit a longer half-life. In one study (Bell and McLellan 2002), alterations in cycling performance were examined 1, 3, and 6 h post-caffeine or placebo ingestion. Data showed increased performance at all time points in the non-users, yet only at 1 and 3 h in the users. These results emphasize that exercise should be completed within 6 h of caffeine ingestion to promote ergogenic effects, with this duration shorter in caffeine users.

Whether repeated doses of caffeine are superior than a single bolus is equivocal. Bell and McLellan (2003) revealed no additional ergogenic effect of a second dose of caffeine ingested 6 h after a morning dose. Repeated small doses of caffeine (1 mg kg^{-1}) ingested 20 and 40 min before a mental challenge enhanced vigilance and alertness (Hewlett and Smith 2007), which may support ingestion of small doses of caffeine in persons whose mood or vigilance may wane during activity, such as soldiers, ultraendurance athletes, hospital personnel, or firefighters.

The widespread availability and popularity of caffeine-containing energy drinks including RedBull^{TM} have initiated research into the ergogenic potential of these beverages. Improved aerobic–anaerobic performance

(Alford *et al* 2001; Ivy *et al* 2009) and muscular endurance (Forbes *et al* 2007) have been demonstrated. Paton *et al* (2010) reported improved sprint cycling performance when chewing gum containing 240 mg of caffeine was ingested during exercise. Because of the limited data pertaining to ergogenic effects of various products containing caffeine, further investigation is merited.

17.4 Tolerance and Withdrawal of Caffeine

Habitual caffeine intake promotes tolerance of its effects by altering individual caffeine sensitivity. Caffeine tolerance is related to an upregulation of adenosine activity and a decrease in adrenergic activity (Latini and Pedata 2001), and requires 5–6 days of repeated use to be established. In one study (Evans and Choice 1992), subjects completed 18 days of caffeine consumption. Results showed that caffeine 'choosers' reported positive effects of caffeine (300 mg per day) and negative effects of placebo. Nevertheless, caffeine-mediated changes in running performance (Wiles *et al* 1992) were similar in users *versus* non-users. Fisher *et al* (1992) reported discrepancies in oxygen uptake, FFA concentration, and respiratory exchange ratio in subjects who consumed and then withdrew from caffeine, although performance was not measured. Currently, it seems that performance is unaltered by caffeine habituation.

Caffeine withdrawal elicits headaches, irritability, and fatigue and can occur within 12–24 h, peaks at 3–4 days, and lasts approximately 7 days (Hughes *et al* 1991). Killgore *et al* (2007) revealed that caffeine withdrawal reduced abstract reasoning and concept formation *versus* placebo. Various lengths of withdrawal (0, 2, and 4 days) did not affect performance at 80% VO_2max compared to no withdrawal (Van Soeren and Graham 1998). As magnitude of dependence varies across individuals, symptoms of withdrawal may vary, but they do not seem to affect resultant exercise performance.

17.5 Effects of Caffeine on Exercise Performance

17.5.1 Caffeine and Endurance Exercise

An early study investigating caffeine's effects on performance was by Costill *et al* (1978), in which cyclists exercised to exhaustion at 80% VO_2max after ingestion of decaffeinated coffee or coffee containing 330 mg of caffeine. Results demonstrated improved endurance due to enhanced fat oxidation as well as lower RPE. Ivy *et al* (1979) revealed increased performance when 250 mg of caffeine was ingested before and during prolonged cycling. In runners, a 10 mg kg^{-1} caffeine dose increased time to exhaustion *versus* placebo and control (French *et al* 1991). Compared to placebo, Pasman *et al* (1995) demonstrated similar increases in performance (11–12 min) when various caffeine doses (5, 9, and 13 mg kg^{-1}) were ingested. Cox *et al* (2002) reported that Coca-Cola containing 1.3 and 1.9 mg kg^{-1} of caffeine augmented cycling performance by 3%. A low dose of caffeine (3 mg kg^{-1}) increased 8 km performance by 1.2% in trained runners (Bridge and Jones 2006), and Jenkins *et al* (2008) revealed that caffeine doses of 2 and 3 mg kg^{-1} increased performance in male cyclists by 4% and 3%, respectively, yet a lower dose (1 mg kg^{-1}) did not alter performance *versus* placebo. However, there was large inter-individual variability in caffeine's ergogenic properties. Overall, data suggest that caffeine doses above 1.3 mg kg^{-1} benefit endurance performance, yet lower doses are ineffective.

Overall, the preponderance of data supports caffeine intake to enhance endurance performance (Table 17.1). However, some studies (Cohen *et al* 1996; Jacobson *et al* 2001; Hunter *et al* 2002) fail to demonstrate improved performance. The Cohen *et al* study required outdoor running in a warm environment, and its lack of ergogenic effect was similar to another study employing outdoor running (Van Nieuwenhoven *et al* 2005). In studies by Hunter *et al* and Jacobson *et al*, cyclists completed prolonged cycling during which CHO was provided, and a fixed end point was the outcome measure. It may be that caffeine is more apt to improve performance during open end point tests rather than when a set distance or pace is used. Overall, there is a large variability in the magnitude of response to caffeine across subjects, which may be related to variations in caffeine metabolism across individuals (Cornelis *et al* 2007). Incorporation of a second caffeine trial within studies would elucidate if caffeine's ergogenic effects are reproducible across days. Overall, further study is merited to better explain these differences.

17.5.2 Caffeine and High-Intensity Exercise

Recently, several review articles (Davis and Green 2009; Astorino and Roberson 2010; Warren *et al* 2010) were published concerning efficacy of caffeine for high-intensity exercise including resistance training, team sports, and sprint-based exercise. During these bouts, performance is dependent upon maximal rates of ATP supply *via* the phosphagen system and–or maximal motor unit recruitment, and not CHO availability or hydration as found with endurance exercise. The following text provides a brief description of caffeine's effects on these activities, with studies thoroughly summarized in Tables 17.2–17.3.

17.5.3 Caffeine and Resistance Training

Resistance training is commonly completed by individuals to promote muscle hypertrophy and increase strength, which enhance exercise performance, or reduce health risks and facilitate completion of day-to-day activities, which aid in quality of life. However, effect of caffeine on resistance training performance is relatively unexplored. In sprinters (Bond *et al* 1986), caffeine ingestion did not alter muscle function *versus* placebo. In football players, Jacobson *et al* (1991) showed significant improvements in muscle torque and

Table 17.1 Change in endurance exercise performance in response to acute caffeine ingestion.

Study	Subjects	Dose	Protocol	Findings
Costill et al (1978)	9 cyclists	330 mg in coffee	Exercise to exhaustion at 80% VO_2max[a]	14.7 min ↑ in performance
Ivy et al (1979)	9 cyclists	250 mg + 250 mg every 15 min of exercise	2 hr of isokinetic cycling	↑ work production by 7.4%
French et al (1991)	6 distance runners	10 mg kg^{-1}	45 min at 75% VO_2max[a] + exercise to exhaustion	↑ time to exhaustion
Graham and Spriet (1992)	7 trained runners	9 mg kg^{-1}	Running and cycling at 85% VO_2max[a]	20–22 min ↑ in time to exhaustion
Wiles et al (1992)	18 middle distance athletes	150–200 mg	1500 m run	4.2 s ↑ in performance
MacIntosh and Wright (1995)	11 swimmers	6 mg kg^{-1}	1500 m swim	23 s ↑ in performance
Pasman et al (1995)	9 cyclists	0, 5, 9, and 13 mg kg^{-1}	Cycling to exhaustion at 80% $Wmax$[b]	11–12 min ↑ duration of exercise
Cohen et al (1996)	7 endurance athletes	0, 5, and 9 mg kg^{-1}	21 km outdoor road race in hot conditions	No effect on performance
Cole et al (1996)	10 endurance athletes	6 mg kg^{-1}	30 min of isokinetic cycling	12% ↑ in total work; ↓ perception of effort
Wemple et al (1997)	6 cyclists	25 mg dl^{-3}	3 h of cycling at 60% VO_2max + time to exhaustion at 85% VO_2max[a]	No effect on performance
Denadai and Denadai (1998)	8 untrained males	5 mg kg^{-1}	Exercise 10% above and below anaerobic threshold	43% ↑ in time to exhaustion at 10% below anaerobic threshold
Graham et al (1998)	9 runners	4.45 mg kg^{-1}	Running until exhaustion at 85% VO_2max[a]	7.5–10 min ↑ in exercise time
Kovacs et al (1998)	15 endurance athletes	150, 225, and 320 mg L^{-1}	1 h cycling time trial	9–17 Watt ↑ power output; ↑ time trial performance
Anderson et al (2000)	8 female rowers	6 and 9 mg kg^{-1}	2000 m rowing trial	1.0% ↑ in performance
Bruce et al (2000)	8 male rowers	6 and 9 mg kg^{-1}	2000 m rowing trial	1.2% ↑ in performance
Jacobson et al (2001)	8 male cyclists	6 mg kg^{-1}	2 h cycling + time trial	No effect on performance
Cox et al (2002)	20 endurance athletes	1.3–1.9 mg kg^{-1}	2 h cycling + time trial	3.1–3.3% ↑ in performance

Table 17.1 (*Continued*)

Study	Subjects	Dose	Protocol	Findings
Hunter et al (2002)	8 cyclists	6 mg kg⁻¹	100 km time trial	No effect on performance
Conway et al (2003)	9 male cyclists and triathletes	6 mg kg⁻¹ or 3 mg kg⁻¹ twice	1.5 h of cycling + time trial	No effect on performance
Van Nieuwenhoven et al (2005)	98 male and female runners	1.3 mg kg⁻¹ in sports drink	18 km outdoor run	No effect on performance
Bridge and Jones (2006)	8 male distance runners	3 mg kg⁻¹	8 km outdoor run	23.8 s ↑ in performance
Cureton et al (2007)	16 cyclists	5.3 mg kg⁻¹	2 h cycling + 15 min time trial	15–23% ↑ in work and RPE[c] and strength loss
Hogervorst et al (2008)	24 endurance trained men	100 mg of caffeine in energy bar	150 min cycling at 60% VO_2max[a] + time to exhaustion at 75% VO_2max[a]	↑ concentration, faster response speed and detection; ↑ time to exhaustion by 27%
Hulston and Juekendrup (2008)	10 male cyclists	CHO[d] + 5.3 mg kg⁻¹	105 min cycling + 45 min time trial	↑ performance by 4.6%
Jenkins et al (2008)	13 cyclists	1, 2, and 3 mg kg⁻¹	15 min performance ride	3–4% ↑ performance with 2 and 3 mg kg⁻¹ dose
Prescino et al (2008)	6 elite swimmers	6.2 mg kg⁻¹	Two 200 m time trials	No effect on performance
Desbrow et al (2009)	9 male cyclists	1.5 and 3 mg kg⁻¹	120 min cycling + time trial	No effect on performance
Ganio et al (2010)	14 male cyclists		2 h of cycling + 15 min time trial	↑ total work accumulated and no decline in MVC[e]
O'Rourke et al (2010)	15 trained runners and 15 recreational runners	5 mg kg⁻¹	Two 5-km time trials	↑ performance in both groups by 1.0 –1.1%
Skinner et al (2010)	10 competitive male rowers	2, 4, or 6 mg kg⁻¹	Four 2000-m time trials during a 6-wk period	No effect on performance
Wallman et al (2010)	10 untrained women	6 mg kg⁻¹	15 min cycling + 10 min maximal effort	No effect on performance
Ganio et al (2011)	11 male cyclists	6 mg kg⁻¹	90 min cycling + 15 min time trial	↑ performance
Irwin et al (2011)	12 male cyclists	3 mg kg⁻¹	1 h time trial	3.0–3.6% ↑ in performance

[a]maximal oxygen uptake. [b]maximal workload. [c]rating of perceived exertion. [d]carbohydrate. [e]maximum voluntary contraction.

Table 17.2 Summary of studies examining effect of caffeine on resistance training performance.

Study	Subjects	Dose	Protocol	Findings
Bond et al (1986)	12 male sprinters	5 mg kg^{-1}	6-RM^a knee extension/flexion	No effect on performance
Jacobson & Edwards (1991)	20 male football players	7 mg kg^{-1}	3–15 reps of knee extension and flexion	4–8% ↑ in peak torque
Jacobson et al (1992)	36 untrained men and women	300 or 600 mg	Peak extension and flexion torque	No effect on performance
Jacobs et al (2003)	13 strength-trained men	4 mg kg^{-1}	Superset of leg press and bench press	No effect on performance
Norager et al (2005)	30 older adults (>75 yr)	6 mg kg^{-1}	Maximal isometric arm flexion strength	No effect on performance
Beck et al (2006)	37 strength-trained men	^{b}2.5 mg kg^{-1}	Wingate test, muscle strength and endurance tests	2.1% ↑ in bench press 1-RM^a; no effects on other parameters
Green et al (2007)	17 strength-trained men	6 mg kg^{-1}	10-RM^a bench press and leg press performance	↑ leg press reps in set 3
Woolf et al (2008)	18 male athletes	5 mg kg^{-1}	Leg–bench press to fatigue + Wingate test	↑ total weight lifted (bench press) and peak power
Astorino et al (2008)	22 strength-trained men	6 mg kg^{-1}	1-RM^a bench/–leg press and reps to fatigue	No effect on 1-RM or muscle endurance
Hoffman et al (2008)	8 strength-trained men	110 mg^b	6 sets of barbell squat at 75% 1-RM^a	↑ total repetitions during set 5
Beck et al (2008)	31 untrained men	2.5 mg kg^{-1b}	1-RM^a bench press + running at 85% VO$_2$peak^c	No effect on performance
Hudson et al (2008)	15 strength-trained men	6 mg kg^{-1}	4 sets of arm curl and leg extensions to fatigue	↑ total reps/number of reps (set 1) of leg extension; no effect on arm performance
Williams et al (2008)	9 strength-trained men	300 mg	1-RM^a bench press, lat-pull down, and Wingate test	No effect on performance
Walter et al (2009)	20 recreationally-active men	200 mg^a	1-RM^a bench press and leg press	No effect on performance
Astorino et al (2010)	14 active men	5 mg kg^{-1}	All-out knee extension–flexion exercise	↑ force production in extension and flexion
Goldstein et al (2010)	15 strength-trained women	6 mg kg^{-1}	1-RM^a bench press + reps to failure at 60% 1-RM	↑ 1-RM^a by 1.8%, but no effect on reps performed
Hendrix et al (2010)	21 untrained men	400 mg^a	1-RM^a keg press and knee extension	No effect on performance
Astorino et al (2011)	14 strength-trained men	6 mg kg^{-1}	Bench, leg press, lat row, and shoulder press reps to fatigue	↑ number of reps of leg press

^{a}repetition maximum. ^{b}caffeine contained in a supplement. ^{c}peak oxygen uptake.

Table 17.3 Summary of studies examining effect of caffeine intake on high intensity, sprint-based exercise.

Study	Subjects	Dose	Protocol	Findings
Williams et al (1988)	9 men	7 mg kg^{-1}	15 s maximal cycling	No effect on performance
Collomp et al (1991)	6 men	5 mg kg^{-1}	Wingate test	No effect on performance
Anselme et al (1992)	14 active men	250 mg	Force: velocity protocol	↑ maximal workload
Collomp et al (1992)	7 trained swimmers; 7 untrained swimmers	250 mg	Two 100 m sprint swims	↑ swimming velocity in bout 1 only in trained swimmers
Stuart et al (1995)	9 male rugby players	6 mg kg^{-1}	Tests of sprint speed, drive power, and passing accuracy	↑ performance in all tests
Jackman et al (1996)	14 endurance athletes	6 mg kg^{-1}	2 min bouts of cycling at 100% VO$_2$max^a	43 s ↑ in endurance
Doherty (1998)	9 trained men	5 mg kg^{-1}	Running at 125% VO$_2$max^a	↑ run time to exhaustion
Greer et al (1998)	9 active men	6 mg kg^{-1}	4 Wingate tests	No effect on performance; ↓ performance in bouts 3 and 4
Bell et al (2001)	24 untrained men	5 mg kg^{-1}	Wingate test + cycling at 125% VO$_2$max^a	↑ time to exhaustion; no effect on Wingate performance
Paton et al (2001)	16 male team-sport athletes	6 mg kg^{-1}	Ten 20 m sprints	No effect on performance
Doherty et al (2004)	11 male cyclists	5 mg kg^{-1}	2 min at 100% Wmax^b + 1 min maximal effort	↑ mean power output
Beck et al (2006)	37 strength-trained men	2.5 mg kg^{-1} in a supplement	Wingate test	No effect on performance
Crowe et al (2006)	17 team sport athletes	6 mg kg^{-1}	Two 60 s maximal cycling bouts	↓ performance
Greer et al. (2006)	18 active men	5 mg kg^{-1}	Wingate test	No effect on performance
Schneiker et al (2006)	10 male team-sport athletes	6 mg kg^{-1}	18 4-s sprints on a cycle ergometer	6.6–8.5% ↑ in total work and mean power
Wiles et al (2006)	8 male cyclists	5 mg kg^{-1}	1 km time-trial	3.1% ↑ in performance; 3.6–8.8% ↑ in power output
Beaven et al (2008)	24 male rugby players	0, 200, 400, or 800 mg	10 m sprints	1.7% ↑ in sprint performance
Glaister et al (2008)	21 active men	5 mg kg^{-1}	12 x 30 m sprint performance	1.4% ↓ in fastest sprint time
Woolf et al (2008)	18 male athletes	5 mg kg^{-1}	Leg-bench press to fatigue + Wingate test	↑ peak power and total weight lifted for bench press
Woolf et al (2009)	17 college football players	5 mg kg^{-1}	40-yd dash + 20 m shuttle run	No effect on performance
Paton et al (2010)	9 male cyclists	240 mg of caffeinated chewing-gum	Four sets of five 30 s-maximal cycling efforts	5.5% ↑ in power output
Roberts et al (2010)	8 male rugby players	4 mg kg^{-1} ingested with CHO^c	15 m sprint + test of speed and agility	↑ sprint performance

^{a}maximal oxygen uptake. ^{b}maximal workload. ^{c}carbohydrate.

Table 17.4   Summary of areas that require investigation of the ergogenic effects of caffeine.

1. To investigate the effect of fed state on ergogenic effects of caffeine.
2. To investigate the effect of training status on ergogenic effects of caffeine.
3. To investigate ergogenic effects of caffeine in persons of various gender, age, and body composition.
4. To investigate ergogenic effects of caffeine during sprint- and power-based exercise in sprint-trained athletes.
5. To investigate efficacy of other items containing caffeine, such as energy drinks, chewing gum, *etc.*, on exercise performance.
6. To investigate if chronic caffeine intake can promote exercise adherence and tolerance to exercise in sedentary populations.
7. To investigate efficacy of acute or repeated caffeine intake on job performance in firefighters, soldiers, and other personnel undergoing sustained operations.
8. To investigate efficacy of acute caffeine intake for resistance training performance.
9. To investigate effects of various length of withdrawal on ergogenic effects of caffeine.
10. To investigate effects of caffeine combined with exercise on weight loss in overweight and obese individuals.
11. To compare ergogenic effects of caffeine across men and women of similar training status.
12. To confirm if adenosine antagonism is the primary mechanism explaining ergogenic effects of caffeine during intense exercise.

power with caffeine. In strength-trained men completing supersets of leg press and bench press, no effect of caffeine on performance was revealed (Jacobs *et al* 2003). Similarly, there was no effect of caffeine on 1-RM bench press and leg press (Astorino *et al* 2008) or bench press and lat pull down in strength-trained men (Williams *et al* 2008). Nevertheless, Beck *et al* (2006) reported a significant increase in 1-RM bench press in men ingesting a caffeine-containing (201 mg) supplement, although the magnitude of this increase was small (+ 2.0%). In elite male athletes, bench press but not leg press performance was augmented with caffeine (Woolf *et al* 2008). However, these data oppose findings (Astorino *et al* 2011) showing improved leg press performance in strength-trained men, yet no change in bench press, shoulder press, or lat row performance. Compared to placebo, Warren *et al* (2010) reported 4 and 14% improvements in muscle strength and endurance with caffeine, with the ergogenic effect typically observed in the knee extensors *versus* other muscle groups.

Overall, caffeine seems to augment muscle strength and endurance only under specific conditions. The majority of data show minimal alterations in 1-RM strength, repetitions completed at a given load, peak torque, or power. Most studies reporting significant increases in performance are characterized by higher doses (5–7 mg kg^{-1}), use of open end point tests, and caffeine combined with other constituents that may alter performance. Despite its purported anti-fatigue properties, caffeine may not be ergogenic during resistance training, especially when fixed end point tests are used.

17.5.4 Caffeine and Team Sport Performance

Several investigations have examined effects of caffeine during protocols that simulate sport such as cycling and team sports. In cyclists, caffeine increased 1 km time-trial performance by 2 s and augmented mean and peak power compared to placebo (Wiles *et al* 2006). In athletes performing 4 s sprints on a cycle ergometer, sprint performance and mean power were enhanced by 6–8% with caffeine (Schneiker *et al* 2006). In rugby players, caffeine increased parameters of intense performance mimicking 80 min of competitive rugby, including sprint speed, power, and passing accuracy, compared to placebo (Stuart *et al* 1995). These data oppose an inability of caffeine to enhance performance of ten 20 m sprints, interspersed with 10 s of recovery, in team-sport athletes (Paton *et al* 2001). The relatively short duration of this bout compared to those used in previous studies may explain the lack of an effect. Caffeine also does not alter agility (Lorino *et al* 2006), so its ability to enhance performance in sports characterized by quick movements, turns, and–or brief sprints of short duration remains equivocal.

17.5.5 Caffeine and Sprint Performance

Early studies failed to reveal benefits of caffeine for short-term exercise. During 15 s of all-out cycling, no change in power, work, or fatigue was revealed (Williams *et al* 1988). Nevertheless, Anselme *et al* (1992) revealed enhanced anaerobic power during a force-velocity test, yet no change in anaerobic capacity. However, Collomp *et al* (1992) demonstrated improved 2 x 100 m swim performance in trained swimmers with 250 mg of caffeine ingested 1 h before exercise *versus* placebo. In adults, cycling time to exhaustion was increased by 20% (4.12 to 4.93 min) with caffeine (Jackman *et al* 1996). In men completing four Wingate tests (Greer *et al* 1998), caffeine did not alter power output *versus* placebo, and in bouts 3 and 4, mean–peak power was lower ($p < 0.05$) with caffeine. A detrimental effect on performance was also revealed during intense cycling (Crowe *et al* 2006). However, Woolf *et al* (2008) revealed significantly higher peak power, but not mean power or fatigue index, in male athletes ingesting caffeine. Collectively, data indicate that caffeine may be ergogenic for high-intensity exercise performed by trained athletes, who provide more consistent performance day-day and thus increase statistical power. For example, improved sprint cycling performance was revealed after ingestion of 240 mg of caffeine (Paton *et al* 2010). An increased sensitivity to caffeine has been revealed in human slow-twitch fibers (Mitsumoto *et al* 1990), and adenosine receptors exist primarily in type I fibers (Latini and Pedata 2001), so a potential ergogenic effect of caffeine in activities depending upon fast-twitch motor unit recruitment may be unexpected. However, further study is needed to elucidate these explanations.

Summary Points

- This chapter reviewed effects of acute caffeine intake on exercise performance.

- Mechanisms of caffeine's actions may reside in the muscle fiber or central nervous system.
- Caffeine is typically ingested 1 h pre-exercise in doses ranging from 3–9 mg kg^{-1} body weight.
- Withdrawal from caffeine tends to promote side effects including headache, anxiety, and lethargy.
- The majority of studies reveal that caffeine enhances endurance performance during cycling, swimming, and running.
- Caffeine seems to increase muscular strength and endurance, but this is not a universal finding of all studies.
- Performance in activities requiring sprinting is frequently enhanced with acute caffeine intake, although this ergogenic effect disappears when these tasks are completed by untrained individuals.

Key Facts of Caffeine's Role as an Ergogenic Aid

- Approximately 90% of adults consume caffeine daily through ingestion of coffee, tea, chocolate, soft drinks, or energy drinks.
- Caffeine was originally thought to enhance performance by improving fat use and sparing muscle glycogen; however, recent studies reveal that caffeine acts as an adenosine antagonist, attenuates sensations of pain and exertion, and improves mood and vigilance.
- Caffeine is ergogenic in coffee, energy drinks, chewing gum, soda, and in pure anhydrous form.
- Ergogenic doses of caffeine range from 1.3–13 mg kg^{-1} body weight, with no dose response evident. Caffeine is typically ingested 1 h pre-exercise to allow its concentration to peak in the blood.
- Caffeine acts upon all bodily tissues including the heart and skeletal muscle, and its effects are maintained in the body for approximately 6 hours post-intake.
- Caffeine withdrawal causes onset of symptoms in most persons including headache, fatigue, anxiety, and lack of focus.
- A variety of caffeine doses has been shown to increase endurance performance in sports including cycling, swimming, and running.
- A 4–14% improvement in strength and endurance has been revealed with caffeine ingestion. These effects are most pronounced in the knee extensors *versus* other muscle groups, and during exercise without a fixed endpoint.
- Caffeine increases performance in team sport athletes including rugby and soccer players participating in repeated sprinting and other tasks that are sport-specific.
- There is clear individual variation in responses to caffeine, which may be due to discrepancies in caffeine metabolism across individuals.

Definitions of Words and Terms

Adenosine: by-product of cellular metabolism that increases with exercise. When it binds to its receptors in the brain, it promotes feelings of fatigue and lethargy.

Adenosine triphosphate (ATP): energy source that is continually made by the muscle cell through metabolism to meet the demands of contracting muscle.

Anaerobic power: maximal rate of ATP generation by skeletal muscle, which is typically assessed during the Wingate test.

Caffeine: methylxanthine compound found naturally in the seeds and leaves of plants that exerts various effects upon the body.

Ergogenic aid: product such as caffeine that enhances exercise performance.

Excitation-contraction coupling: cascade of events starting with neural excitation and subsequent transmission of the impulse along the motor nerve to the muscle fiber, resulting in muscle contraction.

Fatigue: inability to maintain a desired level of performance or intensity.

Glycogen: storage form of carbohydrate located in muscle and liver that is essential for performance in high-intensity exercise, such as endurance sport.

Lactate: product of carbohydrate metabolism whose build-up reflects increasing reliance upon oxygen-independent metabolism.

Lipolysis: degradation of lipid that occurs during exercise to fuel muscle contraction.

One-repetition maximum (1-RM): test employed to assess muscular strength or peak force generation.

Placebo: substance frequently-used during caffeine-based studies, similar in taste, appearance, and volume, to reduce subject bias and learning effects from repeated treatments.

Resistance training: exercise mode characterized by repeated contractions against a given resistance, such as body weight or free weights.

Sprint: exercise mode characterized by high intensity bouts separated by brief rest periods, during which ATP demand is high.

Sympathetic activity: part of the nervous system's 'fight or flight' response that prepares the body for the demands of exercise.

Time trial: cycling-based exercise protocol in which subjects cover a fixed distance in the fastest time possible.

Torque: measure of muscle force in which the force is applied by a joint, such as the knee, over a distance.

Wingate test: a 30-second sprint on a cycle ergometer typically used to assess anaerobic power.

List of Abbreviations

ATP	adenosine triphosphate
CHO	carbohydrate
CNS	central nervous system

FFA	free fatty acid
h	hour
m	meter
mg kg^{-1}	milligram per kilogram body weight
PCr	phosphocreatine
RPE	rating of perceived exertion
VO$_2$max	maximal oxygen uptake

References

Alford, C., Cox, H. and Wescott, R., 2001. The effects of red bull energy drink on human performance and mood. Amino Acids. 21(2): 139–150.

Anderson, M. E., Bruce, C. R., Fraser, S. F., Stepto, N. K., Klein, R., Hopkins, W. G. and Hawley, J. A., 2000. Improved 2000-meter rowing performance in competitive oarswomen after caffeine ingestion. International Journal of Sports Nutrition. 10: 464–475.

Anselme, F., Collomp, K., Mercier, B., Ahmaïdi, S. and Prefaut, C., 1992. Caffeine increases maximal anaerobic power and blood lactate concentration. European Journal of Applied Physiology. 65(2): 188–191.

Astorino, T. A., Firth, K. and Rohmann, R. L., 2008. Effect of caffeine ingestion on one-repetition maximum muscular strength. European Journal of Applied Physiology. 102(2): 127–132.

Astorino, T. A. and Roberson, D. W., 2010. Efficacy of acute caffeine-ingestion for short-term high-intensity exercise: a systematic review. Journal of Strength and Conditioning Research. 24(1): 257–265.

Astorino, T. A., Terzi, M. N., Roberson, D. W. and Burnett, T. R., 2010. Effect of two doses of caffeine on muscular function during knee extension exercise. Medicine and Science in Sports and Exercise. 42(12): 2205–2210.

Astorino, T. A., Martin, B. J., Schachtsiek, L., Wong, K. and Ng, K., 2011. Minimal effect of acute caffeine ingestion on intense resistance training performance. Journal of Strength and Conditioning Research (in press).

Beaven, C. M., Hopkins, W. G., Hansen, K. T., Wood, M. R., Cronin, J. B. and Lowe, T. E., 2008. Dose effect of caffeine on testosterone and cortisol responses to resistance exercise. International Journal of Sports Nutrition and Exercise Metabolism. 18: 131–141.

Beck, T. W., Housh, T. J., Schmidt, R. J., Johnson, G. O., Housh, D. J., Coburn, J. W. and Malek, M. H., 2006. The acute effects of a caffeine-containing supplement on strength, muscular endurance, and anaerobic capabilities. Journal of Strength and Conditioning Research. 20(3): 506–510.

Beck, T. W., Housh, T. J., Malek, M. H., Mielke, M. and Hendrix, R., 2008. The acute effects of a caffeine-containing supplement on bench press strength and time to running exhaustion. Journal of Strength and Conditioning Research. 22(5): 1654–1658.

Bell, D. G., Jacobs, I. and Ellerington, K., 2001. Effect of caffeine and ephedrine ingestion on performance and anaerobic metabolism during the Wingate test. Medicine and Science in Sports and Exercise. 33: 1399–1403.

Bell, D. G. and McLellan, T. M., 2002. Exercise endurance 1, 3, and 6 h after caffeine ingestion. Journal of Applied Physiology. 93(4): 1227–1234.

Bell, D. G. and McLellan, T. M., 2003. Effect of repeated caffeine ingestion on repeated exhaustive exercise endurance. Medicine and Science in Sports and Exercise. 35(8): 1348–1354.

Bond, V., Gresham, K., McRae, J. and Tearney, R. J., 1986. Caffeine ingestion and isokinetic strength. British Journal of Sports Medicine. 20(3): 135–137.

Bridge, C. A. and Jones, M. A., 2006. The effect of caffeine ingestion on 8 km run performance in a field setting. Journal of Sports Science. 24: 433–439.

Bruce, C. R., Anderson, M. E., Fraser, S. F., Stepto, N. K., Klein, R., Hopkins, W. G. and Hawley, J. A., 2000. Enhancement of 2000-m rowing performance after caffeine ingestion. Medicine and Science in Sports and Exercise. 32(11): 1958–1963.

Clausen, T. 2003. Na^+-K^+ pump regulation and skeletal muscle contractility. Physiological Reviews. 83: 1269–1324.

Cohen, B. S., Nelson, A. G., Prevost, M. C., Thompson, G. D., Marx, B. D. and Morris, G. S., 1996. Effects of caffeine ingestion on endurance racing in heat and humidity. European Journal of Applied Physiology. 73(3–4): 358–363.

Cole, K. J., Costill, D. L., Starling, R. D., Goodpaster, B. H., Trappe, S. W. and Fink, W. J., 1996. Effect of caffeine ingestion on perception of effort and subsequent work production. International Journal of Sports Nutrition. 6: 14–23.

Collomp, K., Ahmaidi, S., Audran, M., Chanal, J. L. and Prefaut, C., 1991. Effects of caffeine ingestion on performance and anaerobic metabolism during the Wingate test. International Journal of Sports Medicine. 12: 439–443.

Collomp, K., Ahmaidi, S., Chatard, J. C., Audran, M. and Prefaut, C., 1992. Benefits of caffeine ingestion on sprint performance in trained and untrained swimmers. European Journal of Applied Physiology. 64: 377–380.

Conway, K. J., Orr, J. and Stannard, S. R., 2003. Effect of a divided caffeine dose on endurance exercise performance, urinary caffeine concentration, and plasma paraxanthine. Journal of Applied Physiology. 94(4): 1557–1562.

Cornelis, M. C., El-Sohemy, A. and Campos, H., 2007. Genetic polymorphism of the adenosine A2A receptor is associated with habitual caffeine consumption. American Journal of Clinical Nutrition. 86(1): 240–244.

Costill, D. L., Dalsky, G. P. and Fink, W. J., 1978. Effects of caffeine on metabolism and exercise performance. Medicine and Science in Sports and Exercise. 10: 155–158.

Cox, G. R., Desbrow, B., Montgomery, P. G., Anderson, M. E., Bruce, C. R., Macrides, T. A., Martin, D. T., Moquin, A., Roberts, A., Hawley, J. A. and Burke, L. M., 2002. Effect of different protocols of caffeine intake on metabolism and endurance performance. Journal of Applied Physiology. 93: 990–999.

Crowe, M. J., Leicht, A. S. and Spinks, W. L., 2006. Physiological and cognitive responses to caffeine during repeated, high-intensity exercise. International Journal of Sports Nutrition and Exercise Metabolism. 16: 528–544.

Cureton, K. J., Warren, G. L., Millard-Stafford, M. L., Wingo, J. E., Trilk, J. and Buyckx, M., 2007. Caffeinated sports drink: Ergogenic effects and possible mechanisms. International Journal of Sports Nutrition and Exercise Metabolism. 17: 35–55.

Davis, J. M., Zhao, Z., Stock, H. S., Mehl, K. A., Buggy, J. and Hand, G. A., 2003. Central nervous system effects of caffeine and adenosine on fatigue. American Journal of Physiology. 284: R399–R404.

Davis, J. K. and Green, J. M., 2009. Caffeine and anaerobic performance: ergogenic value and mechanisms of action. Sports Medicine. 39(10): 813–832.

Denadai, B. S. and Denadai, M. L., 1998. Effects of caffeine on time to exhaustion in exercise performed below and above the anaerobic threshold. Brazilian Journal of Medical and Biological Research. 31(4): 581–585.

Desbrow, B., Barrett, C. M., Linahan, C. L., Grant, C. D. and Leveritt, M. D., 2009. Caffeine, cycling performance, and exogenous CHO oxidation: a dose response study. Medicine and Science in Sports and Exercise. 41(9): 1744–1751.

Doherty, M. 1998. The effects of caffeine on the maximally accumulated oxygen deficit and short-term running performance. International Journal of Sports Nutrition. 8: 95–104.

Doherty, M. and Smith, P. M., 2005. Effects of caffeine ingestion on rating of perceived exertion during and after exercise: a meta-analysis. Scandinavian Journal of Medicine and Science in Sports. 15: 69–78.

Doherty, M., Smith, P. M., Hughes, M. and Davison, R., 2004. Caffeine lowers perceptual response and increases power output during high-intensity cycling. Journal of Sports Science. 22(7): 637–643.

Engels, H. J., Wirth, J. C., Celik, S. and Dorsey, J. L., 1999. Influence of caffeine on metabolic and cardiovascular functions during sustained light intensity cycling and at rest. International Journal of Sports Nutrition. 9: 361–370.

Escohotado, A. and Symington, K., 1999. A Brief History of Drugs: From the Stone Age to the Stoned Age. Park Street Press.

Evans, S. M. and Griffiths, R. R., 1992. Caffeine tolerance and choice in humans. Psychopharmacology (Berl.). 108(1–2): 51–59.

Fisher, S. M., McMurray, R. G., Berry, M., Mar, M. H., and Forsythe, W. A., 1986. Influence of caffeine on exercise performance in habitual caffeine users. International Journal of Sports Medicine. 7: 276–280.

Forbes, S. C., Candow, D. G., Little, J. P., Magnus, C. and Chilibeck, P. D., 2007. Effect of Red Bull energy drink on repeated Wingate cycle performance and bench-press muscle endurance. International Journal of Sports Nutrition and Exercise Metabolism. 17(5): 433–444.

Forman, E. S., Dekker, A. H., Javors, J. R. and Davison, D. T., 1995. High-risk behaviors in teenage male athletes. Clinical Journal of Sports Medicine. 5(1): 36–42.

French, C., McNaughton, L. R., Davies, P. and Tristram, S., 1991. Caffeine ingestion during exercise to exhaustion in elite distance runners. Journal of Sports Medicine and Physical Fitness. 31(3): 425–432.

Ganio, M. S., Klau, J. F., Lee, E. C., Yeargin, S. W., McDermott, B. P., Buyckx, M., Maresh, C. M. and Armstrong, L. E., 2010. Effect of various carbohydrate-electrolyte fluids on cycling performance and maximal voluntary contraction. International Journal of Sports Nutrition and Exercise Metabolism. 20(2): 104–114.

Ganio, M. S., Johnson, E. C., Klau, J. F., Anderson, J. M., Casa, D. J., Maresh, C. M., Volek, J. S. and Armstrong, L. E., 2011. Effect of ambient temperature on caffeine ergogenicity during exercise. European Journal of Applied Physiology (in press).

Glaister, M., Howatson, G., Abraham, C., Lockey, R. A., Goodwin, J. E., Foley, P. and McInnes, G., 2008. Caffeine supplementation and multiple sprint performance. Medicine and Science in Sports and Exercise. 40(10): 1835–1840.

Gliottoni, R. C. and Motl, R. W., 2008. Effect of caffeine on leg-muscle pain during intense cycling exercise: possible role of anxiety sensitivity. International Journal of Sports Nutrition and Exercise Metabolism. 18: 103–115.

Goldstein, E., Jacobs, P. L., Whitehurst, M., Penhollow, T. and Antonio, J., 2010. Caffeine enhances upper-body strength in resistance-trained women. Journal of the International Society of Sports Nutrition. 14: 7–18.

Graham, T. E., Helge, J. W., MacLean, D. A., Kiens, B. and Richter, E. A., 2000. Caffeine ingestion does not alter carbohydrate or fat metabolism in human skeletal muscle during exercise. Journal of Physiology. 529: 837–847.

Graham, T. E. 2001. Caffeine and exercise: Metabolism, endurance, and performance. Sports Medicine. 31(11): 785–807.

Graham, T. E., Hibbert, E. and Sathasivam, P., 1998. Metabolic and exercise endurance effects of coffee and caffeine ingestion. Journal of Applied Physiology. 85: 883–889.

Graham, T. E. and Spriet, L. L., 1992. Performance and metabolic responses to a high caffeine dose during prolonged exercise. Journal of Applied Physiology. 71(6): 2292–2298.

Green, J. M., Wickwire, P. J., McLester, J. R., Gendle, S., Hudson, G., Pritchett, R. C. and Laurent, C. M., 2007. Effects of caffeine on repetitions to failure and ratings of perceived exertion during resistance training. International Journal of Sports Physiology and Performance. 2(3): 280–290.

Greer, F., McLean, C. and Graham, T. E., 1998. Caffeine, performance, and metabolism during repeated Wingate exercise tests. Journal of Applied Physiology. 85(4): 1502–1508.

Greer, F., Morales, J. and Coles, M., 2006. Wingate performance and surface EMG frequency variables are not affected by caffeine ingestion. Applied Physiology, Nutrition, and Metabolism. 31: 597–603.

Hendrix, C. R., Housh, T. J., Mielke, M., Zuniga, J. M., Camic, C. L., Johnson, G. O., Schmidt, R. J. and Housh, D. J., 2010. Acute effects of a caffeine-containing supplement on bench press and leg extension strength and time to exhaustion during cycle ergometry. Journal of Strength and Conditioning Research. 24(3): 859–865.

Hewlett, P. and Smith, A., 2007. Effects of repeated doses of caffeine on performance and alertness: new data and secondary analyses. Human Psychopharmacology. 22(6): 339–350.

Hoffman, J. R., Ratamess, N. A., Ross, R., Shanklin, M., Kang, J. and Faigenbaum, A. D., 2008. Effect of a pre-exercise energy supplement on the acute hormonal response to resistance exercise. Journal of Strength and Conditioning Research. 22(3): 874–882.

Hogervorst, E., Bandelow, S., Schmitt, J., Jentjens, R., Oliveira, M., Allgrove, J., Carter, T. and Gleeson, M., 2008. Caffeine improves physical and cognitive performance during exhaustive exercise. Medicine and Science in Sports and Exercise. 40(10): 1841–1851.

Hudson, G. M., Green, J. M., Bishop, P. A. and Richardson, M. T., 2008. Effects of caffeine and aspirin on light resistance training performance, perceived exertion, and pain reception. Journal of Strength and Conditioning Research. 22(6): 1950–1970.

Hughes, J. R., Higgins, S. T., Bickel, W. K., Hunt, W. K., Fenwick, J. W., Gulliver, S. B. and Mireault, S. B., 1991. Caffeine self-administration, withdrawal, and adverse effects among coffee drinkers. Archives of General Psychiatry. 48(7): 611–617.

Hunter, A. M., St Clair Gibson, A., Collins, M., Lambert, M. and Noakes, T. D., 2002. Caffeine ingestion does not alter performance during a 100-km cycling time-trial performance. International Journal of Sports Nutrition and Exercise Metabolism. 12(4): 238–252.

Hulston, C. J. and Juekendrup, A. E., 2008. Substrate metabolism and exercise performance with caffeine and carbohydrate intake. Medicine and Science in Sports and Exercise. 40(12): 2096–2104.

Irwin, C., Desbrow, B., Ellis, A., O'Keeffe, B., Grant, G. and Leveritt, M., 2011. Caffeine withdrawal and high-intensity endurance cycling performance. Journal of Sports Science. 29(5): 509–515.

Ivy, J. L., Costill, D. L., Fink, W. J. and Lower, R. W., 1979. Influence of caffeine and carbohydrate feedings on endurance performance. Medicine and Science in Sports and Exercise. 11: 6–11.

Ivy, J. L., Kammer, L., Ding, Z., Wang, B., Bernard, J. R., Liao, Y. H. and Hwang, J., 2009. Improved cycling time-trial performance after ingestion of a caffeine energy drink. International Journal of Sports Nutrition and Exercise Metabolism. 19(1): 61–78.

Jackman, M., Wendling, P., Friars, D. and Graham, T. E., 1996. Metabolic, catecholamine, and endurance responses to caffeine during intense exercise. Journal of Applied Physiology. 81: 1658–1663.

Jacobs, I., Pasternak, H. and Bell, D. G., 2003. Effects of ephedrine, caffeine, and their combination of muscular endurance. Medicine and Science in Sports and Exercise. 35(6): 987–994.

Jacobson, B. H. and Edwards, S. W., 1991. Influence of two levels of caffeine on maximal torque at selected angular velocities. Journal of Sports Medicine and Physical Fitness. 31(2):147–153.

Jacobson, B. H., Weber, M. D., Claypool, I. and Hunt, L. E., 1992. Effect of caffeine on maximal strength and power in elite male athletes. British Journal of Sports Medicine. 26(4): 276–280.

Jacobson, T. L., Febbraio, M. A., Arkinstall, M. J. and Hawley, J. A., 2001. Effect of caffeine co-ingested with carbohydrate or fat on metabolism and performance in endurance-trained men. Experimental Physiology. 86: 137–144.

Jenkins, N. T., Trilk, J. L., Singhal, A., O'Connor, P. J., Cureton, K. J., 2008. Ergogenic effects of low doses of caffeine on cycling performance. International Journal of Sports Nutrition and Exercise Metabolism. 18(3): 328–342.

Killgore, W. D., Kahn-Greene, E. T., Killgore, D. B., Kamimori, G. H. and Balkin, T. J., 2007. Effects of acute caffeine withdrawal on Short Category Test performance in sleep-deprived individuals. Perception of Motor Skills. 105(3 Pt 2): 1265–1274.

Kovacs, E. M. R., Stegen, J. H. C. H. and Brouns, F., 1998. Effect of caffeinated drinks on substrate metabolism, caffeine excretion, and performance. Journal of Applied Physiology. 85: 709–715.

Lane, J. D. and Philips-Bute, B. G., 1998. Caffeine deprivation affects vigilance, performance, and mood. Physiology and Behavior. 65: 171–175.

Latini, S. and Pedata, F. 2001. Adenosine in the central nervous system: release mechanisms and extracellular concentrations. Journal of Neurochemistry. 79: 463–484.

Lorino, A. J., Lloyd, L. K., Crixell, S. H. and Walker, J. L., 2006. The effects of caffeine on athletic agility. Journal of Strength and Conditioning Research. 20(4): 851–854.

Lovett, R. 2005. Coffee: The demon drink? New Scientist (2518). Retrieved on 2007-11-19.

MacIntosh, B. R. and Wright, B. M., 1995. Caffeine ingestion and performance of a 1500-metre swim. Canadian Journal of Applied Physiology. 20(2): 168–177.

McLellan, T. M., Kamimori, G. H., Voss, D. M., Bell, D. G., Cole, K. G. and Johnson, D., 2005. Caffeine maintains vigilance and improves run times during night operations for Special Forces. Aviation Space and Environmental Medicine. 76: 647–654.

Mitsumoto, H. G., DeBoer, G. E., Bunge, G., Andrish, J. T., Tetzlaff, J. E. and Cruse, R. P., 1990. Fiber-type specific caffeine sensitivity in normal human skinned muscle fibres. Anesthesiology. 72: 50–54.

Motl, R.W., O'Connor, P. J. and Dishman, R. K., 2003. Effect of caffeine on perceptions of leg muscle pain during moderate intensity cycling exercise. Journal of Pain. 4(6): 316–321.

Norager, C. B., Jensen, M. B., Madsen, M. R. and Laurberg, S., 2005. Caffeine improves endurance in 75-year-old citizens: a randomized, double-blind, placebo-controlled crossover study. Journal of Applied Physiology. 99(6): 2302–2306.

O'Rourke, M. P., O'Brien, B. J., Knez, W. L. and Paton, C. D., 2008. Caffeine has a small effect on 5-km running performance of well-trained and recreational runners. Journal of Science and Medicine in Sport. 11: 231–233.

Pasman, W. J., van Baak, M. A., Jeukendrup, A. E., and de Haan, A., 1995. The effects of different dosages of caffeine on endurance performance time. International Journal of Sports Medicine. 16: 225–230.

Paton, C. D., Hopkins, W. G. and Vollebregt, L., 2001. Little effect of caffeine ingestion on repeated sprints in team-sports athletes. Medicine and Science in Sports and Exercise. 33(5): 822–825.

Paton, C. D., Lowe, T. and Irvine, A., 2010. Caffeinated chewing gum increases repeated sprint performance and augments increases in testosterone in competitive cyclists. European Journal of Applied Physiology. 110: 1243–1250.

Prescino, C. L., Ross, M. L. R., Gregory, J. R., Savage, B. and Flanagan, T. R., 2008. Effects of sodium bicarbonate, caffeine, and their combination on repeated 200-m freestyle performance. International Journal of Sports Nutrition and Exercise Metabolism. 16: 116–130.

Porkka-Heiskanen, T. 1999. Adenosine in sleep and wakefulness. Annals in Medicine. 31: 125–129.

Rivers, W. H. R. and Webber, H. N., 1907. The action of caffeine on the capacity for muscular work. Journal of Physiology. 36: 33–47.

Roberts, S. P., Stokes, K. A., Trewartha, G., Doyle, J., Hogben, P. and Thompson, D., 2010. Effects of carbohydrate and caffeine ingestion on performance during a rugby union simulation protocol. Journal of Sports Science. 28(8): 833–842.

Rosser, J. I., Walsh, B. and Hogan, M. C., 2009. Effect of physiological levels of caffeine on Ca2+ handling and fatigue development in Xenopus isolated single muscle fibers. American Journal of Physiology. 296(5): R1512–1517.

Schneiker, K. T., Bishop, D., Dawson, B. and Hackett, L. P., 2006. Effects of caffeine on prolonged intermittent-sprint ability in team-sport athletes. Medicine and Science in Sports and Exercise. 38(3): 578–585.

Smith, A., Sutherland, D. and Christopher, G., 2005. Effects of repeated doses of caffeine on mood and performance of alert and fatigued volunteers. Journal of Psychopharmacology. 19(6): 620–626.

Skinner, T. L., Jenkins, D. G., Coombes, J. S., Taaffe, D. R. and Leveritt, M. D., 2010. Dose response of caffeine on 2000-m rowing performance. Medicine and Science in Sports and Exercise. 42(13): 571–576.

Stuart, G. R., Hopkins, W. G., Cook, C. and Cairns, S. P., 1995. Multiple effects of caffeine on simulated high-intensity team-sport performance. Medicine and Science in Sports and Exercise. 37(11): 1998–2005.

Van Nieuwenhoven, M. A., Brouns, F. and Kovacs, E. M. R., 2005. The effect of two sports drinks and water on GI complaints and performance during an 18-km run. International Journal of Sports Medicine. 26: 281–285.

Van Soeren, M. H. and Graham, T. E., 1998. Effect of caffeine on metabolism, exercise endurance, and catecholamine responses after withdrawal. Journal of Applied Physiology. 85(4): 1493–1501.

Wallman, K. E., Goh, J. W. and Guelfi, K. J., 2010. Effects of caffeine on exercise performance in sedentary females. Journal of Sports Science and Medicine. 9: 183–189.

Walter, A. A., Herda, T. J., Ryan, E. D., Costa, P. B., Hoge, K. M., Beck, T. W., Stout, J. R. and Cramer, J. T., 2009. Acute effects of a thermogenic nutritional supplement on cycling time to exhaustion and muscular strength in college-aged men. Journal of the International Society of Sports Nutrition. 13: 6–15.

Warren, G. L., Park, N. D., Maresca, R. D., McKibans, K. I. and Millard-Stafford, M. L., 2010. Effect of caffeine ingestion on muscular strength and endurance: a meta-analysis. Medicine and Science in Sports and Exercise. 42(7): 1375–1387.

Wemple, R. D., Lamb, D. R. and McKeever, K. H., 1997. Caffeine *vs.* caffeine-free sports drinks: effects on urine production at rest and during prolonged exercise. International Journal of Sports Medicine. 18, 40–46.

Wiles, J. D., Bird, S. R., Hopkins, J. and Riley, M., 1992. Effect of caffeinated coffee on running speed, respiratory factors, blood lactate, and perceived exertion during 1500 m treadmill running. British Journal of Sports Medicine. 26: 116–120.

Wiles, J. D., Coleman, D., Tegerdine, M. and Swaine, I. L., 2006. The effects of caffeine ingestion on performance time, speed, and power during a laboratory-based 1 km cycling time-trial. Journal of Sports Science. 24(11): 1165–1171.

Williams, A. D., Cribb, P. J., Cooke, M. B. and Hayes, A., 2008. The effect of ephedra and caffeine on maximal strength and power in resistance-trained athletes. Journal of Strength and Conditioning Research. 22(2): 464–470.

Williams, J. H., Signorile, J. F., Barnes, W. S. and Henrich, T. W., 1988. Caffeine, maximal power output, and fatigue. British Journal of Sports Medicine. 22: 132–134.

Woolf, K. W., Bidwell, W. K. and Carlson, A. G., 2008. The effect of caffeine as an ergogenic aid in anaerobic exercise. International Journal of Sports Nutrition and Exercise Metabolism. 18(4): 412–429.

Woolf, K. W., Bidwell, W. K. and Carlson, A. G., 2009. Effect of caffeine as an ergogenic aid during anaerobic exercise performance in caffeine naïve collegiate football players. Journal of Strength and Conditioning Research. 23(5): 1363–1369.

CHAPTER 18

The Effects of Caffeine on Ventilation and Pulmonary Function During Exercise

ROBERT F. CHAPMAN*[1], DANIEL P. WILHITE[1] AND
TIMOTHY D. MICKLEBOROUGH[1]

[1] 1025 E. 7th St – HPER 112, Department of Kinesiology, Indiana University,
Bloomington, IN 47405
*E-mail: rfchapma@indiana.edu

18.1 Introduction

In modern times, where athletes from elite to weekend warrior have wide ranging choices of legal and banned substances which can enhance athletic performance, caffeine holds standing as arguably the most utilized ergogenic aid on the planet. What makes caffeine a common and somewhat universal choice is the broad range of exercise modalities where caffeine reportedly improves performance, from activities requiring explosive strength, to short term, high intensity exercise bouts, to exercise activities that necessitate aerobic power and endurance. Interestingly, while many ergogenic substances are considered to be unethical or at least questionable in athletic endeavors, caffeine use in most cultures is typically considered to be socially acceptable. In the United States, we find the marketplace is presently saturated with "energy drinks" featuring substantial quantities of caffeine, matching or even exceeding the "tall" and "grande" sized beverages derived from passing water through coffee grounds that are seemingly available on every street corner.

Food and Nutritional Components in Focus No. 2
Caffeine: Chemistry, Analysis, Function and Effects
Edited by Victor R Preedy
© The Royal Society of Chemistry 2012
Published by the Royal Society of Chemistry, www.rsc.org

A simple search of the literature finds no less than two dozen review articles that have as a primary topic the effects of caffeine on exercise tolerance and muscular performance (for quality examples, see Graham 2001 and Spriet 2000). With the simple fact that caffeine is the most consumed drug in the world by humans, it could be easily argued that this large number of reviews is warranted. When the typical review delves into possible mechanisms for caffeine to affect exercise performance, three are commonly mentioned: 1) changes in substrate utilization resulting in glycogen sparing, 2) alterations in catecholamine release, and 3) central nervous system effects. However, a rarely discussed consequence of caffeine ingestion is the resulting effect on exercise ventilation and pulmonary function (Chapman and Mickleborough 2009). In fact, of the most commonly cited review articles and book chapters in the literature on the subject of caffeine ingestion and exercise, only one (Magkos and Kavouras 2004) briefly mentions the general effects of caffeine on respiration. Within this chapter, the primary purpose is to highlight the effects of caffeine on exercise ventilation and pulmonary function, and to suggest possible effects on exercise performance.

18.2   Effects of Caffeine on Ventilatory Drive

18.2.1   Central and Peripheral Chemosensitivity Changes With Caffeine

Caffeine has been historically cited for its stimulating effects on respiration in both animals and humans (Le Messurier 1936). Clinically, xanthine compounds have been used as safe and effective ventilatory stimulants for the treatment of conditions such as chronic obstructive pulmonary disease and neonatal apnea. Mechanistically, caffeine appears to affect both the central and peripheral chemoreceptors that control ventilation. Looking first at the effects on peripheral chemosensitivity, in moderate doses, caffeine has been demonstrated to significantly raise the resting hypoxic and hypercapnic ventilatory responses in humans (D'Urzo 1990). The hypoxic ventilatory response relates changes in arterial oxyhemoglobin saturation (SaO_2) with changes in minute ventilation during progressive isocapnic hypoxia, thus isolating ventilatory drive with falling inspired PO_2 primarily to the peripheral chemoreceptors located in the carotid bodies and aortic arch. In one study of 7 mostly caffeine habituated, untrained adult males, D'Urzo *et al* (1990) found that 650 mg of orally ingested caffeine resulted in a 135% increase in the slope of the resting hypoxic ventilatory response (Figure 18.1). In other words, the change in resting ventilation per percent change in SaO_2 with hypoxic exposure was nearly doubled with caffeine ingestion. Similarly, the resting hypercapnic ventilatory response (which relates changes in minute ventilation to changes in end tidal PCO_2 with progressive hyperoxic hypercapnia) increased by 31%. Taken together, these data suggest a strong effect of high dose caffeine augmenting the sensitivity of the peripheral chemoreceptors in normal untrained adults.

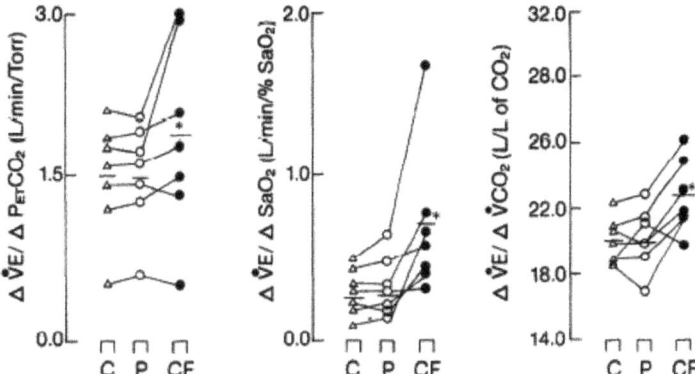

Figure 18.1   Individual hypoxic and hypercapnic ventilatory responses with placebo and 650 mg caffeine ingestion. C (triangles), control; P (open circles), placebo; CF (closed circles), caffeine (650 mg). VE, minute ventilation; PETCO$_2$, end tidal carbon dioxide partial pressure; SaO$_2$, arterial oxyhemoglobin saturation. Horizontal bars represent mean values. * Significantly different from control and placebo, P<0.05. Figure appears in D'Urzo *et al* 1990. Reprinted with permission from the American Physiological Society.

In contrast, how caffeine affects peripheral chemosensitivity in highly trained athletes appears to be another matter entirely. As a group, highly trained endurance athletes commonly show blunted ventilatory responsiveness to hypoxia and hypercapnia (Byrne-Quinn *et al* 1978). Whether this reduced ventilatory responsiveness is an adaptive result to chronic endurance training or is a naturally occurring trait, one which may predispose an individual to success in endurance exercise activities, is not clear. In any case, data suggest that the baseline level of ventilatory responsiveness to hypoxia and hypercapnia may affect the ability of caffeine (and other pharmacological compounds) to augment peripheral chemosensitivity. For example, work by Chapman and Stager (2008) studying highly trained endurance athletes suggests that caffeine does not affect peripheral chemosensitivity in this particular subset of the population. In a group of eight highly trained, caffeine naïve male distance runners, neither the resting hypoxic or hypercapnic ventilatory responses showed a significant change after ingestion of 8 mg kg^{-1} body weight caffeine. All told, it appears that highly trained endurance athletes are generally insensitive to an increased gain in peripheral chemosensitivity with caffeine, likely due to the severely blunted baseline chemoresponsiveness typically seen in this population.

18.2.2   Caffeine and Ventilation During Exercise

In both trained and untrained adults, caffeine has a strong analeptic effect. Peripheral chemosensitivity effects aside, it has been argued that the primary focus behind the ventilatory effect of caffeine is the central stimulation of the respiratory medullary complex. As a result, several studies have demonstrated significant increase in minute ventilation, both during submaximal (Bell *et al* 1999a; Bell *et al* 1999b; Brown *et al* 1991; Chapman and Stager 2008; D'Urzo *et al* 1990; Powers *et al* 1985) and maximal exercise (Chapman *et al* 1998; Chapman and Stager 2008). A listing of these studies with caffeine dosages, habituation and training status, and effects on exercise ventilation can be found in Table 18.1. In the Chapman and Stager (2008) study of highly trained distance runners, who demonstrated no change in resting peripheral chemosensitivity with caffeine, the same 8 mg kg^{-1} body weight dosage resulted in average increases in minute ventilation (VE) of 5–7% during exercise at 75% and 80% of VO$_2$max, compared to placebo (Figure 18.2). Even at VO$_2$max, VE was significantly increased by 5.5 L min^{-1} or nearly 4% with caffeine. It should be noted that some investigations have failed to demonstrate an increase in VE during exercise with caffeine (Sasaki *et al* 1987; Tarnopolsky *et al* 1989; Titlow *et al* 1991). However, many of these studies utilized dosages (<6 mg kg^{-1} body weight) and exercise workloads (<75% of VO$_2$max) far lower than other studies which have shown an increase in VE. Additionally,

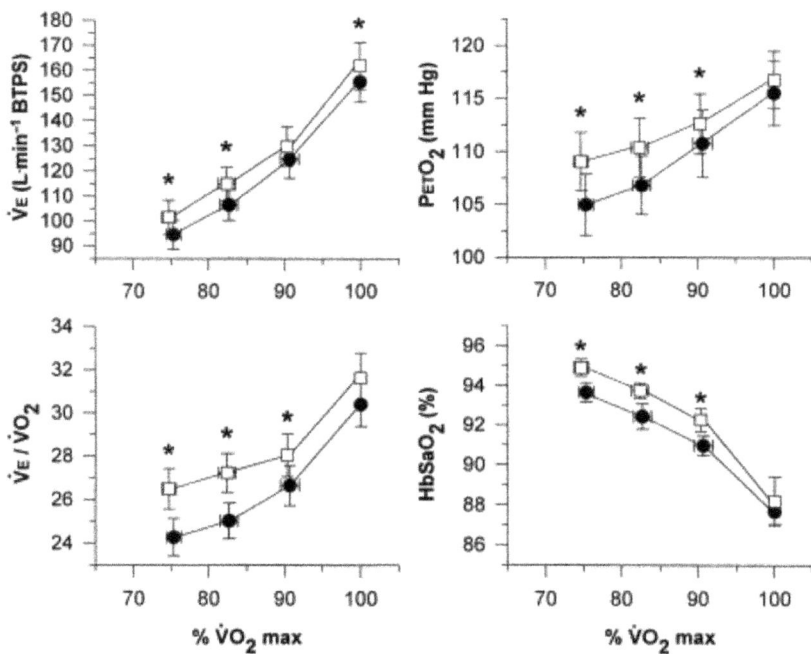

Figure 18.2   Ventilation, ventilatory equivalent for O$_2$, end-tidal PO$_2$, and SaO$_2$ during graded exercise. Filled circles, placebo; empty squares, caffeine. VE, minute ventilation; PETO$_2$, end tidal partial pressure of oxygen; SaO$_2$, arterial oxyhemoglobin saturation; VO$_2$, oxygen uptake. Values are means ± SE. * Significant difference compared with placebo at the same % of VO$_2$max (P<0.05). Figure appears in Chapman and Stager 2008. Reprinted with permission from Wolters Kluwer Health.

Table 18.1 Summary table of select studies reporting effects of caffeine ingestion on exercise ventilation.

Authors	Mode and intensity of exercise	Caffeine dosage and habituation status	Effect on ventilation
Studies demonstrating increases in submaximal exercise ventilation with caffeine			
Brown *et al* 1991	Walking, 50% VO_2max	3.3 mg kg^{-1}, mixed	4% ↑ alveolar ventilation
Chapman and Stager 2008	Running, 75%–80% of VO_2max	8 mg kg^{-1}, naïve	5–7% ↑ VE
D'Urzo *et al* 1990	Cycle, submaximal (<65% VO_2max)	650 mg, mostly habituated	29% ↑ VE
Powers *et al* 1985	Cycling, 80% of ventilatory threshold	7 mg kg^{-1}, NA	9% ↑ VE
Studies demonstrating increases in maximal exercise ventilation with caffeine			
Chapman *et al* 1998	Running, maximal	8 mg kg^{-1}, mixed	8% ↑ VE
Chapman and Stager 2008	Running, maximal	8 mg kg^{-1}, naïve	4% ↑ VE
Studies demonstrating no changes in exercise ventilation with caffeine			
Sasaki *et al* 1987	Running, 80% VO_2max	5 mg kg^{-1}, mixed	No change
Tarnopolski *et al* 1989	Running, 70% VO_2max	6 mg kg^{-1}, habituated	No change
Titlow *et al* 1991	Running, 60% of HRmax	200 mg, NA	No change

VO_2max, maximal oxygen uptake; VE, minute ventilation; NA, caffeine habituation status not reported. Changes reported as increasing all statistically significant compared to placebo or control, P<0.05.

Table 18.2 Effect of increasing and decreasing work of breathing at VO_2max on ventilation, leg blood flow, and leg vascular resistance.

	Inspiratory Assist	Control	Inspiratory Load
W_b, J min^{-1}	194.1 ± 26.3*	527.7 ± 23.2	732.9 ± 52.0*
VE, L min^{-1}	156.7 ± 4.5	145.7 ± 6.2	147.6 ± 6.1
Q_{legs}, L min^{-1}	192.2 ± 0.3*	18.4 ± 0.3	17.1 ± 0.2*
LVR, mm Hg L^{-1} min^{-1}	13.1 ± 0.3*	13.6 ± 0.4	14.6 ± 0.3*
NE spillover, ng min^{-1}	2426.1 ± 191.9	2720.0 ± 183.6	4883.9 ± 233.7*

Values are means ± SE, n = 7. Inspiratory assist trials utilized a pulmonary assist ventilator. Inspiratory load trials had increased resistance in the inspired line. W_b, work of breathing; VE, minute ventilation; Q_{legs}, leg blood flow; LVR, leg vascular resistance; NE, norepinephrine. *Significantly different from Control, P< 0.05. Table adapted from Harms *et al* 1997. Reprinted with permission from the American Physiological Society.

caffeine absorption times and habituation status was widely varied in studies where no increase in exercise ventilation was found.

18.3   Potential Ergogenic Effects of Increased Exercise Ventilation

What are potential positive ergogenic effects on exercise performance from a significant increase in exercise ventilation with caffeine ingestion? An increase in VE could result in a rise in alveolar oxygen partial pressure, improving SaO_2 and ultimately oxygen delivery to the working musculature. In theory, this could be of benefit in unique situations, such as exercise at altitude, or to unique individuals, such as athletes who experience pulmonary gas exchange limitations at sea level. Within the highly endurance trained population, it has been estimated that approximately 50% of athletes experience exercise induced arterial hypoxemia (EIH) during heavy exercise (Powers *et al* 1988). This phenomenon is hallmarked by a significant reduction in arterial PO_2 and SaO_2 during exercise at sea level, with SaO_2 <92% often used as a criterion measure below which VO_2max is impaired. Although originally thought to occur only in male athletes with VO_2max values above 65 mL kg^{-1} min^{-1}, EIH appears to be even more prevalent in female endurance athletes (in part due to smaller lung volumes, compared to men of the same stature) (Richards *et al* 2004) as well as older, masters athletes.

The primary mechanism behind the reduced SaO_2 with EIH is believed to be a reduced red blood cell transit time in the pulmonary capillary, secondary to an enlarged stroke volume and fixed pulmonary capillary blood volume in the athlete (Dempsey *et al* 1984). However, a blunted or "inadequate" hyperventilatory response during exercise in the athlete, resulting in a lower alveolar oxygen partial pressure in the lung, also appears to play a role in the formation of EIH (Harms and Stager 1985). To determine if caffeine could be useful in eliminating the hypoxemia of EIH *via* increasing minute ventilation, 8 caffeine naïve EIH males (VO_2max = 69.2 $\pm$ 4.0 mL kg^{-1} min^{-1}; SaO_2 at VO_2max = 88.0 $\pm$ 1.7%) were given caffeine (8 mg kg^{-1} body weight) or placebo and exercised progressively to VO_2max (Chapman and Stager 2008). During submaximal exercise (75%–90% of VO_2max), caffeine caused a significant increase in SaO_2 in runners with EIH, secondary to increases in minute ventilation and end tidal PO_2 (Figure 18.2). However, at the final minute of exercise corresponding to VO_2max, SaO_2 was unchanged in EIH runners, despite a significant increase in VE. In a follow up experiment, during exercise in simulated altitude (using a hypoxic inspirate), caffeine did not improve SaO_2 during maximal exercise in highly trained male distance runners (despite an increase in VE at VO_2max), but did improve SaO_2 at submaximal workloads (Chapman *et al* 1998). One possible explanation for the lack of improvement of SaO_2 at VO_2max with caffeine is that as exercise intensity approached VO_2max, only about 40% of the increase in VE with caffeine went towards increasing alveolar ventilation, with the remainder going to dead space ventilation. By comparison at 90% of VO_2max, where SaO_2 was significantly improved with caffeine, a much larger 68% of the increase in VE went to increasing alveolar ventilation. In the end, it appears that the stimulatory effects of caffeine on exercise ventilation may be helpful in increasing the alveolar partial pressure and SaO_2 during exercise at submaximal workloads in individuals with impaired pulmonary gas exchange – either physiologic or altitude induced.

18.4   Potential Negative Consequences of Increased Ventilation During Exercise

The above section reads as if increases in exercise ventilation at any given workload with caffeine ingestion may be positive for endurance exercise performance. However, the reality may just be the opposite. One of the potential negative consequences of increased minute ventilation during exercise with caffeine is the increased work of breathing done by the ventilatory muscles. The oxygen consumption of the diaphragm and respiratory musculature is estimated to be 3–6% of whole body VO_2 during moderate exercise, rising to as high as 15% of whole body VO_2 during maximal exercise in highly trained athletes (Figure 18.3; Aaron *et al* 1992). Using previously published data on the metabolic cost of ventilation (Aaron *et al* 1992) the 4–7% increase in VE observed during submaximal and maximal exercise with caffeine (Chapman and Stager 2008) would result in an estimated increase in O_2 consumption by the respiratory muscles of ~ 1–2% of whole body VO_2. For athletes in endurance events, where economy is a significant factor affecting performance, this increase in the metabolic cost of breathing is not trivial.

Perhaps more significantly, an increased work of breathing during exercise holds negative consequences for leg blood flow during endurance exercise. In a study by Harms *et al* (1997), the authors measured blood flow to the legs utilizing a cold saline thermodilution technique in cyclists exercising at VO_2max. In trials where inspiratory resistance was added, increasing the work of breathing, blood flow to the leg was significantly reduced compared to normal breathing trials. Similarly, when ventilatory work was reduced by ~ 60% utilizing a pulmonary assist ventilator, blood flow to the legs increased. Some of the reduced blood flow to the legs with increased ventilatory work is a result of a decreased cardiac output, secondary to a change in pleural pressures. However, it appears that a large portion of the blood flow change is a "metaboreflex" response, where sympathetic outflow to the legs is increased in response to increased ventilatory work (Dempsey *et al* 2008). In other words, during periods of heavy exercise and high ventilatory work, the body responds by "protecting" the diaphragm and accessory respiratory musculature by redirecting blood flow away from locomotor muscles. How much the analeptic effects of caffeine on exercise ventilation increases the overall work of breathing or affects blood flow to the exercising locomotor muscles is not known. During heavy exercise, is a potential gain in SaO_2 and arterial oxygen

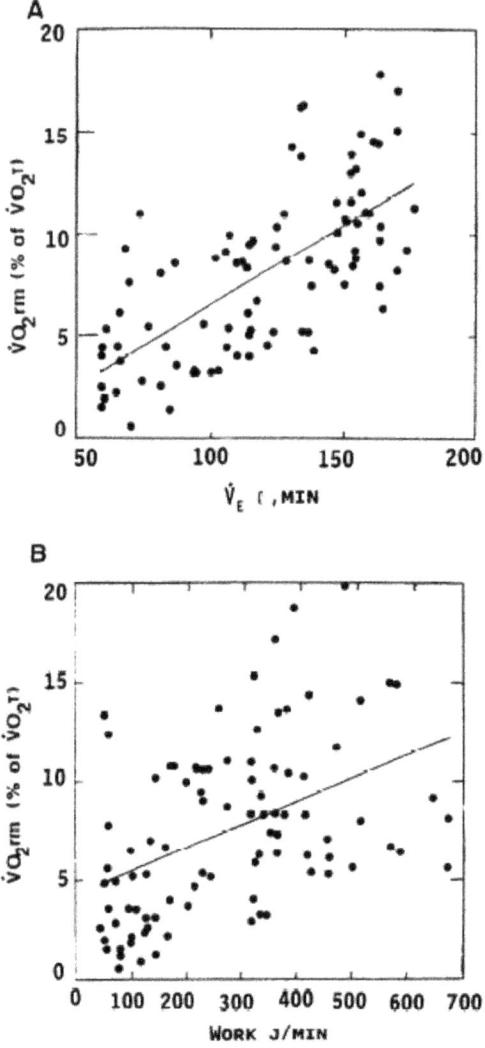

Figure 18.3   Relationship of exercise ventilation and work of breathing to percentage
of total body oxygen uptake required for ventilation alone. $VO_{2\,T}$, total
body oxygen uptake; $VO_{2\,RM}$, oxygen uptake by the respiratory
muscles. Each point represents a single trial in an individual subject.
Figure appears in Aaron *et al* 1992. Reprinted with permission from the
American Physiological Society.

content with caffeine induced increases in ventilation greater than a potential
metaboreflex induced reduction in leg blood flow? The answer to this question
may ultimately determine if the ventilatory effects of moderate to high levels of
caffeine ingestion are ergogenic to endurance exercise performance, from a
ventilatory standpoint.

18.5   Effects of Caffeine on Pulmonary Function

Caffeine is related to theophylline , a methylxanthine drug, which has immunomodulatory, anti-inflammatory, and bronchoprotective effects that potentially contribute to its efficacy as a prophylactic antiasthma drug. Though it has traditionally been classified as a bronchodilator, the ability of theophylline to control chronic asthma is disproportionately greater than is explainable by its relatively small degree of bronchodilator activity. Although at high doses caffeine can inhibit phosphodiesterase activity, dose-response studies have indicated that caffeine produces its stimulant effects by blocking adenosine receptors (Snyder *et al* 1981). It is now established that adenosine receptor antagonism is the most relevant mechanism *in vivo* (Fredholm 1985) because pharmacological doses of methylxanthines (mM) rather than physiological doses (μM) are needed to elicit a Ca^{2+} or phosphodiesterase inhibition effect. The observation that enprophylline selectively inhibits adenosine receptor agonist-induced interleukin-8 secretion by human mast cells *via* A2b receptor antagonism (Feoktistov and Biaggioni 1995), an effect shared by theophylline, raises the possibility that some of the anti-asthmatic effects observed with theophylline and caffeine are *via* A2b receptor antagonism (Feoktistov *et al* 1998).

A number of studies have explored the use of caffeine in asthma and the conclusions from a Cochrane Review (Welch *et al* 2010) suggest that, in the six clinical trials evaluated, caffeine does appear to improve airway caliber modestly for up 4 hours in individuals with asthma. To date, only three studies evaluating the effect of caffeine on pulmonary function in asthmatic individuals with exercise-induced bronchoconstriction (EIB) have been published. Kivity *et al* (1991) studied the effect of two doses of caffeine, 3.5 mg kg^{-1} and 7 mg kg^{-1}, on 10 asthmatics with EIB. Placebo or oral caffeine at 3.5 or 7 mg kg^{-1} was administered 2 hours prior to exercise. On the placebo, post-exercise FEV_1 dropped 25% compared to baseline, while the higher dose of caffeine reduced this decrease to 10% which was statistically significant. The lower dose of caffeine did not demonstrate a statistical effect (with a 14% drop in FEV_1). Although the only statistically significant findings were in doses of 7 mg kg^{-1}, the 3.5 mg kg^{-1} dose also showed a trend towards EIB protection. It should be pointed out that the 7.5 mg kg^{-1} dose of caffeine is equivalent to about 3 cups of coffee, which is an amount unlikely to be consumed by an individual prior to exercise. However, this amount of caffeine can be readily consumed prior to exercise if taken in pill form. Duffy and Phillips (1990), using a randomized double-blind crossover placebo-controlled design, examined the efficacy of caffeine at 5 or 10 mg kg^{-1} or placebo on post-exercise pulmonary function in 11 asthmatics with EIB. Eucapnic voluntary hyperventilation of dry gas was used as a surrogate to exercise as a challenge to the airways. Caffeine or placebo was ingested 90 minutes prior to the dry gas challenge. Similar to the Kivity *et al* study, the higher dose of caffeine, which is equivalent to 5 cups of coffee, reduced the post-exercise fall in FEV_1 significantly from a decrease of 16.7% on placebo to 7.1%. The lower dose of caffeine did not have a significant impact on the post-exercise FEV_1 (-10.2%) compared to placebo (-16.7%). Recently VanHaitsma *et al* (2010) has shown that taking caffeine 1 hour prior to exercise at a dose of 3, 6 and 9 mg kg^{-1}, reduced the post-exercise fall in FEV_1 from a decrease of 18.4% (baseline) to 11.9%, 9.0% and 6.8% respectively (Figure 18.4). In addition, the high dose of caffeine (9 mg kg^{-1}) was shown to be as effective as albuterol, a commonly used asthma medication, in reducing the severity of EIB. Therefore, in the three studies conducted to date examining the efficacy of caffeine ingestion on EIB, post-exercise pulmonary function was significantly improved to subclinical levels in EIB subjects. However, the dose of caffeine required to induce bronchodilation is high compared with usual dietary intakes.

The International Olympic Committee (IOC) has recognized the performance enhancing effects of caffeine, calling it an ergogenic substance and banning the drug from 1962 to 1972. Up through 2004, the IOC set a limit for urinary caffeine levels at 12 mg mL^{-1}, which required approximately 1000 mg of caffeine (about 8 cups of coffee) to exceed that limit. However, it is very important to note that individuals can metabolize caffeine at very different rates. Differences in metabolism, medications, and certain diseases may

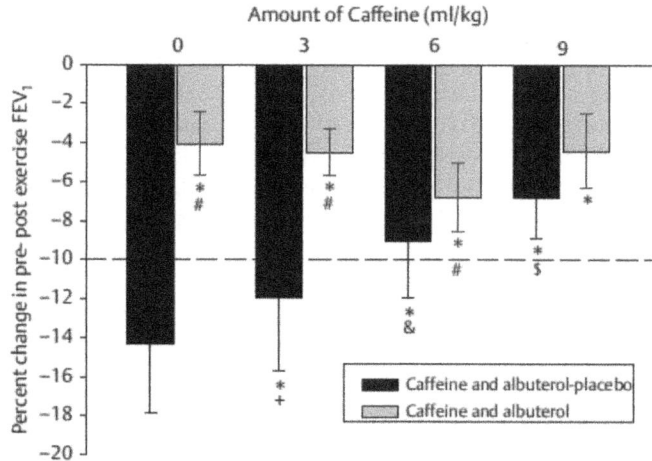

Figure 18.4   The percent change in FEV_1 from pre- to post-exercise for the different caffeine doses with and without albuterol. FEV_1, forced expiratory volume in one second. Dashed line, reductions in post-exercise FEV_1 in excess of 10% represent a positive diagnosis of exercise-induced bronchoconstriction. Values are means $\pm$ SE. * Significantly different from double-placebo, $P<0.05$; $ Significantly different from 3 and 6 mL kg^{-1} caffeine dose, $P<0.05$; & Significantly different from 3 and 9 mL kg^{-1} caffeine dose, $P<0.05$; + Significantly different from 6 and 9 mL kg^{-1} caffeine dose, $P<0.05$. # Albuterol treatment significantly different from albuterol plus caffeine treatment, $P<0.05$. Figure appears in Van Haitsma *et al* 2010. Reprinted with permission from Thieme Medical Publishers.

significantly alter the rate in which caffeine is cleared from the body. Generally caffeine absorption is complete within about 1 hour after ingestion and the plasma concentration peaks after about 90 min. For doses lower than 10 mg kg^{-1}, the half-life of caffeine in the plasma is about 2.5–4.5 hours in healthy adults. Partly due to these individual differences in caffeine clearance, caffeine was removed from the banned list and is currently considered part of a "Monitoring Program" by the World Anti-Doping Agency, with no upper limit for a positive test outcome.

18.6 Summary

In light of caffeine's universal accessibility and increasing social acceptance as an ergogenic aid for exercise and sport, it is important to have an understanding of the ventilatory and pulmonary function effects of caffeine, independent of metabolic, neural, and hormonal responses. While coffee and colas have been the traditional modes of caffeine ingestion, caffeine infused "energy drinks" have dramatically increased in popularity, particularly among younger individuals. For the clinician, coach, and scientist, this increase in caffeine use and acceptance may result in a) an increased reporting of dyspnea during exercise by patients and athletes, secondary to central and possibly peripheral chemoreceptor stimulation increasing ventilatory drive, and b) possible changes in the outcomes of pulmonary function testing.

From an ergogenic standpoint, the significant increase in exercise ventilation seen with high dose (>7 mg kg^{-1}) caffeine ingestion resulted in a rise in end tidal PO_2 and SaO_2 in a) athletes with exercise induced hypoxemia, and b) athletes exercising submaximally at a simulated altitude. Additionally, caffeine has been shown to significantly mitigate the fall in FEV_1 post-exercise in patients with EIB. Therefore, the effect of caffeine specifically on ventilation and pulmonary function may be ergogenic under these circumstances, but this hypothesis has not been examined directly in performance-based testing. Additionally, it is also important to consider that an increase in the work of breathing with a caffeine-mediated increase in VE and a potential metaboreflex effect, diverting blood flow away from the locomotor muscles to the ventilatory muscles, may combine to negate any potential ergogenic advantage from an increase in alveolar oxygenation. These observations on caffeine's effects on ventilation and pulmonary function are separate from the well documented effects of caffeine on metabolism, substrate utilization, catecholamine release, central nervous system function, and exercise performance in general.

Summary Points

- This chapter focuses on the effects of caffeine on exercise ventilation and pulmonary function.

- Caffeine stimulates both the central and peripheral chemoreceptors, which control ventilatory drive.
- In large doses (>6 mg kg^{-1} body weight), caffeine results in significant increases in submaximal and maximal exercise ventilation.
- The increases in exercise ventilation with caffeine may be useful in increasing alveolar oxygen partial pressure and arterial oxygen saturation.
- However, during heavy exercise, the increased work of breathing may cause blood flow to be redirected from the locomotor to the ventilatory muscles *via* a sympathetic "metaboreflex" response, negatively affecting exercise performance.
- Caffeine also demonstrates immunomodulatory, anti-inflammatory, and bronchoprotective effects that contribute to its efficacy as a prophylactic antiasthma drug.
- In three studies conducted to date examining the efficacy of caffeine ingestion on exercise induced bronchoconstriction, pulmonary function after exercise was significantly improved to levels measured before exercise.

Key Facts of Ventilatory Control During Exercise

- The primary purpose of ventilation during exercise is to regulate arterial carbon dioxide content, with arterial oxygen content being a secondary factor.
- Arterial CO_2 content is sensed by both the central chemoreceptors located in the medulla of the brain and the peripheral chemoreceptors located in a) the carotid bodies within the carotid arteries in the neck region, and b) the aortic arch. Arterial O_2 content is sensed solely by the peripheral chemoreceptors.
- The central and peripheral chemoreceptors are normally sensitive to changes in baseline levels, and send a continuous signal to the respiratory musculature which is modulated based on changes in chemical stimuli.
- The airways and lung wall contain mechanoreceptors, which send an afferent signal in response to various stimuli, such as stretch (hyperinflation).
- Ventilation is also affected by temperature (with an increase in temperature resulting in an increase in ventilation).
- Ventilation can be regulated during exercise by modulating either tidal volume (the depth of the breath) or breathing frequency (the number of breaths per minute) or both.
- As exercise progresses from rest through low and moderate workloads, minute ventilation is increased primarily by increasing tidal volume, with modest increases in breathing frequency.
- As exercise approaches heavy or maximal levels, minute ventilation is increased primarily by increasing breathing frequency, with modest increases in tidal volume.

Definition of Words and Terms

Analeptic: A central nervous system stimulant which acts to increase ventilation.

Bronchodilation: A widening of the diameter of the pulmonary airways.

Ergogenic aid: Outside influences on the body which help to enhance performance. Ergogenic aids can be pharmacological, mechanical, psychological, nutritional, or physiological in nature.

Eucapnic voluntary hyperventilation: Technique where the subject mimics exercise ventilation at rest, while breathing gas that has carbon dioxide added (typically 5% CO_2). The added CO_2 allows the subject to hyperventilate at rest without becoming light headed or passing out. This technique is used to create a bronchoconstriction response in subjects known to exhibit EIB, without the stress of actual exercise.

Exercise induced bronchoconstriction (EIB): An abnormal response of airway narrowing during exercise, due to both a constriction response of the airway, as well as an airway inflammatory response, decreasing the diameter of the bronchiole.

Exercise induced hypoxemia (EIH): A physiological condition hallmarked by reduced arterial oxyhemoglobin saturation during heavy exercise at sea level (or low altitudes). Is reported to occur in approximately 50% male endurance athletes with high VO_2max values, a majority of female endurance athletes, and many masters (*i.e.* older) athletes.

Hypercapnic ventilatory response: A measurement of the change in ventilation as the content of inspired carbon dioxide rises.

Hypoxic ventilatory response: A measurement of the change in ventilation as the content of inspired oxygen falls.

Isocapnic: When arterial carbon dioxide levels are maintained at resting levels.

Metaboreflex: A physiological response of increased muscle sympathetic nerve activation during heavy exercise (or heavy ventilatory work), which causes vasoconstriction in the skeletal musculature, redirecting blood flow to the diaphragm and accessory respiratory musculature.

Peripheral chemoreceptors: Specialized cells located in the carotid arteries in the neck and the aorta which are sensitive to changes in the arterial content of oxygen and carbon dioxide, as well as pH. The peripheral chemoreceptors send out a continuous neural signal to the respiratory musculature to control ventilation.

List of Abbreviations

EIB	Exercise induced bronchoconstriction
EIH	Exercise induced hypoxemia
FEV_1	Forced expiratory volume in one second. Commonly reported in units of liters, or as a percent of the lungs' vital capacity.
IOC	International Olympic Committee

PCO$_2$	Partial pressure of carbon dioxide
PO$_2$	Partial pressure of oxygen
SaO$_2$	Arterial oxyhemoglobin saturation, reported as a percent of all of the hemoglobin binding sites that have an oxygen attached.
VO$_2$	The rate of oxygen uptake by the body. Commonly reported in units of liters of oxygen per minute, or milliliters of oxygen per kilogram of body weight per minute.
VO$_2$max	The maximal rate of oxygen uptake by the body. Typical values of VO$_2$max for untrained individuals is approximately 40–45 ml/kg/min, while elite endurance athletes commonly have VO$_2$max values greater than 70 ml/kg/min.

References

Aaron, E. A., Seow, K. C., Johnson, B. D. and Dempsey, J. A., 1992. Oxygen cost of exercise hyperpnea: implications for performance. Journal of Applied Physiology. 72: 1818–1825.

Bell, D. G. and Jacobs, I. 1999. Combined caffeine and ephedrine ingestion improves run times of Canadian Forces Warrior Test. Aviation, Space and Environmental Medicine. 70: 325–329.

Bell, D. G., Jacobs, I., McLellan, T. M., Miyazaki, M. and Sabiston, C. M., 1999. Thermal regulation in the heat during exercise after caffeine and ephedrine ingestion. Aviation, Space and Environmental Medicine. 70: 583–588.

Brown, D. D., Knowlton, R., Sullivan, J. and Sanjabi, P., 1991. Effect of caffeine ingestion on alveolar ventilation during moderate exercise. Aviation, Space and Environmental Medicine. 62: 860–864.

Byrne-Quinn, E., Weil, J. V., Sodal, I., Filley, G. and Grover, R., 1971. Ventilatory control in the athlete. Journal of Applied Physiology. 30: 91–95.

Chapman, R. F., Emery, M. and Stager J. M., 1998. Extent of expiratory flow limitation in normoxia influences the increase in exercise ventilation in mild hypoxia. Respiration Physiology. 113: 65–74.

Chapman, R. F. and Mickleborough T. D., 2009. The effects of caffeine on ventilation and pulmonary function during exercise: an often-overlooked response. The Physician and Sportsmedicine. 37: 97–103.

Chapman, R. F. and Stager, J. M., 2008. Caffeine stimulates ventilation in athletes with exercise-induced hypoxemia. Medicine and Science in Sports and Exercise. 40: 1080–1086.

Dempsey, J. A., Amann, M., Romer, L. M. and Miller, J. D., 2008. Respiratory system determinants of peripheral fatigue and endurance performance. Medicine and Science in Sports and Exercise. 40: 457–461.

Dempsey, J. A., Hanson, P. E. and Henderson, K. S., 1984. Exercise induced arterial hypoxemia in healthy persons at sea level. Journal of Physiology London. 355: 161–175.

Duffy, P. and Phillips, Y. Y., 1991. Caffeine consumption decreases the response to bronchoprovocation challenge with dry gas hyperventilation. Chest. 99: 1374–1377.

D'Urzo, A. D., Jhirad, R., Jenne, H., Avendano, M., Rubinstein, I., D'Costa, M. and Goldstein, R., 1990. Effect of caffeine on ventilatory responses to hypercapnia, hypoxia, and exercise in humans. Journal of Applied Physiology. 68: 322–328.

Feoktistov, I. and Biaggioni, I., 1995. Adenosine A2b receptors evoke interleukin-8 secretion in human mast cells: an enprofylline-sensitive mechanism with implications for asthma. Journal of Clinical Investigation. 96: 1979–1986.

Feoktistov, I., Polosa, R., Holgate, S. T. and Biaggioni, I., 1998. Adenosine A2B receptors: a novel therapeutic target in asthma. Trends in Pharmacological Science. 19: 148–153.

Fredholm, B. 1985. On the mechanism of action of theophylline and caffeine. Acta Medica Scandinavica. 217: 149–153.

Kivity, S., Ben Sharon, Y., Man, A. and Topilsky, M., 1990. The effect of caffeine on exercise-induced bronchoconstriction. Chest. 97: 1083–1085.

Graham, T. E. 2001. Caffeine and exercise: metabolism, endurance and performance. Sports Medicine. 31: 785–807.

Harms, C. A. and Stager, J. M., 1995. Low chemoresponsiveness and inadequate hyperventilation contribute to exercise-induced hypoxemia. Journal of Applied Physiology. 79: 575–580.

Harms, C. A., Wetter, T. J., McClaran, S. R., Pegelow, D. F., Nickele, G. A., Nelson, W. B., Hanson, P. and Dempsey, J. A., 1997. Respiratory muscle work compromises leg blood flow during maximal exercise. Journal of Applied Physiology. 82: 1573–1583.

Le Messurier, D. H. 1936. The site of action of caffeine as a respiratory stimulant. Journal of Pharmacology and Experimental Therapy. 57: 458–463.

Magkos, F. and Kavouras, S. A., 2004. Caffeine and ephedrine: physiological, metabolic and performance enhancing effects. Sports Medicine. 34: 871–889.

Powers, S. K., Dodd, S., Lawler, J., Landry, G., Kirtley, M., McKnight, T. and Grinton, S., 1988. Incidence of exercise induced hypoxemia in elite endurance athletes at sea level. European Journal of Applied Physiology. 58: 298–302.

Powers, S. K., Dodd, S., Woodyard, J. and Magnum, M., 1985. Caffeine alters ventilatory and gas exchange kinetics during exercise. Medicine and Science in Sports and Exercise. 18: 101–106.

Richards, J. C., McKenzie, D. C., Warburton, D. E. and Road, J. D., 2004. Prevalence of exercise-induced arterial hypoxemia in healthy women. Medicine and Science in Sports and Exercise. 36:1514–1521.

Sasaki, H., Maeda, J., Usui, S. and Ishiko, T., 1987. Effect of sucrose and caffeine ingestion on performance of prolonged strenuous running. International Journal of Sports Medicine. 8: 261–265.

Snyder, S. H., Katims, J. J., Annau, A., Bruns, R. F. and Daly, J. W., 1981. Adenosine receptors and behavioral actions of methylxanthines. Proceedings of the National Academy of Sciences of the United States of America. 78: 3260–3264.

Spriet, L. L. 2000. Caffeine and performance. International Journal of Sport Nutrition. 5: S84–S99.

Tarnopolsky, M. A., Atkinson, S. A., MacDougall, J. D., Duncan, J., Sale, D. G. and Sutton, J. R., 1989. Physiological responses to caffeine during endurance running in habitual caffeine users. Medicine and Science in Sports and Exercise. 21: 418–24.

Titlow, L. W., Ishee, J. H. and Riggs, C. E., 1991. Failure of caffeine to affect metabolism during 60 min submaximal exercise. Journal of Sports Science. 9: 15–22.

VanHaitsma, T. A., Mickleborough, T. D., Stager, J. M., Koceja, D., Lindley, M. R., and Chapman, R. F., 2010. Comparative and synergistic effects of caffeine and albuterol on the severity of exercise-induced bronchoconstriction. International Journal of Sports Medicine. 31: 231–236.

Welch, E. J., Bara, A. I., Barley, E. A. and Cates, C. J., 2010. Caffeine for asthma. Cochrane Database Systematic Reviews. 1: CD001112, DOI: 10.1002/14651858.CD001112.pub2.

CHAPTER 19

Dietary Caffeine and Young Children: Implications for Health

WILLIAM J. WARZAK*[1], SHELBY EVANS[1],
LUIS F. MORALES KNIGHT[1], LAURA NEEDELMAN[1]
AND REBECCA K. DOGAN[1]

[1] Department of Psychology, Munroe-Meyer Institute, Department of
Pediatrics, University of Nebraska Medical Center, 985450 Nebraska Medical
Center, Omaha, NE 68198-5450, USA
*E-mail: wwarzak@unmc.edu

19.1 The Prevalence of Caffeine

Caffeine is widely regarded as the most commonly used psychoactive substance
in the world (Fredholm *et al* 1999; James 2011; Miller 2005). It is a naturally
occurring alkaloid that is commonly extracted for human consumption from a
number of plant species, including coffee, tea, and cacao (Thomson and
Schiess 2010). It is artificially added to a wide range of food and medicinal
products because of its CNS arousal, motor-activating and reinforcing effects
(Bramstedt 2007; Ferré *et al* 2011; Nawrot *et al* 2003; Temple 2009).

The caffeine content of beverages and foods varies considerably, even across
products within the same category. For example, a typical 8 oz cup of coffee
contains between 35 and 250 mg of caffeine depending on the brand, the beans,
the roast, the grind and the method of preparation. Tea varies less, with 8 oz
containing between 15 and 35 mg of caffeine. Soft drinks, a preferred item for
children, contain between 36 and 54 mg of caffeine per 12 oz serving. Foods

Food and Nutritional Components in Focus No. 2
Caffeine: Chemistry, Analysis, Function and Effects
Edited by Victor R Preedy

containing chocolate have minimal amounts of caffeine, with few items exceeding 15 mg per serving.

Caffeine is the main active ingredient in energy drinks, many of which contain 70–80 mg per 8 oz serving (Seifert *et al* 2011), and many of these beverages are sold in 16 oz and 24 oz serving sizes. Although the FDA limits caffeine in soft drinks, which are categorized as food, there is no such regulation of energy drinks, which are classified not as food but as dietary supplements. As such, these beverages often contain a number of additional vitamin, mineral, and herbal products that are unregulated (Reissig *et al* 2009).

In countries such as Canada, Australia, and various members of the European Union (EU), the sale of energy drinks is restricted based on age, and, energy drinks are required to carry health warnings and disclose the health risks associated with their consumption. No such regulations are currently in place in the United States (Seifert *et al* 2011).

19.2   Trends in Caffeine Consumption

In the US, individuals under the age of 12 are prohibited from purchasing caffeine in tablet form (*e.g.,* brand names NoDoz, Vivarin, *etc.*); however, no such regulations apply to soda, coffee, tea, chocolate, energy drinks, or other caffeine-added products such as water, gum, and mints (Seifert *et al* 2011). These products are marketed directly to children and adolescents (Bramstedt 2007; Temple 2009) and many of these products are available for purchase by children in schools.

Caffeine consumption in childhood is largely accounted for by soft drinks and chocolate, and consumption of these items has increased dramatically in recent years (Frary *et al* 2005; Temple 2009; Warzak *et al* 2011). Perhaps coincident with this rise in caffeine consumption has been the development of energy drinks, typically marketed to a younger demographic. Half of the energy drink market consists of young children, adolescents and young adults under 25 years of age (Seifert *et al* 2011).

The increase in children's daily caffeine intake over the past several decades raises a number of concerns regarding the long-term impact of caffeine consumption on the developing child and adolescent. It has been estimated that adolescents 12–17 years old consume approximately 70 mg of caffeine per day, which is less than that found in a typical 8 oz cup of coffee. Interestingly, even children two to 11 years old have been recorded as consuming caffeine, although it tends to be much less than that consumed by adults. Precise estimates of caffeine consumption in young children are difficult to obtain and until recently, the caffeine consumption data for children living in the United States was over a decade old (Knight *et al* 2004). Moreover, there are now so many sources of caffeine that it is arguable that many children and adolescents, as well as their parents, may be unaware of, or unable to report their true volume of caffeine consumption.

19.3 The Effects of Caffeine on Children

Caffeine has been found to affect a variety of physiological variables in adults including diuresis, respiration, and heart rate, as well as neurocognitive variables such as attention, reaction time, and psychomotor speed (Nawrot *et al* 2003). The effects of caffeine on children are far less understood. We do know, however, that caffeine has effects on children and that children develop a tolerance to caffeine at doses as low as 1.0 mg kg^{-1} per day (Meltzer *et al* 2008). Withdrawal has been observed in school-aged children, who obtain the majority of their caffeine from soft drinks, and these effects tend to be greater in children with higher rates of consumption (Goldstein and Wallace 1997).

Similar to adults, there may be deleterious effects for children with susceptible medical conditions such as cardiovascular or seizure disorders, among others (Temple 2009). We would note that there are many variables that mediate the effects of caffeine. Not only are many effects dose dependent, often with curvilinear effects, but many also are dependent upon a subject's base level of consumption (*i.e.,* either high or low). There also are potential confounds because the effects of caffeine are often evaluated against a period of deprivation (*i.e.,* caffeine withdrawal). This may not be important for low caffeine consumers, but it is a dilemma for investigators trying to determine the difference between the putative potentiating effects of caffeine, as measured against a baseline of no caffeine, in contrast to the elimination of symptoms that may have occurred as a function of caffeine withdrawal when caffeine is then re-introduced, especially in high caffeine consumers. The implications of these contrasting effects have not been explored in children (Temple 2009). In addition, much of the information regarding children's consumption is reported by their parents or guardians and is subject to inaccuracies inherent in that methodology.

19.4 CNS Effects

At low doses (< 2.5–3 mg kg^{-1}), children have been observed to show increases in attention and manual dexterity (Bernstein *et al* 1994; Rapoport *et al* 1981b). Bernstein *et al* (1994) also found that at least some behavioral effects, such as anxiety, appeared to occur at a dose of 2.5 mg kg^{-1} per day in children under 12; further, all caffeine consuming participants reported higher levels of anxiety, mood swings, crying, and general feelings of restlessness compared to placebo conditions. At higher doses (5–10 mg kg^{-1}), increases in motor activity and speech rate, as well as lowered reaction time and reduced error rates on a measure of sustained attention, have been reported (Durlach 1998; Lieberman *et al* 1987; Smit and Rogers 2000). Deleterious effects on performance have been observed at 10 mg kg^{-1}, including increased fidgeting, restlessness, hyperactivity, and difficulty sleeping (Nawrot *et al* 2003; Temple 2009).

Investigators have differed in administering caffeine by mg or by mg kg^{-1}, which complicates comparison across studies. In addition, many studies used a bolus administration of caffeine that makes generalization to ordinary, daily-dose patterns of use difficult. Some studies have stratified results by daily dose level. For example, Rapoport *et al* (1981a) reported that high-dose consumers (≥300 mg per day) were more easily frustrated, more nervous on tests, and less able to distinguish between high and low doses of caffeine in acute administration than were low-dose consumers (<50 mg per day). High-dose consumers also suffered more apparent effects of withdrawal (Rapoport *et al* 1984). In children, tachycardia, vomiting, and diuresis are among the commonly reported symptoms of high caffeine intake (Nawrot *et al* 2003), but caffeine toxicity is rarely, if ever, fatal. One meta-analysis by Stein *et al* (1996) showed that acute administration of caffeine did not result in significantly negative effects in children. More recently, Meltzer *et al* (2008) concluded that 0.3 mg kg^{-1} per day was the highest level of caffeine ingestion at which no immediate adverse effects were observed in children and adolescents.

In addition, it is known that caffeine has cardiac effects on adults, including arrhythmias and increases in blood pressure. It is not evident, however, whether these effects are similar, more pronounced, or even absent in children and teens. In adults, arrhythmia and palpitations generally follow the consumption of unusually large amounts of caffeine in healthy individuals or result from chronic use in individuals with existing cardiac disorders (Cannon *et al* 2001; Chopra and Morrison 1995; Curatolo and Robertson 1983). The effects of caffeine and cardiac effects in children have yet to be fully investigated. This, combined with the fact that many children and adolescents consume caffeine on a regular basis, suggests that the potential for cardiac events as a function of caffeine consumption should be examined in the pediatric population.

19.5   Sleep Effects

The effects of caffeine on sleep have been documented in a number of studies although not on an experimental basis. Student surveys of middle and high school students have suggested a link between caffeine consumption and sleep reduction (Seicean *et al* 2007; Goldstein and Shapiro 1987). More recently, Warzak *et al* (2011) reported the results of a parent survey of young children 5–12 years old, in which the relationship between caffeine consumption and sleep was significantly negatively correlated. One study (Goldstein and Wallace 1997) suggested that consumption of as little as 50 mg per day could result in higher wakefulness in children. The relationship between sleep and caffeine becomes additionally confounded during puberty, as the circadian effects of melatonin come to be at odds with the common sleep-wake cycle associated with school attendance (Wolfson and Carskadon 2003).

Although insomnia and sleep difficulties have been mentioned as side effects of caffeine in several pediatric studies (Bernstein *et al* 1994, 1998; Elkins *et al* 1981; Rapoport *et al* 1981a; 1981b; 1984), clear data on sleep disruption have not been collected in a systematic manner. In a recent survey of parents of pediatric outpatients, Warzak *et al* (2011) found that the majority of caffeine consuming children under age 8 obtained adequate sleep, but about 25% of these children did not receive the amount of sleep recommended by the Centers for Disease Control (CDC). And, on average, children age 8 to 12 years old who consumed caffeine did not meet the CDC's minimum suggested sleep requirements (Warzak *et al* 2011). Meltzer *et al* (2008), investigating the dose-response relationship between caffeine and various behavioral health concerns, reported that sleep disturbances developed at a caffeine dose of approximately 1.4 mg kg^{-1} per day in children age 12 and under. The evidence collected to date indicates that sleep disruption is common in children who consume caffeine and is perhaps one of the most empirically supported health-related consequences of chronic caffeine use. As a result, health care practitioners have been encouraged to screen children for caffeine use when they present with sleep difficulties.

Not all of caffeine's effects on sleep are negative. The benefits of caffeine in the treatment and prevention of apnea in premature infants have been well documented. In the case of apnea of prematurity and infants of very low birth weight (*i.e.,* 500–1250 g), caffeine has been shown to effectively treat apnea/bradycardia and prevent bronchopulmonary dysplasia (Bancalari 2006; Gray *et al* 2011; Larsen *et al* 1995; Schmidt *et al* 2006). Additionally, intravenous caffeine is known to improve blood pressure and heart rate in these infants (Soloveychik *et al* 2009). Caffeine has therapeutic benefits equal to those of aminophylline and theophylline in the treatment of apnea and bradycardia in preterm infants with fewer harmful side effects, making caffeine the treatment of choice for these conditions (Comer *et al* 2001; Larsen *et al* 1995; Skouroliakou *et al* 2009). Caffeine is now considered one of the safest and most cost-effective procedures in the treatment and prevention of apnea of prematurity (Aranda *et al* 2010).

19.6 Diuretic Effects

The idea that caffeine is a diuretic is a staple of dietary and medical wisdom (Jalkut *et al* 2001) but this has been challenged in recent reviews of empirical literature (Armstrong 2002) which have concluded that an intake of less than 300 mg per day has a diuretic effect no greater than that of water. Rather, the increased urination observed in adult caffeine users appears to be connected to detrusor instability (Edelstein *et al* 1984; James *et al* 1989). Very little research in this vein has been done with children, and what does exist (Warzak *et al* 2011) suggests that caffeine intake is not associated with increased enuresis in children.

19.7 Maternal Consumption

Rates of breast feeding in the US indicate that the majority of infants (63%) are breast-fed at some point during the first year of life, although fewer than one quarter of infants are breast fed for 6 months, and only about 17% of infants are breast fed for a full year (CDC, 2010). Thus, only a relatively small number of infants have the potential to be exposed to caffeine *via* breast milk for an extended period of time.

Nevertheless, it has been reported that withdrawal symptoms have been observed in newborns whose mothers regularly consumed 300–400 mg per day during pregnancy. These infants were reportedly irritable, affectively labile, and had difficulties with sleep, whereas others appeared to be unaffected (Higdon and Frei 2006; Hildebrandt and Gundert-Remy 1983; Nawrot *et al* 2003; Ryu 1985). These symptoms began shortly after birth and spontaneously remitted within a short period of time (McGowan *et al* 1988; Nehlig and Debry 1994).

Conflicting reports of the impact of breast milk is not surprising, given that some variance in the effect of caffeine in very young children (<6 months of age) is to be expected. The half-life of caffeine is extended in neonates; as physiological maturation occurs, the rate of metabolic elimination of caffeine reaches a level similar to that of adults at approximately six months of age (Nawrot *et al* 2003). Therefore, it is likely that the effects of caffeine are more pronounced in younger infants than in older, and may vary within the same infant as she or he matures.

Generally speaking, lower amounts of caffeine do not appear to impact infants, and are considered safe provided the child is healthy. For infants who were premature, low birth weight, or have other health difficulties, the ingestion of even relatively small amounts of caffeine by the mother is hypothesized to cause some negative behavioral and physiological effects (American Academy of Pediatrics Committee on Drugs 2001). Overall, however, the effect of caffeine on infants appears to be time-limited and symptoms remit when caffeine levels decrease. Research to date has not identified a conclusive link between caffeine exposure in infants and long-term health consequences (Nawrot *et al* 2003), and as such, the American Academy of Pediatrics has listed caffeine as a drug that is compatible with breastfeeding when consumed in moderate doses (*i.e.*, ≤300 mg per day).

19.8   Recommended Intake

Although the US Food and Drug Administration has not provided guidelines for caffeine intake, other countries and professional associations have made recommendations for presumably safe levels of caffeine consumption. The American Congress of Obstetricians and Gynecologists (ACOG) concluded that consuming less than 200 mg per day of caffeine during pregnancy was unlikely to cause an adverse effect in the form of miscarriage or preterm birth (ACOG 2010). Similarly, the United Kingdom has also issued a 200 mg per day maximum consumption for pregnant women (Food Standards Agency 2008). Several countries have issued guidelines regarding caffeine consumption and the disclosure of caffeine amounts in beverages and foods. Health Canada

Table 19.1 Recommended intake of caffeine.

Country	Population	*Recommended Maximum Intake*
Canada	Ages 4–6 Years	Less than 45 mg per day
	Ages 7–9 Years	Less than 62 mg per day
	Ages 10–12 Years	Less than 85 mg per day
	Adults	Less than 400 mg per day
	Women of Childbearing Age	Less than 300 mg per day
United Kingdom, United States	Maternal Consumption During Pregnancy	Less than 200 mg per day
United States	Breastfeeding Women	Has an upper limit of 200–300 mg per day

Note. Information obtained from the AAP, ACOG, Pregnant Women Advised to Limit Caffeine Consumption, and Caffeine and Your Health.

(2010) has adopted 2.5 mg kg^{-1} per day as a recommended daily limit on caffeine intake for children under 12. In addition, they recommend no more than 400 mg of caffeine per day in healthy adults, 300 mg per day in women of childbearing age, and no more than 200 mg per day during pregnancy (Health Canada 2010).

Canadian guidelines recommend that children aged 4–6 consume no more than 45 mg per day. This is roughly equivalent to the amount of caffeine found in a 12 oz (355 ml) can of cola, or approximately four 1.5 oz (43 g) milk chocolate bars (Health Canada 2010). Permissible caffeine consumption, per Canadian guidelines, for 7–9 year-old and 10–12 year-old children is higher, with suggested upper limits of 62 mg per day and 85 mg per day, for each group, respectively. Finally, the European Union requires that beverages containing more than 150 mg of caffeine per liter be labeled with the term "high caffeine content." In addition, the label must provide the amount of caffeine per 100 ml of product (Food Standards Agency 2010) to assist consumers in making more informed decisions regarding caffeine intake.

19.9 Future Research

A number of shortcomings in the current knowledge base are readily apparent. (1) Current data on caffeine consumption in children and teens is lacking and does not account for numerous variables of potential importance such as age, developmental stage, or various demographic factors such as gender, socio-economic status and ethnicity. (2) Energy drinks are a new and quite popular product that is marketed directly to children and teenagers. It is known that this population represents approximately one-half of the consumers of such products. Yet, even the most recent data on caffeine consumption predate the introduction of these products into the market. (3) The majority of data collected to date on the effects of caffeine on children and teenagers rely largely on laboratory-based designs using bolus dosing of caffeine with subsequent examination of these effects. This type of design, although informative, does not provide information that can readily be generalized to the day-to-day impact of caffeine on children and teens. (4) No studies to date have examined the long-term health consequences (*e.g.,* cardiac, neurologic, behavioral) of chronic caffeine use by children and teenagers.

What is currently known, based on recent small-scale studies, warrants further examination. Presently, it appears that a number of children who consume caffeine on a regular basis, based on parental report, have some difficulty with sleep. A majority of these children (ages 5–12 years old) fail to get the recommended amount of sleep each night. The impact of sleep deprivation on children and teens is not known; however, it likely has implications for day-to-day performance, behavior, and health. Thus, this relationship requires further explanation and caffeine use should be explored in youngsters presenting with sleep difficulties.

In addition, caffeine appears to play a role in the development of obesity and type 2 diabetes; however, this relationship is clearly moderated by additional dietary factors. Caffeine is often consumed as part of a diet that is high in sugar and lower in milk, vegetables, and other nutritive foods (Temple 2009). This may be due to the mechanisms of caffeine delivery (*e.g.,* in soda) or it may simply reflect poor dietary choices in children and youth. It also has been hypothesized that caffeine might potentiate sensitivity to sugar and, because of its dopaminergic effects, increase the reinforcing properties of sugary foods and drinks (Temple 2009). None of these premises has been sufficiently examined.

Although necessary to fully understand the potential impact of caffeine on health and development, collecting and examining data on caffeine consumption in young children, older children, and teens is not a straightforward undertaking. The primary concern is the collection of accurate data, which is necessary, but also difficult. Most data are collected using a self-report format that is subject to a number of biases. In addition, it is likely that many consumers are unaware of which products, other than the usual coffee, tea, cola, and energy drinks, contain caffeine, and in what amounts, as such information is not readily available on product labeling in the US. Therefore, consumers and parents may be unaware of exactly how much caffeine is in a product and therefore are unable to accurately report quantitative consumption data. Without accurate consumption data, it is difficult to determine the impact of chronic caffeine use on children and teens. There is a plethora of data indicating that caffeine, in moderate doses, is safe in otherwise healthy adults. Nevertheless, it would be ill-advised to assume that caffeine will have the same benign effects in children as it does in adults. The brain undergoes substantial change and maturation between birth and adulthood and it would be unwise to assume that substances with known psychoactive effects have exactly the same effects in the immature, the developing, and the mature brain.

Finally, the EU, New Zealand, and other countries have an upper limit of caffeine content for foods and beverages, yet no such limitations are currently in place in the US. Additionally, the US lacks requirements for labeling and

Table 19.2 Limitations in caffeine literature.

Key Areas	Limitations
Participant information	Current literature does not account for age, developmental stage, gender, SES, ethnicity, and other variables of interest.
Data relevance	Most recent literature predates the widespread availability of energy drinks and caffeine added foods.
Accuracy of variables	Poor ability to collect consumption data as data are recall based and caffeine amounts are not disclosed in food or beverage packaging.
Generalization	Data examining impacts on cognitive function and performance were laboratory based and not representative of daily life. Caffeine naive participants were not used.
Longitudinal data	No long-term data to examine health impacts. Studies doing follow up did so at 1 year or less, which was prior to adulthood.

disclosure of caffeine content in food products. The FDA has yet to provide recommendations for caffeine consumption by children despite the fact that many other countries have done so. The FDA has yet to formally acknowledge recommendations made by the AAP and the ACOG for caffeine consumption in breastfeeding mothers and pregnant women. Clearly, more information needs to be provided to parents and consumers if individuals are to make informed decisions about caffeine consumption for themselves and their children.

Summary Points

- Caffeine is the most commonly used psychoactive substance and studies have indicated that caffeine consumption by children and adolescents is on the rise.
- Special concern has been raised about the consumption of energy drinks, which contain 4–6 times the amount of caffeine in soft drinks, and are frequently marketed directly to children and adolescents.
- Deleterious performance in children has been observed at 10 mg kg^{-1} including increased fidgeting, restlessness, hyperactivity, and difficulty sleeping.
- Recent research concluded that 0.3 mg kg^{-1} per day was the highest level of caffeine ingestion at which no immediate adverse effects were observed in children and adolescents.
- Sleep disruption, common in children who consume caffeine, is perhaps one of the most empirically supported health-related consequences of chronic caffeine use.
- There is little research with caffeine and increased enuresis with children but one recent study suggested that there was not a significant relationship.

- Generally speaking, lower amounts of caffeine consumed by lactating mothers do not appear to affect infants, and are considered safe provided the child is healthy.
- The American Academy of Pediatrics recommends only moderate amounts of caffeine be consumed by breast-feeding women with the upper limit between 200–300 mg per day.
- Withdrawal symptoms have been observed in newborns whose mothers regularly consumed >400 mg per day of caffeine but the effects of caffeine on infants appear to be time-limited and symptoms remit when caffeine levels decrease.
- Caffeine is considered one of the safest and most cost-effective interventions in the treatment and prevention of apnea of prematurity.
- Several countries have issued guidelines regarding caffeine consumption and the European Union requires that beverages containing more than 150 mg of caffeine be labeled "high caffeine."
- Current data regarding caffeine consumption in children and teens is lacking and does not sufficiently account for numerous variables such as developmental stage, gender, socio-economic status, or ethnicity and no studies have examined the long-term health consequences of chronic caffeine use by children and adolescents.

Key Facts

Trends in Caffeine Consumption

- Caffeine is the most commonly used psychoactive substance in the world. It is most commonly consumed as coffee, tea, and cola drinks; however, a new class of beverages called "energy drinks," that contain up to three times the caffeine in a typical cola drink, are becoming increasingly popular among children. Caffeine is not a heavily regulated substance, and although some governments have promulgated guidelines for maximum daily intake, the US Food and Drug Administration has not done so.
- Caffeine drinks are heavily marketed to children and adolescents, who have radically increased their average daily intake of caffeine over the last three decades. Younger people (under 25 years old) currently comprise half the market for energy drinks. In the US, there has recently been an effort to limit the availability of caffeine drinks in schools, but this has had only a limited affect on caffeine consumption among this population.

Effects on children

- The effects of caffeine on the central nervous system have been well investigated in adults and adolescents, but much less so in young children. Low to moderate doses seem to improve attention, memory, and reaction time performance in children, but higher doses are associated with deleterious effects including increased fidgeting, restlessness, hyperactivity, and difficulty sleeping. Very high doses can result in tachycardia, vomiting, and diuresis, but caffeine toxicity is rarely fatal. It is known that caffeine can cause arrhythmias and changes in blood pressure in adults, but these effects have not been studied in children.

- Dose-dependent performance improvement in tests of attention was reported in children receiving caffeine versus a placebo. However, studies have not shown caffeine to be useful in moderating symptoms of ADHD. Caffeine has also been shown to have deleterious effects on mood and affect.

- The American Academy of Pediatrics recommends only moderate levels of caffeine be consumed by breast-feeding women with the upper limit between 200–300 mg per day. In these amounts, caffeine *via* breast milk does not appear to affect healthy infants.

Definitions of Words and Terms

AAP (American Academy of Pediatrics): A professional organization in the US, representing *pediatricians* (physicians specializing in the health of children).

Aminophylline: A drug typically used as a *bronchodilator* (to open up narrowed or obstructed *bronchi* and *bronchioles*, the tubes through which air passes from the throat to the lungs). Aminophylline is a combination of *theophylline* and ethylenediamine.

ACOG (American College of Obstetricians and Gynecologists): A professional organization in the US, representing *obstetricians* (physicians specializing in gestation and birth) and *gynecologists* (physicians specializing in the female reproductive system).

ADHD (Attention Deficit / Hyperactivity Disorder): A mental health diagnosis, most typically made in childhood, that denotes difficulties with focusing and/or sustaining attention, and/or difficulties related to an unusually high level of motor activity. It is critical to note that a diagnosis of ADHD may denote problems with attention *or* problems with hyperactivity *or* both.

Anxiety: A psychological state characterized mainly by a subjective sense of fear and/or worry about an unpleasant future event. Anxiety also has physiological symptoms, such as muscle tension, increased heart rate, and/or increased blood pressure.

Apnea: The stoppage of breathing. Typically the term *apnea* refers to involuntary stoppage of breathing, as in *sleep apnea*, which occurs during sleep, or *apnea of prematurity*, which occurs in premature infants with poor muscle or nervous system development.

Arrhythmia: Any abnormal pattern of heartbeat, typically due to problems with the electrical signals that regulate heartbeat.

Bradycardia: An abnormally slow heart rate, typically defined as less than 60 beats per minute in an adult patient who is resting.

Bronchopulmonary dysplasia: A type of lung injury most common among premature infants. It is characterized by inflammation and swelling in the lungs.

Caffeine: An alkaloid, most typically found in plants, that is used as a stimulant drug. It is the world's most widely used *psychoactive drug* (drug with effects on thinking, emotions, and/or behavior).

CDC (Centers for Disease Control and Prevention): A US government agency charged with protecting public health and safety by providing information to the public about disease prevention and by coordinating efforts to contain outbreaks of contagious disease.

Circadian: Relating to the natural daily cycle (roughly 24 hours long) of biochemical processes that is present in many life forms.

CNS (Central nervous system): The brain and spinal cord.

CNS arousal: A condition involving increased blood pressure, increased heart rate, and an increased ability to perceive and react to stimuli.

Depression: A psychological state characterized mainly by negative emotions (*e.g.*, sadness), negative thoughts (*e.g.*, guilt, thoughts of worthlessness, thoughts of suicide), and a slowing down or absence of behaviors (*e.g.*, not participating in activities that were formerly found to be pleasant.)

Detrusor instability: An instability of the *detrusor urinae muscle*, which is the primary muscle in the bladder that contracts in order to expel urine and relaxes in order to allow the bladder to fill with urine.

Diuresis: Increased production of urine, typically accompanied by increased urination.

Dopaminergic effects: Effects in the CNS focused on receptors sensitive to the neurotransmitter *dopamine*. Dopaminergic effects often involve learning and reward-seeking behavior.

Energy drink: A broad class of beverages sold on the basis of claims about their ability to promote energy and wakefulness. Energy drinks typically have large amounts of caffeine and sugars, but also typically include other additives which are promoted as energy-producing.

Food Standards Agency (UK): A department of the government of the United Kingdom, charged with regulating products sold as food in order to protect public health and safety.

Food and Drug Administration (US): An agency of the United States federal government, charged with regulating products sold as food and as drugs in order to protect public health and safety.

Health Canada: A department of the federal government of Canada, charged with protecting the public health.

Melatonin: A chemical that is naturally present in the brain and which is involved in the regulation of the *circadian* rhythm.

Motor effects: Effects on movement of skeletal muscles, typically mediated through the motor cortex of the brain.

Psychomotor speed: The relationship between perceiving a stimulus and reacting to it via motor movement.

Temperament: The aspects of personality that are innate, *i.e.*, not learned.

Theophylline: A drug typically used as a *bronchodilator* (to open up narrowed or obstructed *bronchi* and *bronchioles*, the tubes through which air passes from the throat to the lungs). See also *Aminophylline*.

List of Abbreviations

AAP	American Academy of Pediatrics
ACOG	American College of Obstetricians and Gynecologists
CNS	central nervous system
CDC	Centers for Disease Control and Prevention (USA)
EU	European Union
FDA	Food and Drug Administration (US)
mg	milligrams
mg kg^{-1}	milligrams per kilogram (of body weight)
mg kg^{-1} per day	milligrams per kilogram per day
oz	ounce

Acknowledgements

The authors gratefully acknowledge Christine Majors for her editorial contributions to this manuscript.

References

American Academy of Pediatrics Committee on Drugs. 2001. The transfer of drugs and other chemicals into human milk. Journal of Pediatrics. 108: 776–789.

American Congress of Obstetricians and Gynecologists (ACOG) Committee Opinion No. 462. 2010. Moderate caffeine consumption during pregnancy. American Journal of Obstetrics and Gynecology. 116: 467–468.

Aranda, J., Beharry, K., Valencia, G., Natarajan, G. and Davis, J., 2010. Caffeine impact on neonatal morbidities. Journal of Maternal-Fetal and Neonatal Medicine. 23: 20–23.

Armstrong, L. 2002. Caffeine, body fluid-electrolyte balance, and exercise performance. International Journal of Sport Nutrition and Exercise Metabolism. 12: 189–206.

Bancalari, E. 2006. Caffeine for apnea of prematurity. The New England Journal of Medicine. 354: 2179–2181.

Bernstein, G., Carroll, M., Crosby, R., Perwien, A., Go, F. and Benowitz, N., 1994. Caffeine effects on learning, performance, and anxiety in normal school-age children. Journal of the American Academy of Child and Adolescent Psychiatry. 33: 407–415.

Bernstein, G., Carroll, M., Dean, N., Crosby, R., Perwien, A. and Benowitz, N., 1998. Caffeine withdrawal in normal school-age children. Journal of the American Academy of Child and Adolescent Psychiatry. 37: 858–865.

Bramstedt, K. 2007. Caffeine use by children: the quest for enhancement. Substance Use and Misuse. 42: 1237–1251.

Cannon, M., Cooke, C. and McCarthy, J., 2001. Caffeine-induced cardiac arrhythmia: an unrecognised danger of healthfood products. Medical Journal of Australia. 174: 520–521.

Centers for Disease Control and Prevention (CDC). 2010. Sleep and sleep disorders: how much sleep do I need? Available at: http://www.cdc.gov/sleep/how_much_sleep.htm. Accessed 8 June 2010.

Chopra, A. and Morrison, L., 1995. Resolution of caffeine-induced complex dysrhythmia with procainamide therapy. Journal of Emergency Medicine. 13: 113–117.

Comer, A., Perry, C. and Figgitt, D., 2001. Caffeine citrate: a review of its use in apnoea of prematurity. Paediatric Drugs. 3: 61–79.

Curatolo, P. and Robertson, D., 1983. The health consequences of caffeine. Annals of Internal Medicine. 98: 641–653.

Durlach, P. 1998. The effects of a low dose of caffeine on cognitive performance. Psychopharmacology. 140: 116–119.

Edelstein, B., Keaton-Brasted, C. and Burg, M., 1984. Effects of caffeine withdrawal on nocturnal enuresis, insomnia, and behavior restraints. Journal of Consulting and Clinical Psychology. 152: 857–862.

Elkins, R., Rapoport, J., Zahn, T., Buchsbaum, M., Weingartner, H., Kopin, I., Langer, D. and Johnson, C., 1981. Acute effects of caffeine in normal prepubertal boys. American Journal of Psychiatry. 138: 178–183.

Ferré, S., Jensen, M., Kempf, K., Martin, S., Kolb, H., Temple, J., Nardi, A., O'Connor, P., Simola, N., Morelli, M., Syikis, D., Lara, D., Hughes, R. and Killgore, W., 2011. What do you see as the main priorities, opportunities, and challenges in caffeine research in the next five years? Journal of Caffeine Research. 1: 5–12.

Food Standards Agency. 2010. High caffeine energy drinks. Available at: http://www.food.gov.uk/safereating/chemsafe/energydrinks. Accessed 12 October 2011.

Food Standards Agency. 2008. Pregnant women advised to limit caffeine consumption. Available at: http://www.food.gov.uk/news/newsarchive/2008/nov/caffeinenov08. Accessed 12 October 2011.

Frary, C., Johnson, R. and Wang, M., 2005. Food sources and intakes of caffeine in the diets of persons in the United States. Journal of the American Dietetic Association. 105: 110–113.

Fredholm, B., Bättig, K., Holmén, J., Nehlig, A. and Zvartau, E., 1999. Actions of caffeine in the brain with special reference to factors that contribute to its widespread use. Pharmacological Reviews. 51: 83–133.

Goldstein, I. and Shapiro, D., 1987. The effects of stress and caffeine on hypertensives. Psychosomatic Medicine. 49: 226–235.

Goldstein, A. and Wallace, M., 1997. Caffeine dependence in schoolchildren? Experimental Clinical Psychopharmacology. 5: 388–392.

Gray, P., Flenady, V., Charles, B. and Steer, P., 2011. Caffeine citrate for very preterm infants: Effects on development, temperament and behaviour. Journal of Paediatrics and Child Health. 47: 167–172.

Health Canada. 2010. Caffeine and your health. Available at: http://www.hc-sc.gc.ca/fn-an/securit/factsfaits/caffeine-eng.php. Accessed 9 December 2009.

Higdon, J. and Frei, B., 2006. Coffee and health: a review of recent human research. Critical Reviews in Food Science and Nutrition. 46: 101–123.

Hildebrandt, R. and Gundert-Remy, U., 1983. Lack of pharmacological active saliva levels of caffeine in breast-fed infants. Pediatric Pharmacology. 3: 237–244.

Jalkut, M., Lerman, S. and Churchill, B., 2001. Enuresis. Pediatric Clinics of North America. 48: 1461–1488.

James, J. 2011. A new journal to advance caffeine research. Journal of Caffeine Research. 1: 1–3.

James, J., Sawczuk, D. and Merrett, S., 1989. The effect of chronic caffeine consumption on urinary incontinence in psychogeriatric inpatients. Psychology & Health. 3: 297–306.

Knight, C., Knight, I., Mitchell, D. and Zepp, J., 2004. Beverage caffeine intake in US consumers and subpopulations of interests: Estimates from the share of intake panel survey. Food and Chemical Toxicology. 42: 1923–1930.

Larsen, P., Brendstrup, L., Skov, L. and Flachs, H., 1995. Aminphylline *versus* caffeine citrate for apnea and bradycardia prophylaxis in premature neonates. Acta Paediatrica. 84: 360–364.

Lieberman, H., Wurtman, R., Emde, G., Roberts, C. and Coviella, I., 1987. The effects of low doses of caffeine on human performance and mood. Psychopharmacology (Berl). 92: 308–312.

McGowan, J., Altman, R. and Kanto, W., 1988. Neonatal withdrawal symptoms after chronic maternal ingestion of caffeine. South Medical Journal. 81: 1092–1094.

Meltzer, H., Fotland, T., Alexander, J., Elind, E., Hallstrom, H., Lam, H., Liukkonen, K., Petersen, M. and Solbergsdottir, E., 2008. Risk assessment of caffeine among children and adolescents in the Nordic countries. TemaNord. 551.

Miller, M. 2005. What are caffeine's psychological benefits and risks? The Harvard Mental Health Letter / from Harvard Medical School. 22: 8.

Nawrot, P., Jordan, S., Eastwood, J., Rotstein, J., Hugenholtz, A. and Feeley, M., 2003. Effects of caffeine on human health. Food Additives and Contaminants. 20: 1–30.

Nehlig, A. and Debry, G., 1994. Consequences on the newborn of chronic maternal consumption of coffee during gestation and lactation: a review. Journal of the American College of Nutrition. 13: 6–21.

Rapoport, J., Berg, C., Ismond, D., Zahn, T. and Neims, A., 1984. Behavioral effects of caffeine in children. Relationship between dietary choice and effects of caffeine challenge. Archives of General Psychiatry. 41: 1073–1079.

Rapoport, J., Elkins, R., Neims, A., Zahn, T. and Berg, C., 1981a. Behavioral and autonomic effects of caffeine in normal boys. Developmental Pharmacology and Therapeutics. 3: 74–82.

Rapoport, J., Jensvold, M., Elkins, R., Buchsbaum, M., Weingartner, H., Ludlow, C., Zahn, T., Berg, C. and Neims, A., 1981b. Behavioral and cognitive effects of caffeine in boys and adult males. Journal of Nervous and Mental Disease. 169: 726–732.

Reissig, C., Strain, E. and Griffiths, R., 2009. Caffeinated energy drinks: a growing problem. Drug and Alcohol Dependence. 99: 1–10.

Ryu, J. 1985. Effect of maternal caffeine consumption on heart rate and sleep time of breast-fed infants. Developmental Pharmacology and Therapeutics. 8: 355–363.

Schmidt, B., Roberts, R., Davis, P., Doyle, L., Barrington, K., Ohlsson, A., Solimano, A. and Tin, W., 2006. Caffeine therapy for apnea of prematurity. New England Journal of Medicine. 354: 2112–2121.

Seicean, A., Redline, S., Seicean, S., Kirchner, H., Gao, Y., Sekine, M., Zhu, X. and Storfer-Isser, A., 2007. Association between short sleeping hours and overweight in adolescents: results from a US Suburban High School survey. Sleep and Breathing. 11: 285–293.

Seifert, S., Schaechter, J., Hershorin, E. and Lipshultz, S., 2011. Health effects of energy drinks on children, adolescents, and young adults. Journal of Pediatrics. 127: 511–528.

Skouroliakou, M., Bacopoulou, F. and Markantonis, S., 2009. Caffeine *versus* theophylline for apnea of prematurity: a randomised controlled trial. Journal of Paediatrics and Child Health. 45: 587–592.

Smit, H. and Rogers, P., 2000. Effects of low doses of caffeine on cognitive performance, mood and thirst in low and higher caffeine consumers. Psychopharmacology (Berl). 152: 167–173.

Soloveychik, V., Bin-Nun, A., Ionchev, A., Sriram, S. and Meadow, W., 2009. Acute hemodynamic effects of caffeine administration in premature infants. Journal of Perinatology. 29: 205–208.

Stein, M., Krasowski, M., Leventhal, B., Philips, W. and Bender, B., 1996. Behavioral and cognitive effects of methylxanthines. A meta-analysis of theophylline and caffeine. Archives of Pediatric and Adolescent Medicine. 150: 284–288.

Temple, J. 2009. Caffeine use in children: what we know, what we have left to learn, and why we should worry. Neuroscience and Biobehavioral Reviews. 33: 793–806.

Thomson, B. and Schiess, S., 2010. Risk profile: caffeine in energy drinks and energy shots. Institute of Environmental Science & Research Limited.

Warzak, W., Evans, S., Floress, M., Gross, A. and Stoolman, S., 2011. Caffeine consumption in young children. Journal of Pediatrics. 158: 508–509.

Wolfson, A. and Carskadon, M., 2003. Understanding adolescents' sleep patterns and school performance: a critical appraisal. Sleep Medicine Reviews. 7: 491–506.

Caffeine and Type 2 Diabetes

ATSUSHI GOTO[1] AND SIMIN LIU*[2]

[1] Program on Genomics and Nutrition, Department of Epidemiology, UCLA School of Public Health, Los Angeles, CA, USA; Center for Metabolic Disease Prevention, UCLA School of Public Health, Los Angeles, CA, USA; Department of Diabetes and Metabolic Medicine, National Center for Global Health and Medicine, Tokyo, Japan, 1-21-1 Toyama, Shinjuku-ku, Tokyo 162-8655, Japan; [2] Center for Metabolic Disease Prevention, University of California, Los Angeles, 650 Charles E. Young Drive South, Box 951772, Los Angeles, CA 90095-1772
*E-mail: siminliu@ucla.edu

20.1 Introduction

Type 2 diabetes (T2D) is a chronic disease with severe complications, such as neuropathies, blindness, renal failure, amputations, stroke, heart disease, and premature death (Joslin and Kahn 2005). Its prevalence is now reaching epidemic proportion worldwide particularly among minorities and those from developing countries (Shaw *et al* 2009). A large body of experimental and observational evidence indicates a role of caffeine in T2D development. This chapter provides an overview of the available evidence relating coffee intake and caffeine to T2D risk.

20.2 Type 2 Diabetes: Pathophysiology and Risk Factors

Impaired insulin action and secretion play fundamental roles in the pathogenesis of T2D. Both genetic and environmental factors are considered

Food and Nutritional Components in Focus No. 2
Caffeine: Chemistry, Analysis, Function and Effects
Edited by Victor R Preedy
© The Royal Society of Chemistry 2012
Published by the Royal Society of Chemistry, www.rsc.org

Table 20.1   Risk factors of type 2 diabetes.

- obesity
- physical inactivity
- older age
- race–ethnicity
- history of gestational diabetes
- family history of diabetes

to modulate insulin action and secretion (Joslin and Kahn 2005). Because environmental factors are modifiable, considerable attention has been paid to their potential for T2D prevention. Obesity is the most important risk factor in the development of T2D that can be modified with comprehensive lifestyle interventions (Table 20.1). Nonetheless, potential beneficial effects of specific dietary components on diabetes risk are still not fully understood. A number of recent observational studies have indicated that promising modifiable lifestyle factors may include intake of caffeinated-coffee and/or caffeine. Coffee is one of the most frequently consumed beverages worldwide (FDA 2007), perhaps due in large part to the psychosomatic effects of caffeine and it has been reported that about 500 billion cups of coffee are consumed annually (Clarke and Vitzthum 2001).

20.3   Coffee Intake and Type 2 Diabetes

20.3.1   Epidemiologic Evidence

To date, a number of prospective cohort studies have examined the association between coffee and the risk of T2D (Goto *et al* 2011, Salazar-Martinez *et al* 2004, Tuomilehto *et al* 2004, van Dam *et al* 2006). In the most recent meta-analysis and systematic review that summarizes findings from 18 studies involving 457 922 men and women, those who drank 3–4 cups per day had ∼25% lower T2D risk compared to those drinking 2 or fewer cups per day (relative risk [RR], 0.76; 95% confidence interval [CI], 0.69–0.82). Further, an inverse log-linear association (RR, 0.93 [0.91–0.95] per every additional cup intake of coffee) was observed, suggesting a dose-response relationship between coffee consumption and the risk of T2D (Huxley *et al* 2009).

20.3.2   Potential Mechanisms

Although the exact biological mechanism responsible for the inverse association between coffee consumption and T2D remains to be established, several mechanistic explanations have been proposed. Coffee contains a myriad of potentially beneficial components, including caffeine, cafestol, kahweol, potassium, niacin, magnesium, tocopherols, and phenol chlorogenic acid, many of which are biologically active (Table 20.2) (Ranheim and

Table 20.2   Major components of coffee.

- caffeine
- cafestol
- kahweol
- potassium
- niacin
- magnesium
- tocopherols
- phenol chlorogenic acid

Halvorsen 2005). One cup (8 fluid ounces) of regular coffee contains approximately 7 mg of magnesium (USDA 2011) which by acting as a cofactor in many enzymatic reactions may lower T2D risk (Song *et al* 2004). However, several prospective studies found that further adjusting for magnesium did not materially change the relative risk (RR) estimates for coffee consumption and T2D risk (Iso *et al* 2006), suggesting that magnesium alone cannot completely explain the coffee-diabetes relation. Coffee also contains phytoestrogens, which has been suggested to explain the possible protective effect of coffee (Tuomilehto *et al* 2004). Further, coffee is a rich source of chlorogenic acid, a biologically active phenol. Chlorogenic acid may have beneficial effects on glucose metabolism through its anti-oxidant property, and inhibition of hepatic glucose-6-phosphatase and intestinal glucose absorption (van Dam and Hu 2005). However, its biological effects in the human body are not well-understood, and chlorogenic acid may not fully explain the possible protective effect of coffee on T2D risk.

   Caffeine is another component of coffee which has been proposed to have potential protective effect on glucose metabolism. Caffeine (1,3,7-trimethyl-xanthine) is a xanthine alkaloid rich in coffee, tea, and chocolates derived from cocoa beans. Given its well-defined chemistry and ease of implementation in a clinical setting, we herein provide an assessment of the potential role of caffeine intake in the development of T2D risk and glucose metabolism.

20.4   Caffeine Intake and Type 2 Diabetes Risk

20.4.1   Epidemiologic Evidence

Several prospective cohort studies have examined the association between intake of caffeine and T2D risk. Data from prospective cohort studies are summarized in Table 20.3. Most but not all studies have found an inverse association between caffeine intake and T2D risk. In a study conducted in United States, Salazar-Martinez and colleagues showed that adults in the highest quintile of caffeine intake (median, 708 mg per day among women and 566 mg per day among men) had a 17–30% lower risk of diabetes than those in the lowest quintile of intake (median, 13 mg per day among women and 69 mg per day among men) (Salazar-Martinez *et al* 2004). A prospective study conducted in France also reported that women in the highest quintile of caffeine intake (mean, 397 mg per day) compared with the lowest quintile (mean, 48 mg per day) had 33% lower risk of developing T2D (Sartorelli *et al* 2010). A study from Japan also found an inverse association between caffeine intake and T2D risk (Iso *et al* 2006). Interestingly, the authors speculated that the observed inverse associations of coffee and green tea with T2D risk were mainly due to the association between caffeine intake and T2D risk because green tea and coffee are both major sources (approximately 45% for each) of caffeine in Japan.

These observations should be interpreted with the caution that residual confounding by unmeasured or unknown confounders could explain the inverse relation between caffeine or coffee consumption and T2D risk. For example, unhealthy lifestyle is a potential confounding factor that is often difficult to measure and accounted for in observational epidemiology. However, caffeine or coffee consumption is usually associated with unhealthy life style. Therefore, further adjusting for unhealthy lifestyle would strengthen the association between caffeine or coffee and T2D risk.

Of note, as shown in Table 20.3, the inverse association between caffeine intake and T2D risk appears to be stronger in women than men. Further, caffeine intake has been positively associated with circulating levels of sex hormone binding globulin (SHBG) in several studies (Goto *et al* 2011). Circulating SHBG levels have been reported to be higher in women than men, and strongly and positively associated with T2D risk (Ding *et al* 2009). In a recent study of post-menopausal women enrolled in the US, women consuming more than 500 mg per day had 44% lower risk of T2D compared with 50 mg or less per day of caffeine (Goto *et al* 2011). Additionally, the inverse association of caffeine intake with risk of T2D was markedly attenuated after adjusting for SHBG levels (Goto *et al* 2011). Also, in recent two nested case-control studies using a Mendelian randomization analysis, it has been demonstrated that genetically determined lower SHBG levels were associated with higher T2D risk (Ding *et al* 2009). Using two single nucleotide polymorphisms (SNPs) in the SHBG gene (rs6257 and rs6259) as instruments, the Mendelian randomization analysis revealed that RRs for T2D per standard-deviation increase in the plasma levels of SHBG were 0.28 (95% CI, 0.13–0.58) among women and 0.29 (95% CI, 0.15–0.58) among men (Ding *et al* 2009). The findings have been subsequently confirmed in a large consortium of case-control studies (Perry *et al* 2010). Taken together, these findings support the notion that caffeine intake may affect T2D risk *via* alteration of SHBG metabolism, and SHBG may account for the sex-difference in the inverse association between caffeine intake and T2D risk.

20.4.2   Experimental Evidence

The presence of adenosine A1-receptors in the INS-1 cells, a pancreatic β-cell line, has been demonstrated by Western blotting (Topfer *et al* 2008). It is

Table 20.3 Cohort studies of caffeine consumption and type 2 diabetes risk.

Source	Country	Sex	Age (years)	N/n cases	Diabetes ascertainment	Follow-up (years)
Goto *et al* 2011 (WHS)	United States	F	≥45	718/359	Self-report	10 (median)
Sartorelli *et al* 2010 (E3N/EPIC)	France	F	41–71	1415/69532	Self-report	11 (mean)
Oba *et al* 2010 (Takayama study)	Japan	F	35–70	7643/175	Self-report	11 years (1992–2002)
Oba *et al* 2010 (Takayama study)	Japan	M	35–70	5897 /278	Self-report	11 years (1992–2002)
van Dam *et al* 2006 (NHSII)	United states	F	26–46	88 259/1,263	Self-report	10 years (1991–2001)
Iso *et al* 2006 (JACC study)	Japan	F	40–65	10 686/213	Self-report	5 years
Iso *et al* 2006 (JACC study)	Japan	M	40–65	6727/231	Self-report	5 years
Greenberg *et al* 2005 (NHANES)	United states	F and M	32–88	7006/309	Self-report	8.4 years
Salazar-Martinez *et al* 2004 (NHS)	United states	F	30–55	84 276/ 4085	Self-report	18 years (1980–1998)
Salazar-Martinez *et al* 2004 (HPFS)	United states	M	40–75	41 934/ 1333	Self-report	12 years (1986–1998)

F female; M male; RR relative risk; WHS Women's Health Study; E3N/EPIC Etude Epidémiologique auprès de Femmes de la Mutuelle Générale de l'Education Nationale/ European Prospective Investigation into Cancer and Nutrition; NHSII Nurses' Health Study II; JACC Japan Collaborative Cohort; NHANES National Health and Nutrition Examination Survey; NHS Nurses' Health Study; HPFS Health Professionals Follow-Up Study.

Table 20.3 (*Continued*)

Adjustments	Caffeine consumption (mg)	Multivariable adjusted RR (95% CI)
age, race, duration of follow-up, time of blood draw, smoking, physical activity, family history of diabetes, alcohol, total calories, and BMI	≤50	1.00 (Reference)
	51–250	1.00 (0.56–1.82)
	251–500	1.26 (0.68–2.32)
	500	0.56 (0.27–1.15)
age, family history of diabetes, physical activity, alcohol, education, hypercholesterolemia, hypertension, smoking, fiber, saturated fat, total energy, menopausal status, hormone replacement therapy, oral contraceptives use, and BMI	Quartiles	
	48 ± 25	1.00 (Reference)
	126 ± 21	0.84 (0.72, 0.96)
	209 ± 28	0.77 (0.66, 0.89)
	397 ± 150 (mean ± SD)	0.67 (0.58, 0.78)
age, smoking status, BMI, physical activity, education, alcohol, total energy intake, fat intake and menopausal status	Tertiles	
	1st tertile	1.00 (Reference)
	2nd tertile	1.26 (0.88–1.82)
	3rd tertile	0.95 (0.63–1.43)
age, smoking status, BMI, physical activity, education, alcohol, total energy intake, and fat intake	Tertiles	Tertiles
	1st tertile	1.00 (Reference)
	2nd tertile	0.98 (0.60–1.10)
	3rd tertile	0.95(0.69–1.30)
age, smoking status, BMI, physical activity, alcohol, hormone replacement therapy, oral contraceptive use, family history of type 2 diabetes, history of hypertension, history of hypercholesterolemia, consumption of sugar-sweetened soft drinks and punch, processed meat consumption, polyunsaturated-to-saturated fat intake ratio, total energy intake, the glycemic index, and cereal fiber intake	Quintiles	
	22	1.00 (Reference)
	93	0.88 (0.75–1.05)
	180	0.89 (0.75–1.05)
	341	0.74 (0.62–0.89)
	528 (median)	0.55 (0.45–0.67)
age, body mass index, family history of diabetes, smoking status, alcohol intake, magnesium intake, hours of walking and exercise	Quintiles	
	57	1.00 (Reference)
	137	0.86 (0.56–1.32)
	196	0.81 (0.54–1.22)
	267	0.67 (0.43–1.05)
	399 (median)	0.52 (0.32–0.84)
age, body mass index, family history of diabetes, smoking status, alcohol intake, magnesium intake, hours of walking and exercise	Quintiles	
	57	1.00 (Reference)
	137	1.23 (0.77–1.95)
	203	1.22 (0.77–1.91)
	274	1.47 (0.94–2.31)
	480 (median)	0.85 (0.51–1.42)
per-capita income, education, race, gender, physical activity, smoking status, alcohol consumption, BMI, age, and type of diet.	≤150	
	150–300	1.00 (Reference)
	300–450	0.69 (0.42–1.11)
	450–600	0.73 (0.38–1.41)
	>600	0.65 (0.30–1.42)
age, total caloric intake, family history of diabetes, alcohol consumption, smoking status, menopausal status and postmenopausal hormone use, intakes of glycemic load, trans fat, polyunsaturated fatty acid, cereal fiber, and magnesium, BMI, and physical activity	Quintiles	
	69	1.00 (Reference)
	193	0.98 (0.90–1.07)
	328	0.86 (0.78–0.94)
	432	0.81 (0.73–0.89)
	708 (median)	0.70 (0.63–0.77)
age, total caloric intake, family history of diabetes, alcohol consumption, smoking status, intakes of glycemic load, trans fat, polyunsaturated fatty acid, cereal fiber, and magnesium, BMI, and physical activity	Quintiles	
	13	1.00 (Reference)
	74	1.12 (0.94–1.33)
	172	1.09 (0.92–1.29)
	323	0.95 (0.80–1.13)
	566 (median)	0.83 (0.69–0.99)

possible that caffeine may directly stimulate insulin secretion *via* the antagonism of adenosine A1-receptors in the pancreatic β-cells. Several *in-vitro* studies and studies in rodent have documented that caffeine is a stimulant of pancreatic β-cells (Kagami *et al* 2008). One recent study in rats reported that caffeine may protect pancreatic β-cells against natural toxins such as streptozotocin (Kagami *et al* 2008). Moreover, several studies have shown that caffeine increases resting metabolic rate (Astrup *et al* 1990) and decreases body weight (Boozer *et al* 2002). In addition, studies have shown that after ingestion of caffeine (10mg kg^{-1}), lipid turnover increases by 2-fold. Also, thermic changes (13.3% increase), oxidative free fatty acids (FFA) disposal (44% increase), and non-oxidative FFA disposal (2.3-fold increase) were observed (Acheson *et al* 2004). Interestingly, it has been reported that the inverse association of coffee intake with T2D risk only applied to those who had previously lost weight (Greenberg *et al* 2005). Further, a prospective cohort study conducted in United States examined a relation between caffeine intake and 12-year weight change. The authors reported that participants who increased their caffeine consumption had lower mean body gain than those who decreased their caffeine consumption in both men and women (Lopez-Garcia *et al* 2006). These findings suggest the crucial role of weight loss in relating coffee intake to T2D risk.

It is important to note that several experimental studies indicated that caffeine may acutely impair insulin sensitivity. In a cross-over study, caffeinated-coffee consumption for 4 weeks increased fasting insulin levels compared with coffee abstinence, but did not reduce fasting glucose levels, raising concern of the acute impairment of insulin sensitivity (van Dam *et al* 2004). It is considered that caffeine acutely decreases insulin sensitivity in skeletal muscle *via* antagonism of adenosine A1 receptors in skeletal muscle and increasing sympathetic activity.

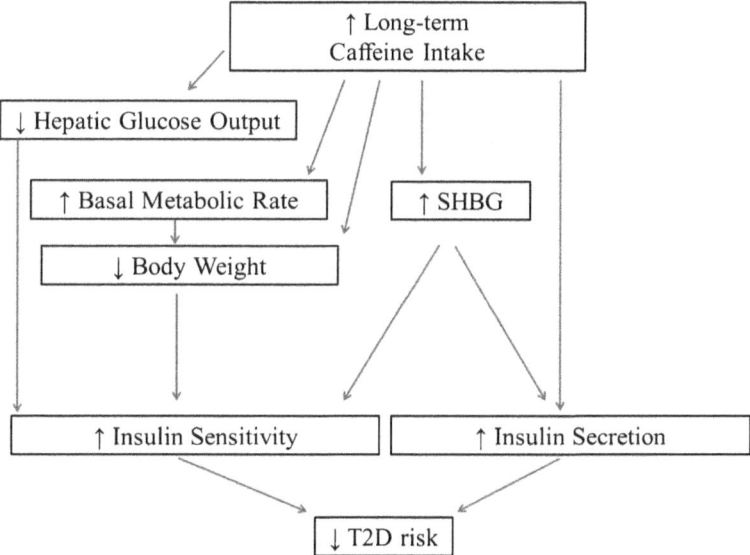

Figure 20.1 Potential mechanisms for the relation between long-term caffeine intake and type 2 diabetes risk.

However, its acute effect of caffeine on insulin sensitivity is considered to be tolerated after several days of caffeine use (Robertson *et al* 1981). Further, adenosine perfusion has been demonstrated to increase hepatic glucose production in dogs (McLane *et al* 1990), suggesting that antagonism of adenosine A1 receptors in the liver by caffeine can reduce hepatic glucose production. In a non-randomized intervention of 47 participants, caffeinated-coffee intake of 8 cups per day for 4 weeks led to favorable changes in inflammatory markers, including serum adiponectin and serum interleukin-18 (IL-18), although no changes were seen for markers of glucose metabolism (Kempf *et al* 2010). Taken together, these findings suggest that long-term caffeine intake may exert beneficial effects on insulin secretion and sensitivity through several mechanisms, including increasing resting metabolic rate, losing body weight, reducing hepatic glucose production, improving chronic inflammation, increasing insulin secretion, and increasing SHBG levels (Figure 20.1).

20.5 Conclusion

A large number of prospective cohort studies have reported that higher intakes of caffeinated-coffee and caffeine were consistently associated with reduced risk of T2D. However, much uncertainty exists concerning whether or not caffeine consumption can be used as a safe agent for T2D prevention. Recent molecular epidemiologic work has linked SHBG genotype and phenotype to the relation between coffee intake and T2D risk. This supports the view that caffeine does play a role in the development of T2D *via* altering SHBG metabolism. In some individuals, caffeine intake has also been linked to arrhythmias, dyslipidemia, pregnancy complications, and drug interactions (Higdon and Frei 2006). Currently, some authoritative body such as Health Canada reported that moderate caffeine intakes ($\sim$400 mg per day) were not associated with adverse effects such as general toxicity, cardiovascular effects, effects on bone status, changes in adult behavior, increased incidence of cancer, and effects on male fertility in healthy adults (Nawrot *et al* 2003). On balance, modest amount of caffeine intake ($\sim$300 mg per day for healthy adults) can be incorporated into a healthy lifestyle that includes exercise and weight reduction for diabetes prevention. Clearly, further mechanistic and intervention studies are needed to determine the long-term effect of caffeine intake on T2D risk.

Summary Points

- T2D is a chronic disease whose pathogenesis involves impaired insulin action and secretion.
- Both genetic and environmental factors are considered to modulate insulin action and secretion and thus diabetes risk.

- Large prospective studies of apparently healthy men and women have consistently shown that intake of caffeinated-coffee and/or caffeine is inversely related to T2D risk in a dose response manner.
- Coffee is one of the most frequently consumed beverages worldwide and about 500 billion cups of coffee are consumed annually.
- Coffee contains a myriad of components including caffeine, cafestol, kahweol, potassium, niacin, magnesium, tocopherols, and phenol chlorogenic acid.
- The inverse association between caffeine intake and T2D risk appears to be stronger in women than men.
- SHBG may account for caffeine's potential protective effect of caffeine on glucose metabolism.
- Caffeine may directly stimulate insulin secretion *via* the antagonism of adenosine A1-receptors in the pancreatic β-cells.
- Caffeine increases resting metabolic rate and decreases body weight.
- Caffeine might acutely decrease insulin resistance, whereas long-term caffeine intake could exert beneficial effects on insulin sensitivity through several mechanisms, including increasing resting metabolic rate, losing body weight, reducing hepatic glucose production, and improving chronic inflammation.

Key Facts

1. Key Facts of Insulin

- Insulin is a 51-amino acid peptide hormone that is produced and secreted by pancreatic β-cells.
- In the early 20th century, insulin was discovered by Dr. John Macleod, Dr. Frederick Banting, and Charles Herbert Best. Dr. Macleod and Dr. Banting received a Nobel prize.
- Insulin lowers blood glucose levels through facilitation of glucose absorption by the skeletal muscle cells, adipose tissue, and liver.

2. Key Facts of SHBG

- Sex hormone-binding globulin (SHBG) is a circulating glycoprotein synthesized and secreted by liver hepatocytes.
- SHBG has been thought to modulate the bioavailability of sex hormones *via* binding sex hormones with high affinity and regulating their accessibility to target cells.
- It has been recently discovered that the plasma membranes of various cell types are able to bind SHBG specifically and with high affinity, and SHBG mediates sex hormones signaling at the cell membrane through the SHBG receptors.
- This recognition of the function of SHBG as a mediator of a novel steroid-signaling system has drawn much interest to biologic effects of SHBG.

3. Key Facts of Sex-Differences in T2D Risk

- It has been found that the strong, positive association between adiposity and T2D risk is stronger in women than men.
- T2D is a stronger risk factor for coronary heart disease mortality in women than in men.
- It has been proposed that circulating levels sex-steroid hormones or SHBG may account for the sex-differences in T2D risk and coronary heart disease risk.

Definitions of Words and Terms

Type 2 diabetes: Type 2 diabetes (T2D) is a chronic disease with high blood glucose levels, characterized by impaired insulin secretion and decreased insulin sensitivity. T2D is accompanied by severe complications, such as neuropathies, blindness, renal failure, amputations, stroke, heart disease, and premature death. The most important risk factors for T2D include obesity and physical inactivity. There are a number of drugs available for treating T2D. However, given the global diabetes epidemic, non-pharmacologic strategies to prevent T2D are warranted.

Insulin sensitivity: Insulin sensitivity is the ability of endogenous or exogenous insulin to promote glucose clearance. In contrast, insulin resistance is defined as a subnormal response to endogenous or exogenous insulin.

Insulin secretion: Insulin secretion, or β-cell function, is the ability of pancreatic β-cell to produce and secrete insulin.

Adenosine: Adenosine is an endogenous molecule that modulates tissue function by activating four adenosine receptor subtypes (A1, A2A, A2B and A3). The effects of caffeine on the human body are primarily through the antagonism of adenosine receptors.

Prospective cohort study: Prospective cohort study is an observational epidemiologic study in which a defined group of individuals (the cohort) is prospectively followed over time.

Meta-analysis: Meta-analysis is a quantitative analysis of multiple studies, typically using statistical methods to combine results.

Systematic review: A systematic review is an overview of multiple studies with clear objectives, methods, and results.

Confounding: Confounding occurs when the exposure-outcome association is distorted simply because a distinct third factor (*i.e.,* confounder) is associated with both the exposure and the outcome. In this case, the exposure is merely a marker of the third factor but does not directly cause disease. Residual confounding occurs when confounder is misclassified, or there are unmeasured–unknown confounders.

Mendelian randomization analysis: Mendelian randomization analysis is an epidemiologic observational study designed to estimate a phenotype-disease association, using genetic variants (*e.g.,* single nucleotide polymorphisms

[SNPs]), phenotype (*e.g.,* blood glucose levels), and outcome (*e.g.,* developed T2D or not). Under several assumptions, Mendelian randomization analysis estimates a causal association between the phenotype and disease.

SNPs: SNPs (single nucleotide polymorphisms) are small genetic changes, or genetic variations that occur when a single nucleotide (A, T, C, or G) in the genome sequence is altered.

Adiponectin: Adiponectin is an inflammatory cytokine primarily synthesized and secreted by adipose tissue. While most inflammatory cytokines are considered to exert detrimental effects on the development of T2D, cardiovascular diseases, and cancer, adiponectin is believed to have beneficial health effects.

IL-18: IL-18 (interleukin-18) is a cytokine which is produced by inflammatory cells such as macrophages. IL-18 plays an active role in inflammatory reactions including acute and chronic inflammation. Because chronic inflammation impairs insulin sensitivity, higher levels of IL-18 might exert deleterious effects on glucose metabolism.

List of Abbreviations

T2D	Type 2 diabetes
RR	Relative risk
CI	Confidence interval
SHBG	Sex hormone binding globulin
FFA	Free fatty acids
SNPs	Single nucleotide polymorphisms
IL-18	Interleukin-18

References

Joslin, E. P. and Kahn, C. R., 2005. Joslin's diabetes mellitus. Lippincott Williams & Willkins, Philadelphia, Pa.

Shaw, J. E., Sicree, R. A. and Zimmet, P. Z., 2009. Global estimates of the prevalence of diabetes for 2010 and 2030. Diabetes Res. Clin. Pract.

US Food and Drug Administration (FDA). Medicines in my home: caffeine and your Body. Available at:http://www.fda.gov/downloads/Drugs/ResourcesForYou/Consumers/BuyingUsingMedicineSafely/UnderstandingOver-the-CounterMedicines/UCM205286.pdf. Accessed 23 Oct 2011.

Clarke, R. J. and Vitzthum, O. G., 2001. Coffee : recent developments. Blackwell Science, Oxford; Malden, MA.

Goto, A., Song, Y., Chen, B. H., Manson, J. E., Buring, J. E. and Liu, S., 2011. Coffee and caffeine consumption in relation to sex hormone-binding globulin and risk of type 2 diabetes in postmenopausal women. Diabetes. 60: 269–275.

Salazar-Martinez, E., Willett, W. C., Ascherio, A., Manson, J. E., Leitzmann, M. F., Stampfer, M. J. and Hu, F. B., 2004. Coffee consumption and risk for type 2 diabetes mellitus. Ann. Intern. Med. 140: 1–8.

Tuomilehto, J., Hu, G., Bidel, S., Lindstrom, J. and Jousilahti, P., 2004. Coffee consumption and risk of type 2 diabetes mellitus among middle-aged Finnish men and women. J. Am. Med. Assoc. 291: 1213–1219.

van Dam, R. M., Willett, W. C., Manson, J. E. and Hu, F. B., 2006. Coffee, caffeine, and risk of type 2 diabetes: a prospective cohort study in younger and middle-aged US women. Diabetes Care. 29: 398–403.

Huxley, R., Lee, C. M., Barzi, F., Timmermeister, L., Czernichow, S., Perkovic, V., Grobbee, D. E., Batty, D. and Woodward, M., 2009. Coffee, decaffeinated coffee, and tea consumption in relation to incident type 2 diabetes mellitus: a systematic review with meta-analysis. Arch. Intern. Med. 169: 2053–2063.

Ranheim, T. and Halvorsen, B., 2005. Coffee consumption and human health–beneficial or detrimental? –Mechanisms for effects of coffee consumption on different risk factors for cardiovascular disease and type 2 diabetes mellitus. Mol. Nutr. Food Res. 49: 274–284.

US Department of Agriculture. USDA National Nutrient Database for Standard Reference, Release 24. Available at:http://www.ars.usda.gov/SP2UserFiles/Place/12354500/Data/SR24/reports/sr24fg14.pdf. Accessed 23 Dec 2011.

Song, Y., Manson, J. E., Buring, J. E. and Liu, S., 2004. Dietary magnesium intake in relation to plasma insulin levels and risk of type 2 diabetes in women. Diabetes Care. 27: 59–65.

Iso, H., Date, C., Wakai, K., Fukui, M. and Tamakoshi, A., 2006. The relationship between green tea and total caffeine intake and risk for self-reported type 2 diabetes among Japanese adults. Ann. Intern. Med. 144: 554–562.

van Dam, R. M. and Hu, F. B., 2005. Coffee consumption and risk of type 2 diabetes: a systematic review. J. Am. Med. Assoc., 294: 97–104.

Sartorelli, D. S., Fagherazzi, G., Balkau, B., Touillaud, M. S., Boutron-Ruault, M. C., de Lauzon-Guillain, B. and Clavel-Chapelon, F., 2010. Differential effects of coffee on the risk of type 2 diabetes according to meal consumption in a French cohort of women: the E3N/EPIC cohort study. Am. J. Clin. Nutr. 91: 1002–1012.

Ding, E. L., Song, Y., Manson, J. E., Hunter, D. J., Lee, C. C., Rifai, N., Buring, J. E., Gaziano, J. M. and Liu, S., 2009. Sex hormone-binding globulin and risk of type 2 diabetes in women and men. N. Engl. J. Med. 361: 1152–1163.

Perry, J. R., Weedon, M. N., Langenberg, C., Jackson, A. U., Lyssenko, V., Sparso, T., Thorleifsson, G., Grallert, H., Ferrucci, L., Maggio, M., Paolisso, G., Walker, M., Palmer, C. N., Payne, F., Young, E., Herder, C., Narisu, N., Morken, M. A., Bonnycastle, L. L., Owen, K. R., Shields, B., Knight, B., Bennett, A., Groves, C. J., Ruokonen, A., Jarvelin, M. R., Pearson, E., Pascoe, L., Ferrannini, E., Bornstein, S. R., Stringham, H. M., Scott, L. J., Kuusisto, J., Nilsson, P., Neptin, M., Gjesing, A. P., Pisinger, C., Lauritzen, T., Sandbaek, A., Sampson, M., Zeggini, M. E., Lindgren, C. M., Steinthorsdottir, V., Thorsteinsdottir, U., Hansen, T., Schwarz, P., Illig, T., Laakso, M., Stefansson, K., Morris, A. D., Groop, L., Pedersen, O., Boehnke, M., Barroso, I., Wareham, N. J., Hattersley, A. T., McCarthy, M. I. and

Frayling, T. M., 2010. Genetic evidence that raised sex hormone binding globulin (SHBG) levels reduce the risk of type 2 diabetes. Hum. Mol. Genet. 19: 535–544.

Topfer, M., Burbiel, C. E., Muller, C. E., Knittel, J. and Verspohl, E. J., 2008. Modulation of insulin release by adenosine A1 receptor agonists and antagonists in INS-1 cells: the possible contribution of 86Rb+ efflux and 45Ca2+ uptake. Cell Biochem. Funct. 26: 833–843.

Kagami, K., Morita, H., Onda, K., Hirano, T. and Oka, K., 2008. Protective effect of caffeine on streptozotocin-induced beta-cell damage in rats. J. Pharm. Pharmacol. 60: 1161–1165.

Astrup, A., Toubro, S., Cannon, S., Hein, P., Breum, L. and Madsen, J., 1990. Caffeine: a double-blind, placebo-controlled study of its thermogenic, metabolic, and cardiovascular effects in healthy volunteers. Am. J. Clin. Nutr. 51: 759–767.

Boozer, C. N., Daly, P. A., Homel, P., Solomon, J. L., Blanchard, D., Nasser, J. A., Strauss, R. and Meredith, T., 2002. Herbal ephedra/caffeine for weight loss: a 6-month randomized safety and efficacy trial. Int. J. Obes. Relat. Metab. Disord. 26: 593–604.

Acheson, K. J., Gremaud, G., Meirim, I., Montigon, F., Krebs, Y., Fay, L. B., Gay, L. J., Schneiter, P., Schindler, C. and Tappy, L., 2004. Metabolic effects of caffeine in humans: lipid oxidation or futile cycling? Am. J. Clin. Nutr. 79: 40–46.

Greenberg, J. A., Axen, K. V., Schnoll, R. and Boozer, C. N., 2005. Coffee, tea and diabetes: the role of weight loss and caffeine. Int. J. Obes. 29: 1121–1129.

Lopez-Garcia, E., van Dam, R. M., Rajpathak, S., Willett, W. C., Manson, J. E. and Hu, F. B., 2006. Changes in caffeine intake and long-term weight change in men and women. Am. J. Clin. Nutr. 83: 674–680.

van Dam, R. M., Pasman, W. J. and Verhoef, P., 2004. Effects of coffee consumption on fasting blood glucose and insulin concentrations: randomized controlled trials in healthy volunteers. Diabetes Care. 27: 2990–2992.

Robertson, D., Wade, D., Workman, R., Woosley, R. L. and Oates, J. A., 1981. Tolerance to the humoral and hemodynamic effects of caffeine in man. J. Clin. Invest. 67: 1111–1117.

McLane, M. P., Black, P. R., Law, W. R. and Raymond, R. M., 1990. Adenosine reversal of *in vivo* hepatic responsiveness to insulin. Diabetes. 39: 62–69.

Kempf, K., Herder, C., Erlund, I., Kolb, H., Martin, S., Carstensen, M., Koenig, W., Sundvall, J., Bidel, S., Kuha, S. and Tuomilehto, J., 2010. Effects of coffee consumption on subclinical inflammation and other risk factors for type 2 diabetes: a clinical trial. Am. J. Clin. Nutr. 91: 950–957.

Higdon, J. V. and Frei, B., 2006. Coffee and health: a review of recent human research. Crit. Rev. Food Sci. Nutr. 46: 101–123.

Nawrot, P., Jordan, S., Eastwood, J., Rotstein, J., Hugenholtz, A. and Feeley, M., 2003. Effects of caffeine on human health. Food Addit. Contam. 20: 1–30.

CHAPTER 21
Caffeine and Apoptosis

WEINONG HAN[1] AND YU-YING HE*[1]

[1] Section of Dermatology, Department of Medicine, University of Chicago, 5841 S. Maryland Ave, MC5067, Chicago, IL 60637, USA
*E-mail: yyhe@medicine.bsd.uchicago.edu

21.1 Introduction

Caffeine (methylxanthine), a purine alkaloid, is a key component of many popular drinks, most notably tea and coffee. Therefore, caffeine is the most frequently ingested neuroactive substance in the world. For many years, caffeine has been generally believed to induce cell apoptosis, suppress cell proliferation, abolish chemical- or radiation-induced delays in cell cycle progression and enhance the toxicity of radiation and anti-cancer agents. Caffeine potentiates the lethal effects of toxic agents such as irradiation or alkylating agents (Busse *et al* 1978; Fingert *et al* 1986; Musk and Steel 1990). Caffeine has also been shown to inhibit ultraviolet B (UVB)-induced skin cancer in mice (Lu *et al* 2000; Lu *et al* 2002a). In contrast to the pro-apoptotic effect, caffeine has been found to be anti-apoptotic in neuronal cells (Nakaso *et al* 2008). These results suggest that the cytoprotective effect of caffeine is likely to be specific for neuroblastoma cells.

There are several critical pathways that are vital for effective tumor suppression. One of them is the tumor suppressor gene p53, which functions as a key component of a cellular emergency response system to induce cell growth arrest or apoptosis in response to DNA damage (el-Deiry *et al* 1993; Hartwell and Kastan 1994). Activation of p53 can result in either G1 cell cycle arrest or apoptosis, which contribute to suppression of the malignant transformation. The maintenance of genomic integrity and inactivation of p53 in cancers may

Food and Nutritional Components in Focus No. 2
Caffeine: Chemistry, Analysis, Function and Effects
Edited by Victor R Preedy
© The Royal Society of Chemistry 2012
Published by the Royal Society of Chemistry, www.rsc.org

be responsible for accelerated cell growth, increased cell survival, and resistance to genotoxic anticancer agents.

Following DNA damage, p53 has a major influence on whether a cell will live or die. Cells exposed to DNA-damaging agents such as γ-radiation or many chemotherapeutic agents pause in their cell cycle progression to allow time for DNA repair. If the damage is too extensive, the affected cells will undergo apoptosis. Both cell cycle arrest and apoptosis are normally mediated by p53 activities (Canman *et al* 1994; Enoch and Norbury 1995). One of the primary p53 targets is the Bax protein, a member of the Bcl-2 family that controls cell death through its participation in disruption of mitochondria and subsequent cytochrome c release (Marzo *et al* 1998).

In addition, the Ataxia-telangiectasia mutated (ATM) and ATM and RAD3-related (ATR) pathways play important roles in DNA damage response and tumor suppression. ATM and ATR belong to the phosphoinositide 3-kinase (PI3K)-related protein kinases (PIKKs). ATM responds primarily to double-strand breaks (DSBs) induced by ionizing radiation, while ATR reacts to UV or stalled replication forks. The list of ATR and ATM substrates is rapidly expanding; however, the best studied is the Ser–Thr kinase checkpoint kinase-1 (Chk1). Chk1 kinase is an important effector and substrate of ATR; it plays a key role in mediating replication fork stability following DNA damage and is also required for cell cycle checkpoint activation (Chen and Sanchez 2004). ATM activates another checkpoint protein, checkpoint kinase-2 (Chk2). These pathways coordinate the DNA damage checkpoint function and apoptosis. Defects in the ATR–Chk1 and ATM–Chk2 pathways increase cancer risk.

Another protein that is critical for tumorigenesis through regulating cell survival and growth is the serine–threonine kinase AKT, also known as protein kinase B. AKT is a downstream effector of phosphatidylinositol 3-kinase (PI3K) that has recently been a focus of intense research. It appears that AKT lies at the cross-roads of multiple cellular signaling pathways and acts as a transducer of many functions initiated by growth factors and other receptors that activate PI3K. One of the major activities of AKT is to promote cell survival. AKT is activated in response to UVB irradiation and is frequently activated in human cancers (Bode and Dong 2003; Bowden 2004; Ming *et al* 2010).

In this review, we will focus on the mechanistic action of caffeine on apoptosis *via* the p53, ATR–Chk1 and AKT–COX-2 pathways, respectively. We hope to establish the relationship between apoptosis and caffeine as a framework for a more general understanding of the function of caffeine.

21.2 Caffeine and p53

In serum-starved JB6 CL41 cells containing wild-type p53, caffeine induces apoptosis through p53, Bax, and caspase 3; in p53(-/-) fibroblast cells, caffeine did not enhance apoptosis (He *et al* 2003) (Figure 21.1). Both phosphorylation of p53 at Ser15 and p53 activity increased in JB6 Cl41 cells after cells were treated with increasing concentrations of caffeine. Phosphorylation of p53 at Ser15 is regarded as a critical step for p53-dependent transactivation. In addition, p53 is upstream of Bax, and Bax is a p53 downstream effector (Marzo *et al* 1998). p53 is not only a functional regulator of Bax but also can cause Bax overexpression in JB6 Cl41 and p53-/- fibroblast cells. Caspase 3 activation is also involved in the process of apoptosis induced by caffeine (He *et al* 2003). In a separate study, caffeine was reported to sensitize the human non-small cell lung cancer H358 cell line to p53-mediated apoptosis by inducing mitochondrial translocation and a conformational change in the Bax protein (Dubrez *et al* 2001).

The pro-apoptotic effect of caffeine can be p53-independent. The resistance of MCF-7 cells to ionizing radiation-induced apoptosis is caused by two independent events; one of them is a caffeine- or UCN-01-inhibitable event that does not depend on p53 or the release of a G(2)-M cell cycle arrest (Essmann *et al* 2004). The second event is the loss of caspase-3, which, surprisingly, seems essential for a fully functional caspase-9 pathway, even in the face of a previous release of mitochondrial proapoptotic proteins.

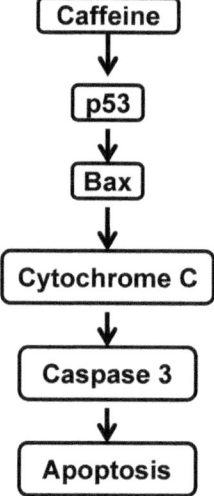

Figure 21.1 p53-dependent apoptosis induced by caffeine through the Bax and caspase 3 pathways. Caffeine can induce p53-dependent apoptosis through the Bax and caspase 3 pathways. The tumor suppressor gene, *p53*, functions as a key component of a cellular emergency response system to induce cell growth arrest or apoptosis. Caffeine induces p53 phosphorylation. p53 can cause overexpression of Bax. Bax can promote the cytosolic release of cytochrome *c*, which in turn, activates caspase 3, one of the key executioners of apoptosis. Activated caspase 3 cleaves its substrate and marks the beginning of DNA cleavage.

Studies on the mechanism(s) of the inhibitory effect of caffeine on UVB-induced carcinogenesis indicated a stimulatory effect on UVB-induced apoptosis in the epidermis (Lu *et al* 2000; Lu *et al* 2002a) and apoptosis in tumors (Lu *et al* 2002b) as well as a sunscreen effect (Lu *et al* 2007). Mechanistic studies indicate that administration of caffeine enhances UVB-induced apoptosis in mouse epidermis by increasing wild-type p53 (Lu *et al* 2000) as well as by a p53- and Bax-independent mechanism (Lu *et al* 2004).

Oral administration of caffeine to mice during chronic UVB irradiation inhibited the formation of patches with p53 mutation (Lu *et al* 2005). In addition, topical application of caffeine after stopping UVB in UVB-pretreated high risk mice decreased the number of patches when compared with control mice treated with acetone. It was of interest that patches from caffeine-treated mice had fewer types of mutations than patches from mice treated with water or tea (Kramata *et al* 2005). Whether there is a difference among cells harboring various homozygous p53 mutations in their sensitivity to caffeine remains to be investigated.

21.3 Caffeine and ATR–Chk1

Caffeine exposure sensitizes tumor cells to ionizing radiation and other genotoxic agents by inhibiting both ATR and ATR (Sarkaria *et al* 1999; Cortez 2003; Alao and Sunnerhagen 2009). Caffeine (in amounts in the millimolar range) can inhibit both ATR and the related ataxia-telangiectasia-mutated (ATM) kinase, both of which are central components of the DNA damage response. ATM is activated by DNA double-strand breaks, whereas ATR is preferentially activated by single-stranded regions of DNA, which can occur at stalled DNA replication forks or as intermediates during nucleotide excision repair of UV photoproducts (O'Driscoll *et al* 2003).

Caffeine is reported to inhibit the catalytic activity of both ATM and ATR at drug concentrations similar to those that induce radiosensitization (Hall-Jackson *et al* 1999). The kinase activity of ATR in the presence and absence of DNA is suppressed by caffeine. Cortez suggests that although caffeine is an inhibitor of ATM-ATR kinase activity *in vitro*, it can block checkpoints without inhibiting ATM-ATR activation *in vivo* (Cortez 2003).

The combined impact of caffeine and UVB radiation on keratinocytes has attracted great attention. Administration of caffeine enhances the removal of DNA-damaged cells by inhibiting the ATR-mediated phosphorylation of Chk1 and prematurely increasing the number of cyclin B1-containing cells that undergo lethal mitosis (Lu *et al* 2008b). UVB irradiation of mouse skin activated the ataxia-telangiectasia mutated- and Rad3-related (ATR) pathway, causing a several-fold increase in keratinocytes with phospho-Chk1 (Ser345) and a marked decrease in mitotic keratinocytes with cyclin B1. Caffeine administration reduced this mitotic delay to only 4 h and caused a marked increase in apoptosis by 6 to 10 h after UVB (Lu *et al* 2008b). Chk1-345 phosphorylation activated by UVB was attenuated by caffeine (Figure 21.2).

It appears that the ATR–Chk1 pathway may represent the main biologically relevant target of caffeine in normal keratinocytes (Heffernan *et al* 2009). siRNA of ATR, but not that of the related kinase ATM, augmented UVB-induced apoptosis, mimicking the effect of caffeine in these cells. Importantly, caffeine did not further enhance UVB-induced apoptosis following siRNA of ATR in normal keratinocytes, indicating that caffeine's effects are epistatic to those of ATR-pathway inhibition. Furthermore, treatment of normal keratinocytes with PF610666, a novel Chk1 kinase inhibitor that appears to be more selective than UCN-01, the archetypal Chk1 inhibitor, similarly potentiated UVB-induced apoptosis (Blasina *et al* 2008).

Caffeine also increases apoptosis in tumors long after all UVB treatment has ceased (Lu *et al* 2011). In the absence of caffeine, UVB-induced tumors often had islands of phospho-Chk1 (Ser317)-staining cells that were not present in non-tumor areas of the epidermis. Treatment of mice with topical caffeine significantly diminished phospho-Chk1 (Ser317) staining and increased the number of mitotic cells that expressed cyclin B1 and caspase 3 in tumors, consistent with caffeine-induced lethal mitosis selectively in tumors. It was proposed that as compared with adjacent uninvolved skin, UVB-induced skin tumors have elevated activation of and dependence on the ATR–Chk1 pathway long after UVB exposure has ceased, and that caffeine can induce apoptosis selectively in tumors by inhibiting this pathway and promoting lethal mitosis.

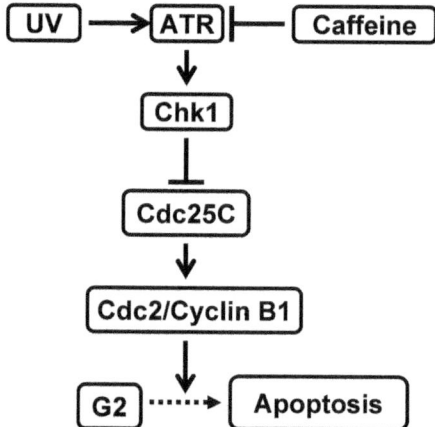

Figure 21.2   p53-independent apoptosis induced by caffeine: ATR mechanism. UVB causes DNA damage, resulting in activation of ataxia-telangiectasia-mutated-Rad3 related (ATR) kinase. ATR-dependent activation leads to Chk1 phosphorylation and to an inactivation of Cdc25c phosphatase by increased phosphorylation of Cdc25C. Cdc25C phosphorylation has been shown to increase both its turnover and cytoplasmic retention. Consequently, activation of the Cdc2/cyclin B1 complex is prevented and mitosis entry is inhibited. Caffeine may abrogate the G_2 checkpoint and enhance apoptosis inhibiting ATR-mediated phosphorylation of Chk1.

However, our recent studies have shown that caffeine promotes UVB-induced apoptosis independent of the ATR pathway (Han *et al* 2011). ATR knockdown did not resemble the proapoptotic effect of caffeine in human HaCaT cells and HeLa cells. These results further support the notion that caffeine has multiple molecular targets along the signaling pathways that regulate cell survival and apoptosis.

21.4 Caffeine and AKT–COX-2

Caffeine clearly has a role in the AKT pathway. In JB6 cells, caffeine was found to significantly inhibit epidermal growth factor (EGF)- and 12-O-tetradecanoylphorbol-13-acetate (TPA)-induced cell transformation in the JB6 mouse epidermal cell line (Nomura *et al* 2005). Pretreatment with caffeine suppressed EGF-induced phosphorylation and activation of Akt and ribosomal p70 S6 protein kinase (p70 S6K), a target of Akt, without inhibiting phosphatidylinositol 3-kinase (PI3K) activation. The inhibition of Akt activation of caffeine was not a result of its adenosine receptor antagonism. Inhibition of PI3K (p110 delta in particular) can likely explain some of the physiological and pharmacological properties of caffeine (Foukas *et al* 2002).

Caffeine reduces mitochondrial membrane potentials and apoptosis in a dose-dependent manner. Caffeine-induced apoptosis is also associated with other cell death mechanisms, such as autophagy, where it increases both the levels of microtubule-associated protein 1 light chain 3-II and the number of autophagosomes. There were fewer early apoptotic cells among autophagy-deficient mouse embryonic fibroblasts treated with caffeine than in their wild-type counterparts (Saiki *et al* 2011).

Furthermore, caffeine was shown to interact with the tumor suppressor phosphatase and tensin homologue deleted from chromosome 10 (PTEN), a protein that is a negative regulator of the phosphatidylinositol-3-kinase (PI3K)–AKT signaling pathway (Davies *et al* 1999; Stambolic *et al* 1998; Stocker *et al* 2002) . Overexpression of PTEN in cancer cells results in cell-cycle arrest and cell death through inhibition of PI3K (Saito *et al* 2003). A combination of adenovirus-mediated transfer of PTEN (Ad-PTEN) and caffeine synergistically suppressed cell growth and induced apoptosis in colorectal cancer cells but not in normal colorectal fibroblast cells through abrogation of G(2)–M arrest, downregulation of the AKT pathway, and modulation of the p44–42MAPK pathway. Caffeine-mediated radiosensitization of glioma cells was associated with the inhibition of AKT hyperphosphorylation in PTEN-deficient cells to a level comparable with PTEN-proficient cells and was significantly correlated with the activation of the G(1) DNA damage checkpoint independent of the *de novo* synthesis of p53 and p21 (Sinn *et al* 2010).

Our recent studies have shown that in HaCaT keratinocytes, inhibiting the AKT–COX-2 pathway is a critical molecular mechanism by which caffeine promotes UVB-induced apoptosis of unrepaired keratinocytes for elimination (Han *et al* 2011) (Figure 21.3). Caffeine has a proapoptotic effect similar to

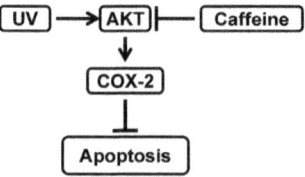

Figure 21.3 p53-independent apoptosis induced by Caffeine: ATR-independent AKT/COX-2 mechanism. UVB induces anti-apoptotic AKT and COX-2 signaling. Caffeine promotes UVB-induced apoptosis by inhibiting the AKT/COX-2 pathways independent of ATR.

either AKT inhibition or COX-2 inhibition. The effect of caffeine on AKT and COX-2 has been shown in other cells. In primary rat microglial cells caffeine was shown to inhibit the protein synthesis of COX-2 (Fiebich *et al* 2000). Caffeine also inactivated the survival signal including the AKT and ERK pathways to trigger apoptosis in osteoblasts (Lu *et al* 2008a).

21.5   Conclusion and Perspective

Caffeine is the most frequently ingested neuroactive agent in the world. In multiple cell types and in mouse models, caffeine has been shown to increase apoptosis. p53, ATR, ATM, AKT, and COX-2 have been found to play vital roles in caffeine's pro-apoptotic action. Considering the essential roles of p53, ATR, ATM, AKT and COX-2 in tumorigenesis, the effect of caffeine on these pathways may serve as the fundamental mechanism by which caffeine prevents cancer development. In the near future the development of caffeine analogs with biologic activity similar to the parent compound, but with less toxicity, could be very useful in combination with DNA-damaging agents for preventing and treating cancers through enhancing apoptotic cell death.

Summary Points

- This chapter focuses on caffeine and apoptosis.
- Caffeine is the most widely consumed psychoactive substance.
- Caffeine has been reported to induce or enhance apoptosis.
- Apoptosis is a form of cell death in which a programmed sequence of events leads to the elimination of cells without releasing harmful substances into the surrounding area. Apoptosis plays a crucial role in developing and maintaining health by eliminating old cells, unnecessary cells, and unhealthy cells.
- p53, ATR, ATM, AKT, and COX-2 have been found to play vital roles in caffeine's pro-apoptotic action.
- Considering the essential role of p53, ATR, ATM, AKT and COX-2 in tumorigenesis, the inhibitory effect of caffeine on these pathways may

serve as the fundamental mechanism by which caffeine prevents cancer development.

• In the near future, the development of caffeine analogs with biologic activity similar to the parent compound, but with less toxicity, could be very useful in combination with DNA-damaging agents for preventing and treating cancers through enhancing apoptotic cell death.

Key Facts

Key Facts of AKT

• The first indication that AKT plays a role in oncogenesis was provided by the isolation of the transforming retrovirus from an *AKR* mouse *T*-cell lymphoma in 1977.
• Dr Testa first began working on AKT in the mid-1980s, when he mapped the chromosomal location of the human *AKT1* gene by an *in situ* hybridization technique.
• In 1991, three independent research groups cloned and characterized Akt kinases.
• In 1992, the first recurrent alterations of an *AKT* gene were identified in human cancer, with the demonstration of amplification and overexpression of *AKT2* in a subset of ovarian carcinomas.
• In 1995, AKT was shown to be the target of platelet-derived growth factor-activated phosphatidylinositol 3-kinase (PI3K), and AKT kinases are now known to be key mediators of signal transduction pathways downstream of activated growth factor and cytokine receptors and PI3K.
• Many human cancers, including carcinomas, glioblastoma multiforme, and various hematological malignancies, exhibit frequent activation of AKT.
• The importance of the Akt pathway in human cancers is highlighted by the mutation of various components of the pathway such as the *PTEN* and PI3-kinase (P110α), which occur in more than 30% of human tumors.

Key Facts of ATR

• The first human ATR cDNA full-length clone (originally named FRP1, FRAP-related protein 1) was isolated from a Jurkat T-cell cDNA library and identified by its significant homology to other members of the phosphatidylinositol kinase-related kinase (PIKK) family.
• ATR's function is essential for cell viability.
• ATR-deficiency at the organismal level affects normal development, tissue homeostasis, and aging.
• ATR haploinsufficiency has been found in mismatch repair (MMR)-deficient cancers.

Key Facts of ATM

- Ataxia telangiectasia is caused by mutations in the ATM gene located on chromosome 11q22.23 in 1988 and it positional cloning by Savitsky *et al* in 1995.
- Ataxia telangiectasia usually runs in families. The mode of inheritance is autosomal recessive, so in a family with two parents who are carriers of the A-T allele, there is 1 chance in 4 for each child born to the parents to have the disorder. Prenatal diagnosis can be carried out in most families, but this is complex and must be arranged before conception.
- Cloning of full-length ATM cDNA was first accomplished in 1997.
- c-Abl was identified as the ATM interacting protein in 1997.
- p53 was identified as a substrate for ATM in radiation signal transduction by different investigators.
- The ATM gene encodes a 13kb mature transcript with an open reading frame of 9168 nucleotides.
- The ATM protein is about 370kDa, is ubiquitously expressed, and is localized to the cell nucleus.
- The ATM protein is a large serine-threonine kinase thought to play a role in regulating cell cycle checkpoints, repair of double stranded DNA and meiosis.
- ATM is also known to play a role in regulating p53, BRCA1 and Chk2. Part of ATM's role in DNA repair is known to be that of telomere repair as telomeres degrade more rapidly in people affected with Ataxia telangiectasia.
- Atm-deficient mice exhibit growth retardation.
- Both male and female Atm-deficient mice are infertile.
- Atm -/- mice develop aggressive malignant thymic lymphomas and usually succumb to the disease by 4 months of age.
- Mice homozygous for Atm disruption display acute radiation sensitivity, which is manifested selectively in certain tissues, including the gastro-intestinal tract, salivary glands and skin.

Key Facts of Apoptosis

- In addition to cell-cycle arrest and repair machinery, the damaged cells, where damage is beyond repair, may induce an apoptotic (programmed cell death; PCD) response that is highly cell-specific and is the most common form of physiologic cell death in multicellular forms.
- Apoptosis is described by its morphological characteristics, including cell shrinkage, membrane blebbing, chromatin condensation and nuclear fragmentation.
- The concept of PCD was first developed by plant biologists in 1920s.

- The term programmed cell death was introduced in 1964, proposing that cell death during development is not accidental in nature but follows a sequence of controlled steps leading to locally and temporally defined self-destruction.
- The term apoptosis had been coined in order to describe the morphological processes leading to controlled cellular self-destruction and was first introduced in a publication by Kerr, Wyllie and Currie in 1972.
- Apoptosis plays an essential role in survival of the organism and is considered to be an imperative component of various processes including normal cell turnover, proper development and functioning of the immune system, multiplication of mutated chromosomes, hormone-dependent atrophy, normal embryonic development, elimination of indisposed cells and maintenance of cell homeostasis.
- DNA damage and production of the predominant lesions such as cyclobutane pyrimidine dimers (CPDs), 6–4 photoproducts (6–4PPs) and certain other lesions as a result of UV radiation (UVR), ionizing radiation (IR), oxidative stress, replication or recombination errors as well as from environmental and therapeutic genotoxins is one of the principal triggers of apoptosis.
- There are at least two broad pathways that lead to apoptosis, an "Extrinsic" and an "Intrinsic" Pathway. In both pathways, signaling results in the activation of a family of Cys (Cysteine) Proteases, named Caspases, that act in a proteolytic cascade to dismantle and remove the dying cell.
- Improper apoptosis or malfunctioning of individual apoptotic machinery may cause several human diseases such as cancer and neurodegenerative disorders as well as several types of autoimmune disorder.
- Unnecessary cell death and unsound regulation of caspase activity are associated with certain diseases such as Alzheimer's disease, Parkinson's disease and Huntington's disease.

Key Facts of UV

- Ultraviolet (UV) radiation is a component of solar radiation.
- UV radiation levels are influenced by a number of factors: Sun elevation, latitude, cloud cover, altitude, ozone and ground reflection.
- Small amounts of UV radiation are beneficial to health, and play an essential role in the production of vitamin D.
- However, excessive exposure to UV radiation is associated with different types of skin cancer, sunburn, accelerated skin aging, cataract and other eye diseases. There is also evidence that UV radiation reduces the effectiveness of the immune system.
- Excessive UV exposure results in a number of chronic skin changes:

  a. Cutaneous malignant melanoma: a life-threatening malignant skin cancer.

b. Squamous cell carcinoma of the skin: a malignant cancer, which generally progresses less rapidly than melanoma and is less likely to cause death.
c. Basal cell carcinoma of the skin: a slow-growing skin cancer appearing predominantly in older people.
d. Photoaging: a loss of skin tightness and the development of solar keratoses.

- UV radiation appears to diminish the effectiveness of the immune system by changing the activity and distribution of the cells responsible for triggering immune responses. Immunosuppression can cause reactivation of the herpes simplex virus in the lip ("cold sores").

Key Facts of PTEN

- *PTEN* (phosphatase and tensin homolog deleted on chromosome 10) is one of the most frequently mutated tumor suppressor genes in human cancer.
- In 1997, *PTEN* was first discovered by independent groups (Li group and Steck *et al*) and recognized as the frequently lost tumor suppressor gene on human chromosome 10q23, a locus that is highly susceptible to mutation in primary human cancers.
- The frequency of monoallelic mutations at this locus has been estimated at 50–80% in sporadic tumors (including endometrial carcinoma, glioblastoma, and prostate cancer) and at 30–50% in breast, colon, and lung tumors.
- Complete loss of *PTEN* is observed at the highest frequencies in endometrial cancer and glioblastoma and is generally associated with advanced cancers and metastases.
- A recent landmark study reveals that *PTEN* loss is a common event in breast cancers caused by *BRCA1* deficiency.

Definitions of Words and Terms

Apoptosis: A form of cell death in which a programmed sequence of events leads to the elimination of cells without releasing harmful substances into the surrounding area. Apoptosis plays a crucial role in developing and maintaining health by eliminating old cells, unnecessary cells, and unhealthy cells. The human body replaces about a million cells a second. Too little or too much apoptosis plays a role in many diseases. Reduced apoptosis can lead to cancer, while increased apoptosis contributes to neurodegenerative diseases such as Parkinson's disease and Alzheimer's.

COX-2: Cyclooxygenase-2, a protein, acts as an enzyme and specifically catalyzes the production of certain chemical messengers called prostaglandins. Some of these messengers are responsible for promoting inflammation.

When COX-2 activity is blocked, inflammation is reduced. Cyclooxygenase-2 (COX-2) is up-regulated in murine and human NMSC. Overexpression of COX-2 promotes UV-induced skin tumorigenesis in mice. Inhibition of COX-2 by biochemical inhibitors or genetic deletion decreases chemical- or UV-induced skin tumor development. In addition, UVB radiation induces COX-2 up-regulation in human skin, mouse skin and keratinocytes *in vitro*. Extensive studies have identified multiple mechanisms mediating such COX-2 induction.

PTEN: PTEN (phosphatase and tensin homolog deleted on chromosome 10) functions as a highly effective tumor suppressor in a wide variety of tumor tissues by negatively regulating the phosphatidylinositol 3-kinase (PI3K)–protein kinase B (AKT) pathway. The loss of PTEN's function through deletion, mutation, and–or decreased expression has been found both in human sporadic cancers and in hereditary cancer syndromes. This ubiquitous and evolutionarily conserved signaling cascade influences many functions, including cell growth, survival, proliferation, migration, and metabolism.

DNA damage response pathway ATM–ATR: the Ataxia-telangiectasia mutated (ATM) and ATM and RAD3-related (ATR) pathways play important roles in DNA damage response and tumor suppression. ATM and ATR belong to the phosphoinositide 3-kinase (PI3K)-related protein kinases (PIKKs). ATM responds primarily to double-strand breaks (DSBs) induced by ionizing radiation, while ATR reacts to UV or stalled replication forks. ATM activates checkpoint kinase-2 (Chk2). Chk1 kinase is an important effector and substrate of ATR. Chk1 plays a key role in mediating replication fork stability following DNA damage and is also required for cell cycle checkpoint activation. ATR phosphorylates several proteins, including the histone H2A variant H2AX, p53, and BRCA1.These pathways coordinate the DNA damage checkpoint function. Defects in the ATR–Chk1 and ATM–Chk2 pathways increase cancer risk. Mutations in ATM predispose carriers to cancer and are found in approximately 0.5–1.0% of the population. People with mutations in both alleles of ATM suffer from the neurodegenerative and cancer predisposition disorder ataxia-telangiectasia8. Mutations in ATR in humans are associated with Seckel syndrome, a developmental disorder characterized by microcephaly and proportionate growth retardation. ATR-mutated Seckel syndrome primary skin fibroblasts exhibit decreased survival following UV irradiation.

UVB: ultraviolet (UV) radiation with wavelengths between 290 and 320 nm, comprising less than 1 per cent of the ultraviolet radiation that reaches the earth's surface. Ultraviolet B causes sunburn and a number of damaging photochemical changes within cells, including damage to DNA, leading to premature aging of the skin, premalignant and malignant changes, and a variety of photosensitivity reactions; it is also used therapeutically for treatment of skin disorders.

AKT: the serine–threonine kinase AKT, also known as protein kinase B, is a downstream effector of phosphatidylinositol 3-kinase (PI3K) that has recently been a focus of intense research. It appears that AKT lies at the cross-roads of multiple cellular signaling pathways and acts as a transducer of many functions initiated by growth factors and other receptors that activate PI3K. One of the major activities of AKT is to promote cell survival. AKT is activated in response to UVB irradiation and is frequently activated in human cancers.

p53: also known as protein 53 or tumor protein 53, is a tumor suppressor protein that in humans is encoded by the TP53 gene. p53 is crucial in multicellular organisms, where it regulates the cell cycle and apoptosis, and thus functions as a tumor suppressor that is involved in preventing cancer. As such, p53 has been described as "the guardian of the genome", the "guardian angel gene", and the "master watchman", referring to its role in conserving stability by preventing genome mutation.

Caffeine: Caffeine is a drug that stimulates the central nervous system. Caffeine is found naturally in coffee, tea, and chocolate. For many years, caffeine has been generally believed to induce cell apoptosis, suppress cell proliferation, abolish chemical- or radiation-induced delays in cell cycle progression and enhance the toxicity of radiation and anti-cancer agents. Caffeine has also been shown to inhibit ultraviolet B (UVB)-induced skin cancer in mice.

Cell cycle: The sequence of events within the cell during cell divisions. The cell cycle is traditionally divided into the following phases:

- G0 (G zero, the G standing for gap)
- G1, (G one, the first gap)
- S (synthesis phase during which the DNA is synthesized, replicated)
- G2 (G two, the second gap)
- M (standing for mitosis, cell division).

Cells that are not destined to divide again are considered to be in G0. The transition from G0 to G1 is thought to commit the cell to completing the cell cycle and dividing.

DNA damage: Drug- or radiation-induced injuries in DNA that introduce deviations from its normal double-helical conformation. These changes include structural distortions which interfere with replication and transcription, as well as point mutations which disrupt base pairs and wield damaging effects on prospective generations through changes in the DNA sequence. If the damage is insignificant, it can often be repaired (DNA REPAIR). If the damage is extensive, it can induce apoptosis.

List of Abbreviations

1. 293T cell	a cell line derived from human embryonic kidney cells
2. E-BP1	eukaryotic initiation factor 4E binding protein 1
3. -OHDA	6-Hydroxydopamine
4. ATM	Ataxia-telangiectasia mutated
5. ATR	ATM and RAD3-related

6. BL-13 cells and BL-28 cells — bladder cancer cell lines expressing p53 with a wild-type sequence

7. BL-17/2 cells — bladder cancer cell lines expressing p53 with a point mutation in codon 280

8. BPDE — benzo[a]pyrene dihydrodiol epoxide

9. BRCA1 — breast cancer 1, early onset

10. Chk1 — checkpoint kinase-1

11. Chk2 — checkpoint kinase-2

12. COX-2 — Cyclooxygenase-2

13. DSBs — double-strand breaks

14. EGF — epidermal growth factor

15. ERK1–2 — extracellular signal-regulated kinases 1 and 2

16. H2AX — histone H2A variant

17. H358 cell — the p53-null non-small cell lung carcinoma cell line

18. HaCaT cell — Human keratinocyte cell line

19. HeLa cell — Human epithelial carcinoma cell line

20. JB6 CL41 cells — a post-initiated mouse epidermal cell line

21. LY294002 — PI3K inhibitor

22. MAPK — mitogen-activated protein kinase

23. MCF-07 cell — human breast adenocarcinoma cell line

24. MPP+ — 1-methyl-4-phenylpyridinium

25. mTOR — the murine target of rapamycin

26. p70S6K — 70-kDa ribosomal protein S6 kinase

27. PCC — premature chromosome condensation

28. PD — Parkinson's disease

29. PF610666 — a novel Chk1 kinase inhibitor

30. PI3K — phosphatidylinositol 3-kinase

31. PIKKs — phosphoinositide 3-kinase (PI3K)-related protein kinases

32. PTEN — phosphatase and tensin homolog deleted from chromosome 10

33. ROS — reactive oxygen species

34. RT112 cells — bladder cancer cell lines expressing a functionally defective p53

35. SH-SY5Y — human neuroblastoma cell line

36. TPA — 12-O-tetradecanoyl-phorbol-13-acetate

37. U2OS cell — human osteosarcoma cell line expressing wild type p53 and Rb, but lacking p16

38. UCN-01 — 7-Hydroxystaurosporine

39. UCN-01 — archetypal Chk1 inhibitor

40. US8-93 cell and LMS6-93 cell — human sarcoma cell lines, both with a p53 mutation

41. UVB — ultraviolet B

Acknowledgement

We apologize to those investigators whose work could not be directly referenced owing to space limitations. Work in the authors' laboratory was supported by NIH grant ES016936, the University of Chicago Comprehensive Cancer Center (P30 CA014599), and the CTSA (NIH UL1RR024999). The authors are grateful to Dr. Ann Motten for her critical reading of this manuscript.

References

Alao, J. P. and Sunnerhagen, P., 2009. The ATM and ATR inhibitors CGK733 and caffeine suppress cyclin D1 levels and inhibit cell proliferation. Radiat. Oncol. 4: 51.

Blasina, A., Hallin, J., Chen, E., Arango, M. E., Kraynov, E., Register, J., *et al.* 2008. Breaching the DNA damage checkpoint via PF-00477736, a novel small-molecule inhibitor of checkpoint kinase 1. Mol. Cancer Ther. 7: 2394–2404.

Bode, A. M. and Dong, Z., 2003. Mitogen-activated protein kinase activation in UV-induced signal transduction. Sci. STKE. 2003: RE2.

Bowden, G. T. 2004. Prevention of non-melanoma skin cancer by targeting ultraviolet-B-light signalling. Nat. Rev. Cancer. 4: 23–35.

Busse, P. M., Bose, S. K., Jones, R. W. and Tolmach, L. J., 1978. The action of caffeine on X-irradiated HeLa cells. III. Enhancement of X-ray-induced killing during G2 arrest. Radiat. Res. 76: 292–307.

Canman, C. E., Chen, C. Y., Lee, M. H. and Kastan, M. B., 1994. DNA damage responses: p53 induction, cell cycle perturbations, and apoptosis. Cold Spring Harb. Symp. Quant. Biol. 59: 277–286.

Chen, Y. and Sanchez, Y., 2004. Chk1 in the DNA damage response: conserved roles from yeasts to mammals. DNA Repair. 3: 1025–1032.

Cortez, D. 2003. Caffeine inhibits checkpoint responses without inhibiting the ataxia-telangiectasia-mutated (ATM) and ATM- and Rad3-related (ATR) protein kinases. J. Biol. Chem. 278: 37139–37145.

Davies, M. A., Koul, D., Dhesi, H., Berman, R., McDonnell, T. J., McConkey, D., *et al.* 1999. Regulation of Akt/PKB activity, cellular growth, and apoptosis in prostate carcinoma cells by MMAC/PTEN. Cancer Res. 59: 2551–2556.

Dubrez, L., Coll, J. L., Hurbin, A., Solary, E. and Favrot, M. C., 2001. Caffeine sensitizes human H358 cell line to p53-mediated apoptosis by inducing mitochondrial translocation and conformational change of BAX protein. J. Biol. Chem. 276: 38980–38987.

el-Deiry, W. S., Tokino, T., Velculescu, V. E., Levy, D. B., Parsons, R., Trent, J. M., *et al.* 1993. WAF1, a potential mediator of p53 tumor suppression. Cell. 75: 817–825.

Enoch, T. and Norbury, C. 1995. Cellular responses to DNA damage: cell-cycle checkpoints, apoptosis and the roles of p53 and ATM. Trends Biochem. Sci. 20: 426–430.

Essmann, F., Engels, I. H., Totzke, G., Schulze-Osthoff, K. and Janicke, R. U., 2004. Apoptosis resistance of MCF-7 breast carcinoma cells to ionizing radiation is independent of p53 and cell cycle control but caused by the lack of caspase-3 and a caffeine-inhibitable event. Cancer Res. 64: 7065–7072.

Fiebich, B. L., Lieb, K., Hull, M., Aicher, B., van Ryn, J., Pairet, M., *et al.* 2000. Effects of caffeine and paracetamol alone or in combination with acetylsalicylic acid on prostaglandin E(2) synthesis in rat microglial cells. Neuropharmacology. 39: 2205–2213.

Fingert, H. J., Chang, J. D. and Pardee, A. B., 1986. Cytotoxic, cell cycle, and chromosomal effects of methylxanthines in human tumor cells treated with alkylating agents. Cancer Res. 46: 2463–2467.

Foukas, L. C., Daniele, N., Ktori, C., Anderson, K. E., Jensen, J. and Shepherd, P. R., 2002. Direct effects of caffeine and theophylline on p110 delta and other phosphoinositide 3-kinases. Differential effects on lipid kinase and protein kinase activities. J. Biol. Chem. 277: 37124–37130.

Hall-Jackson, C. A., Cross, D. A., Morrice, N. and Smythe, C., 1999. ATR is a caffeine-sensitive, DNA-activated protein kinase with a substrate specificity distinct from DNA-PK. Oncogene. 18: 6707–6713.

Han, W., Ming, M. and He, Y. Y., 2011. Caffeine promotes ultraviolet B-induced apoptosis in human keratinocytes without complete DNA repair. J. Biol. Chem. 286: 22825–22832.

Hartwell, L. H. and Kastan, M. B., 1994. Cell cycle control and cancer. Science. 266: 1821–1828.

He, Z., Ma, W. Y., Hashimoto, T., Bode, A. M., Yang, C. S. and Dong, Z., 2003. Induction of apoptosis by caffeine is mediated by the p53, Bax, and caspase 3 pathways. Cancer Res. 63: 4396–4401.

Heffernan, T. P., Kawasumi, M., Blasina, A., Anderes, K., Conney, A. H. and Nghiem, P., 2009. ATR-Chk1 pathway inhibition promotes apoptosis after UV treatment in primary human keratinocytes: potential basis for the UV protective effects of caffeine. J. Invest. Dermatol. 129: 1805–1815.

Kramata, P., Lu, Y. P., Lou, Y. R., Cohen, J. L., Olcha, M., Liu, S., *et al.* 2005. Effect of administration of caffeine or green tea on the mutation profile in the p53 gene in early mutant p53-positive patches of epidermal cells induced by chronic UVB-irradiation of hairless SKH-1 mice. Carcinogenesis. 26: 1965–1974.

Lu, P. Z., Lai, C. Y. and Chan, W. H., 2008a. Caffeine induces cell death *via* activation of apoptotic signal and inactivation of survival signal in human osteoblasts. Int. J. Mol. Sci. 9: 698–718.

Lu, Y. P., Lou, Y. R., Li, X. H., Xie, J. G., Brash, D., Huang, M. T., *et al.* 2000. Stimulatory effect of oral administration of green tea or caffeine on ultraviolet light-induced increases in epidermal wild-type p53, p21(WAF1/CIP1), and apoptotic sunburn cells in SKH-1 mice. Cancer Res. 60: 4785–4791.

Lu, Y. P., Lou, Y. R., Li, X. H., Xie, J. G., Lin, Y., Shih, W. J., *et al.* 2002a. Stimulatory effect of topical application of caffeine on UVB-induced apoptosis in mouse skin. Oncol Res. 13: 61–70.

Lu, Y. P., Lou, Y. R., Liao, J., Xie, J. G., Peng, Q. Y., Yang, C. S., *et al.* 2005. Administration of green tea or caffeine enhances the disappearance of UVB-induced patches of mutant p53 positive epidermal cells in SKH-1 mice. Carcinogenesis. 26: 1465–1472.

Lu, Y. P., Lou, Y. R., Peng, Q. Y., Nghiem, P. and Conney, A. H., 2011. Caffeine decreases phospho-Chk1 (Ser317) and increases mitotic cells with cyclin B1 and caspase 3 in tumors from UVB-treated mice. Cancer Prev. Res. 4: 1118–1125.

Lu, Y. P., Lou, Y. R., Peng, Q. Y., Xie, J. G. and Conney, A. H., 2004. Stimulatory effect of topical application of caffeine on UVB-induced apoptosis in the epidermis of p53 and Bax knockout mice. Cancer Res. 64: 5020–5027.

Lu, Y. P., Lou, Y. R., Peng, Q. Y., Xie, J. G., Nghiem, P. and Conney, A. H., 2008b. Effect of caffeine on the ATR/Chk1 pathway in the epidermis of UVB-irradiated mice. Cancer Res. 68: 2523–2529.

Lu, Y. P., Lou, Y. R., Xie, J. G., Peng, Q. Y., Liao, J., Yang, C. S., *et al.* 2002b. Topical applications of caffeine or (–)-epigallocatechin gallate (EGCG) inhibit carcinogenesis and selectively increase apoptosis in UVB-induced skin tumors in mice. Proc. Natl. Acad. Sci. U.S.A. 99: 12455–12460.

Lu, Y. P., Lou, Y. R., Xie, J. G., Peng, Q. Y., Zhou, S., Lin, Y., *et al.* 2007. Caffeine and caffeine sodium benzoate have a sunscreen effect, enhance UVB-induced apoptosis, and inhibit UVB-induced skin carcinogenesis in SKH-1 mice. Carcinogenesis. 28: 199–206.

Marzo, I., Brenner, C., Zamzami, N., Jurgensmeier, J. M., Susin, S. A., Vieira, H. L., *et al.* 1998. Bax and adenine nucleotide translocator cooperate in the mitochondrial control of apoptosis. Science. 281: 2027–2031.

Ming, M., Han, W., Maddox, J., Soltani, K., Shea, C. R., Freeman, D. M., *et al.* 2010. UVB-induced ERK/AKT-dependent PTEN suppression promotes survival of epidermal keratinocytes. Oncogene. 29: 492–502.

Musk, S. R. and Steel, G. G., 1990, Override of the radiation-induced mitotic block in human tumour cells by methylxanthines and its relationship to the potentiation of cytotoxicity. Int. J. Radiat. Biol. 57: 1105–1112.

Nakaso, K., Ito, S. and Nakashima, K. 2008. Caffeine activates the PI3K/Akt pathway and prevents apoptotic cell death in a Parkinson's disease model of SH-SY5Y cells. Neurosci. Lett. 432: 146–150.

Nomura, M., Ichimatsu, D., Moritani, S., Koyama, I., Dong, Z., Yokogawa, K., *et al.* 2005. Inhibition of epidermal growth factor-induced cell transformation and Akt activation by caffeine. Mol. Carcinog. 44: 67–76.

O'Driscoll, M., Ruiz-Perez, V. L., Woods, C. G., Jeggo, P. A. and Goodship, J. A., 2003. A splicing mutation affecting expression of ataxia-telangiectasia and Rad3-related protein (ATR) results in Seckel syndrome. Nat. Genet. 33: 497–501.

Saiki, S., Sasazawa, Y., Imamichi, Y., Kawajiri, S., Fujimaki, T., Tanida, I., *et al.* 2011. Caffeine induces apoptosis by enhancement of autophagy via PI3K/Akt/mTOR/p70S6K inhibition. Autophagy. 7: 176–187.

Saito, Y., Gopalan, B., Mhashilkar, A. M., Roth, J. A., Chada, S., Zumstein, L., *et al.* 2003. Adenovirus-mediated PTEN treatment combined with caffeine produces a synergistic therapeutic effect in colorectal cancer cells. Cancer Gene Ther. 10: 803–813.

Sarkaria, J. N., Busby, E. C., Tibbetts, R. S., Roos, P., Taya, Y., Karnitz, L. M., *et al.* 1999. Inhibition of ATM and ATR kinase activities by the radiosensitizing agent, caffeine. Cancer Res. 59: 4375–4382.

Sinn, B., Tallen, G., Schroeder, G., Grassl, B., Schulze, J., Budach, V., *et al.* 2010. Caffeine confers radiosensitization of PTEN-deficient malignant glioma cells by enhancing ionizing radiation-induced G1 arrest and negatively regulating Akt phosphorylation. Mol. Cancer Ther. 9: 480–488.

Stambolic, V., Suzuki, A., de la Pompa, J. L., Brothers, G. M., Mirtsos, C., Sasaki, T., *et al.* 1998. Negative regulation of PKB/Akt-dependent cell survival by the tumor suppressor PTEN. Cell. 95: 29–39.

Stocker, H., Andjelkovic, M., Oldham, S., Laffargue, M., Wymann, M. P., Hemmings, B. A., *et al.* 2002. Living with lethal PIP3 levels: viability of flies lacking PTEN restored by a PH domain mutation in Akt/PKB. Science. 295: 2088–2091.

Subject Index

Illustration and figures are in **bold**. Tables are in *italics*.